pushing rubber tubing over a glass tube; a severe explosion and fire may result from attempting to distill a substance in a completely closed system.

Procedures to be followed in case of accidents are given in the Appendix and also are printed on the inside back cover of this manual.

Toxic Chemicals and Mutagens

Many common organic chemicals, particularly those containing nitrogen, are toxic or even lethal when ingested in amounts as small as a few tenths of a gram. A major rule of safety laboratory practice is *never* taste any compound.

Certain organic compounds, when ingested or inhaled over a long period of time, even in minute doses, may eventually induce tumors—some benign and some cancerous. An attempt has been made in this manual to avoid use of such compounds, but this is difficult because of the limited information available in most cases. It seems probable that many relatively safe compounds have been included on the "cancer suspect" list, whereas many others that should be included are not. Unfortunately, simplicity of structure or lack of active functional groups does not guarantee safety. In the absence of specific information, it is wise to minimize exposure to the vapors of all organic solvents; even pleasant-smelling vapors may be quite toxic.

Laboratory Apparel

It is desirable to protect street clothing from being soiled and damaged by chemicals and accidents of various sorts by wearing an inexpensive laboratory coat or a rubber apron. For freedom of arm movement, tight sleeves should be avoided; loose and bulky sleeves may cause overturning of fragile apparatus. Unprotected light-weight, flammable apparel constitutes a serious fire hazard in the organic laboratory. Many synthetic fabrics are soluble in acetone and other common organic solvents.

Shoes that cover at least the toes and instep must be worn in the laboratory.

Shoulder-length (or longer) hair should be tied back, especially if a burner is being used.

In experiments requiring transfer of corrosive chemicals, it is desirable to wear some type of resistant gloves. Inexpensive, disposable gloves made from polyethylene are available. Rubber gloves afford better protection, although they make laboratory operations more cumbersome.

Experimental Organic CHEMISTRY
A Small-Scale Approach

Experimental Organic
CHEMISTRY

A Small-Scale Approach

Charles F. Wilcox, Jr.

Cornell University

MACMILLAN PUBLISHING COMPANY
NEW YORK
Collier Macmillan Publishing
LONDON

Macmillan Publishing Company
866 Third Avenue, New York, New York 10022

Collier Macmillan Canada, Inc.

Library of Congress Cataloging in Publication Data
Wilcox, Charles F., (date)
  Experimental organic chemistry.

  Includes index.
  1. Chemistry, Organic—Laboratory manuals.   I. Title.
QD261.W497 1988   547'.0028   87-14625
ISBN 0-02-427620-0

Printing: 4 5 6 7 8   Year: 0 1 2 3 4 5 6 7

Preface

In recent years the undergraduate organic laboratory has had to cope with both large student populations and an increased sensitivity to the hazards of organic chemicals. The first has placed a severe burden on available space, and the second has mandated costly waste disposal. At Cornell, at least, it now costs us more to remove a pound of product than it does to bring in a pound of starting material.

A partial solution to both problems is to reduce sharply the scale of the experiments—to use smaller amounts in a smaller apparatus. The practical question is how small. Some schools have gone over to micro-scale experiments, but we at Cornell felt that this was too small, that it lost an important visual contact between the student and the materials being used. Another consideration against such a small scale was the necessity to use analytical balances and the inability to carry out ordinary distillations on products. Our compromise was to run experiments mostly on the 0.3–3-g scale. Corresponding to about one fifth of the scale used previously, this represents a significant reduction in volumes and a welcome increase in safety factors. At this scale we are able to use the less expensive 10-mg sensitivity electronic top-loading balances (very fast weighings) and to carry out ordinary distillations by going over to a one-piece distillation head. These heads are moderately expensive (about $65, if purchased), but they substitute for an equally expensive array of distillation adapter, vacuum adapter, and one of two water-cooled condensers previously used. At this scale, too, we are able to use a *single* Thermowell heating mantle as the heat source for temperatures above 90° (we have retained the use of steam baths).

The present text is a rewriting of the author's *Experimental Organic Chemistry: Theory and Practice* using most of the same experiments reduced to about one-fifth scale. However, there is more to scaling down experiments than just dividing by five. In general, the number of transfers between vessels must be minimized to reduce losses due to surface adhesion; and, where the quantities are particularly small, the conventional extraction procedure is better replaced by what I refer to as the "squirting" technique—two test tubes and a Pasteur pipet to forcibly mix the layers and transfer one of them. In some cases reaction times should be changed. The organization of the previous text has been largely maintained except that some of the old experiments have been moved to a new section called "Compounds of Medicinal and Biological Interest." Several preparations have been added: cyclohexyl chloride, *p*-methoxytetraphenylmethane, *o*- and *p*-nitrophenol, indigo, deuterotoluene, and binaphthol. Certain preparations have been deleted—*sec*-butyl chloride, *m*-dinitrobenzene, *p*-bromonitrobenzene, 3-phenylsydnone, and triphenylpropionic acid—either because they were judged unacceptable on the new scale or for safety reasons. Almost all of the experiments, new and old, have been tested, some by classes of students, some by small groups of undergraduate and graduate students, and others by an experienced organic teacher. As a result, the description of virtually every experiment has been changed. Where practical, the experimental procedures have been simplified. An effort has been made to make the mechanistic and experimental descriptions clearer, and many new questions have been added. The previous edition has been well accepted largely because the experiments work in the hands of students; with this text they should work even better.

The text has been divided into four parts. Part I describes the common separation methods and the theories underlying them. As before, the concept of vapor pressure serves as a unifying theme to correlate the various separation methods. Part II presents the identification of organic compounds by both classical wet chemical methods and spectrometric methods. Part III offers a variety of preparations that illustrate much basic functional group chemistry. Part IV is concerned with the preparation and reactions of compounds of medicinal and biological interest. All of the experiments are carried out on a scale that significantly reduces the hazards and the disposal problem. The author is well aware of the fear many neophytes have of chemicals. Rather than hide the chemical hazards, the approach taken here is to try to identify them, especially those that students might not be aware of, and indicate how to cope with particular hazards and prevent them from becoming dangers.

The author appreciates the helpful suggestions from Macmillan's reviewers: Professors Donald C. Berndt, Western Michigan University; Kenneth L. Johnson, Emporia State University; Sheue L. Keenen, University of Wisconsin–River Falls; Russell C. Petter, University of Pittsburgh; and R. Schoffstall, University of Colorado.

The author gratefully acknowledges the assistance of his wife Mary, who helped both to develop the new experiments and to test the old ones.

C. F. W., Jr.

Contents

I SEPARATION AND PURIFICATION OF ORGANIC COMPOUNDS

II IDENTIFICATION OF ORGANIC COMPOUNDS

III PREPARATIONS AND REACTIONS OF TYPICAL ORGANIC COMPOUNDS

IV COMPOUNDS OF MEDICINAL AND BIOLOGICAL INTEREST

1 Introduction

The first organic laboratory course is primarily intended to acquaint you with the principles and practice of organic laboratory operations, as well as to reinforce the theoretical aspects of the subject. The experience of working with a variety of typical organic materials and observing their characteristic properties and transformations gives a sense of reality to organic structural formulas and their frequently strange names. If approached seriously, the laboratory experiments will stimulate your intellectual curiosity and develop your powers of observation, in addition to giving training in careful and skillful manipulation. It is important to realize that none of this will be achieved if the experiments are performed in an unenlightened routine fashion, reading one sentence and proceeding to the next without having beforehand a general knowledge of the whole sequence of laboratory operations and of the underlying principles.

This chapter reviews the laboratory setting—and the important safety concerns—and provides a discussion of several basic laboratory operations that will be used throughout the course.

1.1 General Precautions

The organic laboratory contains many hazards. Fortunately, the risks can be minimized if certain basic safety practices, described in the following paragraphs, are followed faithfully.

Safety Glasses. The eyes are particularly vulnerable to injury by splashing droplets of corrosive chemicals and flying particles of glass or other solid fragments. You *must* wear safety glasses with side-shields (or goggles) *at all times*

1

in the laboratory. Prescription glasses are not a substitute for safety glasses and if they must be worn, they should be supplemented with a pair of plastic goggles that fit over them. Goggles provide even more protection than safety glasses and some laboratories require them to be worn instead.

Contact lenses should *never* be worn in the laboratory unless safety glasses or goggles are also worn because contact lenses cannot be removed rapidly enough to prevent damage from reagents splashed in the eyes. Soft contact lenses offer no protection against shrapnel, and they can capture organic vapors that might lead to irritation or damage.

Fire Hazards. One of the chief dangers of organic laboratory work is the fire hazard associated with the manipulation of volatile, flammable organic liquids. With few exceptions, organic liquids and vapors catch fire readily, and many organic vapors form explosive mixtures with air. Because of the fire hazard, many laboratories have banned the use of burners except under special, carefully controlled conditions. If you are going to use a burner, it is important to keep the following points in mind. Organic liquids must not be manipulated near an open flame, and precautions must be taken to avoid the escape of organic vapors into the laboratory. For general safety, you should form the habit of scanning the adjacent laboratory bench space for lighted burners before working with flammable solvents, and it is good practice to look around for fire hazards to yourself and to adjacent workers before lighting a match or a burner.

The degree of flammability of organic compounds varies widely. The vapors of diethyl ether, petroleum ether, acetone, and ethanol catch fire quite readily and the manipulation of these liquids requires careful attention at all times to fire hazards. Methylene chloride (bp 40°)[1] is a much safer solvent. Carbon disulfide is so readily ignited (even by a hot steam pipe) that it should never be used by an inexperienced worker.

Chemical Burns and Cuts. Specific precautions for handling particularly dangerous chemicals are noted in the directions for the experiment where they are used, but any ordinary chemical or piece of apparatus can be dangerous if manipulated carelessly. It is important to develop a general awareness of hazards and accidents that arise from carelessness in simple routine operations. To cite two examples, a severe cut or laceration may result from carelessly pushing rubber tubing over a glass tube; a severe explosion and fire may result from attempting to distill a substance in a completely closed system. Your instructors will be on the lookout for potential problems, but you cannot depend on them to see everything. *You must think for yourself.*

Procedures to be followed in case of accidents are given in the Appendix and also are printed on the inside back cover of this manual.

---

[1] Temperatures in this book are in degrees Celsius unless otherwise specified.

⫸ **Toxic Chemicals and Mutagens.** Many common organic chemicals, particularly those containing nitrogen, are toxic or even lethal when ingested in amounts as small as a few tenths of a gram. A major rule of safe laboratory practice is *never* to taste anything.

Some organic compounds, when ingested or inhaled over a long period of time, even in minute doses, may eventually induce tumors, some of which are benign and some cancerous. An attempt has been made in this manual to avoid such compounds, but because of the limited information available in most cases, you should not assume that all is well. It seems probable that many relatively safe compounds have been included on the "cancer suspect" list, while many others that should be included have not. Unfortunately, simplicity of structure or lack of active functional groups does not guarantee safety. In the absence of specific information it is wise to minimize exposure to the vapors of all organic solvents; even pleasant-smelling vapors may be quite toxic.

Federal law now requires that purchasers of a chemical product be provided with a "material safety data sheet" (MSDS) that, in addition to listing physical and chemical data, summarizes any hazardous components, health hazard data, spill and leak procedures, and special precautions and protective equipment to be used in handling the chemical.

⫸ **Laboratory Apparel.** It is desirable to protect street clothing from soiling and damage from chemicals and accidents of various sorts by wearing an inexpensive laboratory coat or a rubber apron. For freedom of arm movement tight sleeves should be avoided; loose and bulky sleeves may cause overturning of fragile apparatus. Unprotected wearing apparel of lightweight, flammable fabrics constitutes a serious fire hazard in the organic laboratory. Many synthetic fabrics are soluble in acetone and many other common organic solvents.

Shoes that cover at least the toes and instep must be worn in the laboratory; sandals *are not allowed*. Shoulder-length (or longer) hair should be tied back, especially if a burner is being used. In experiments requiring transfer of corrosive chemicals it is desirable to wear some type of resistant glove. An inexpensive, disposable glove made from polyethylene is available.[2] Rubber gloves afford better protection, although they make laboratory operations more cumbersome.

1.2 Apparatus

One of your first activities in the laboratory will be to check in and determine if your assigned complement of glassware and hardware is present and in undamaged condition. The rules of the laboratory governing the degree of financial responsibility for the equipment at the end of the term will vary from school

---

[2] Polygloves, distributed by Cole-Parmer, come in large and medium sizes: similar gloves are available from Will Scientific, Inc. in large, medium, and small sizes.

Distillation head

Vigreux column

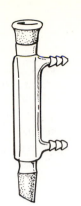
Reflux column

Separatory funnel

Claisen adapter

Graduated cylinder

Round-bottomed flask

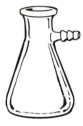
Erlenmeyer flask

Suction flask

Beaker

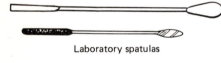
Laboratory spatulas

Forceps

Powder funnel

Buchner funnel

Hirsch funnel

Pasteur pipet

Filter funnel

Three-finger clamp

Clamp holder

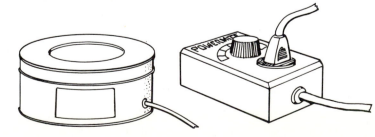
Thermowell heater

Heat controller

FIGURE 1.1 *Common Organic Apparatus*

to school, and you should understand your liability clearly before you begin. Carefully check the glassware for cracks and chips. It is a good idea to clean any dirty glassware to be certain that it really can pass inspection. Determine that the thermometers are not broken and that the mercury thread is continuous.

Drawings of some of the more common organic laboratory apparatus are given in Figure 1.1.

Cleaning and Drying Glassware. It is advantageous to clean laboratory glassware immediately after use, since tars and gummy matter are most easily removed before they harden. Much time is saved by having glassware clean and dry, ready for use at once. Many water-insoluble organic compounds and gums can be removed quickly and economically with scouring powder, a brush, and warm water. The use of strong acids, such as concentrated sulfuric or sulfuric–chromic acid cleaning solution, is dangerous and messy. *Nitric acid is particularly dangerous* as a cleaning agent because it reacts explosively with many organic compounds.

To remove resins and gummy material from glassware, first pour or scrape out as much material as possible directly into a labeled waste container; *never put organic tars, paper, or other solid wastes into the sink*. Next, try to remove or loosen the resin by using a small amount of acetone (1–2 mL) and allowing the solvent to stand in contact with the material for 5 or 10 min. The solvent action may be hastened by warming the glassware on a steam bath (not over a flame) with care to avoid accidental ignition of the flammable solvent vapor. Do not expect tars and gums to dissolve quickly; allow ample time for the organic solvent to act. Acetone is usually a good solvent for tars, but ethanol is generally not effective.

To remove the remaining small amounts of tars and dirt, use scouring powder and a large test tube brush. If the brush is bent properly, it will reach the inner surfaces of flasks. The use of a little washing powder or liquid detergent followed by a good water rinse (preferably distilled water) will give glassware a clean, brilliant sparkle when it dries. The best way to dry apparatus is to allow it to stand overnight on the laboratory desk. Beakers and flasks should be inverted to permit drainage; test tubes and small funnels may be inverted over crumpled paper placed in the bottom of a large beaker.

If wet glassware must be dried quickly for immediate use, it may be rinsed with one or two small portions, *not over 2 mL*, of acetone and allowed to drain; the last traces of acetone are removed by drawing or passing a current of dry air through the apparatus. Methanol or ethanol may be used instead of acetone, but they evaporate less quickly. Ordinary compressed air is not suitable for drying since it is apt to be saturated with water and may even contain suspended droplets of water or oil. A better alternative is to draw a stream of air through the apparatus by means of a glass tube connected to a suction pump. With the suction pump on and drawing air, insert the tube into the apparatus and leave it there until the apparatus is dry.

Apparatus with Interchangeable Ground-Glass Joints. Most contemporary organic laboratory work uses apparatus having interchangeable ground-glass joints (standard taper joints, ⑤). The principal advantage of ground-jointed apparatus is that the joints are not affected by corrosive liquids and vapors that attack corks and rubber stoppers (chlorosulfonic acid, phosphorus trichloride, bromine, nitric acid, etc.); reaction mixtures containing such corrosive materials may be distilled or refluxed without contamination of the product or loss of material through leakage at the joints. Dimensions of the joints have been standardized[3] (⑤ 12/18, 14/20, etc.), so that a variety of assemblies can be set up from a small stock of standard taper flasks, condensers, and adapters.

The foremost rule for assembling ground-jointed apparatus is that the ground surfaces *must be free* of any gritty material that might score them when they are mated. It is good practice to wipe each surface gently with a lint-free cloth or tissue before assembling them.

Because of the highly precise grinding of standard taper apparatus, it is not essential in most ordinary laboratory work to apply grease to the ground surfaces before assembly (this does not apply to stopcocks or other apparatus with ground surfaces that must be rotated during use). Grease must be applied, however,

1. When the apparatus will be heated above 150°.
2. When there is any possibility that the joint will come in contact with strongly alkaline solutions.
3. When the apparatus will be required to hold a vacuum.
4. When the surfaces will be rotated during use.

As an additional precaution, joints that have been exposed to base or basic solutions should be disassembled and cleared immediately after use, especially if they have been heated.

Although careful application of the foregoing rules will prevent mated joints from fusing together (becoming "frozen"), many workers prefer to apply grease to the ground surfaces in every situation. The trade-off is the risk of losing the apparatus versus the risk of contaminating the reaction with grease leached from the joint during the reaction.

A good method for applying grease is to place several small dabs on one of the two surfaces to be mated. The joint members are then placed together and rotated back and forth gently until the grease has formed a thin, continuous, film between the two ground surfaces. Care should be taken to avoid excess grease, since the excess will gradually flow out of the bottom of the joint and contaminate any material with which it comes in contact. It should be remem-

---

[3] The joints are ground to a standard taper (⑤) of 1-mm taper for a length of 10 mm; the maximum diameter and length in millimeters are given in that order by the designation 12/18, etc.

bered also that if a liquid is to be poured out of a flask having a greased joint, the surface must first be wiped clean.

Greased joints should always be cleaned thoroughly when the apparatus is disassembled. Hydrocarbon greases are readily removed by acetone and many other organic solvents. Silicone grease has the advantage over hydrocarbon greases of lower vapor pressure and lower solubility in most organic solvents, but if left on glass surfaces it tends to oxidize to an unsightly white film that cannot be removed and gives the appearance of a very dirty flask. At check-out time, an apparatus with such a film is likely to be rejected. The problem can be avoided by using the minimum amount of silicone grease and removing it from the apparatus as soon as the experiment is completed; methylene chloride, CH_2Cl_2, is a good solvent.

A limitation of ground-jointed apparatus is that, unlike glassware assembled with corks or rubber stoppers, it has no mechanical flexibility. Special care must be exercised when a clamp is tightened onto a portion of an assembly that is jointed to another clamped member.

1.3 Weighing and Measuring Reagents

In performing laboratory experiments, it is important to weigh or measure the amounts of materials carefully and to use the exact quantities called for in the directions. When reagents need not be accurately measured (as in certain tests or in drying operations), the laboratory directions indicate approximate quantities. When an approximate quantity is indicated (e.g., 1–2 mL) you should use a quantity within the specified limits. At first it will be advisable to actually measure the quantity used in order to learn to judge such quantities, but after some experience you should be able to make acceptable approximations.

In many other cases an operation's success depends on using the starting materials and other reagents in the definite, prescribed amounts. A careful laboratory worker will acquire a habit of using solutions of known strength and of weighing or measuring the reagents and solutions used. The strengths of the common laboratory desk reagents are listed in Table 1.1.

TABLE 1.1
Desk Reagents

Reagent	Density g/mL	Reagent concentration		
		g/100 g	g/100 mL	mole/L (approx.)
Acetic acid (glacial)	1.06	99.5	105.5	17.5
Hydrochloric acid (concd)	1.18	35.4	42	12
Nitric acid (concd)	1.42	70	100	17
Sulfuric acid (concd)	1.84	96	176	18
Sodium hydroxide solution (dil)	1.11	10	11.1	3
Ammonia solution (concd)	0.90	29	26	15

Several types of balances are used in undergraduate organic laboratories. For the experiments in this book, the principal requirement is that masses as large as 100 g can be measured with an accuracy of about 0.01–0.001 g (1–10 mg). All balances should be treated as delicate instruments, and anything spilled on one should be removed immediately. Many balances use knife edges to support the balance beam, and these can be damaged easily by rough use.

Needless accuracy in weighing should be avoided, since preparative organic laboratory work is done with a precision of only about 1%. If you are using a swing balance, it is helpful to know how much weight is required to displace the balance beam pointer by a small amount from the null mark. A few minutes spent in gaining familiarity with the balance and its responsiveness can save much time later.

Many modern electronic balances have a "tare" button that, when pressed, automatically adjusts subsequent weighings to be displayed relative to that "tare" value. The tare value can be reset to the current weight by simply pushing the button again.

If the balance is of the mechanical kind that has a zero adjustment screw, the screw setting should not be altered for the purpose of taring. With such balances, the difference between two weighings must be determined.

Interconversion of Weights and Volumes. In recent years there has been an effort to develop and use a common, international system of units. From a long series of meetings the "SI units" (Le Système International d'Unités) was evolved. Unfortunately, in the SI system the base unit of length is the meter (m) and the derived unit of volume is a "cubic meter" (m^3); the base unit of mass is a "kilogram" (kg). The liter and gram are not part of the SI system. Thus a flask that in traditional units had a volume of 25 milliliters would have in strict SI units a volume of 25 micro cubic meters and a 2.0-gram sample would have a mass of 2.0 milli kilograms. Because of the inconvenient sizes of SI units for mass and volume, I have decided to use the traditional gram and liter units in this book. The common laboratory volume unit is the milliliter (mL), which is 1/1000 of a liter (L). A liter is the volume of a cube 10 cm on a side. Thus the milliliter and cubic centimeter (cm^3 or cc) are identical.

In laboratory practice it is often necessary or desirable to convert weight measures into volume measures, and vice versa. These conversions may be made by use of the following relationships.

$$\text{Weight (g)} = \text{volume (mL) at } T° \times \text{density (g/mL) at } T°$$

$$\text{Volume (mL) at } T° = \frac{\text{weight (g)}}{\text{density (g/mL) at } T°}$$

The numerical value of the density and the specific gravity of a particular liquid (at a given temperature) are usually so nearly equal that they may be used interchangeably for organic work. Nevertheless, you should be aware of the following accurate definitions.

The density of a liquid is equal to the mass of a unit volume of the substance. An accurate density value includes a statement of the temperature and the units. For example, the density of water at 20° is 0.9982 g/mL, which is commonly expressed by the notation $d_4^{20} = 0.9982$.

When solutions such as hydrochloric acid, concentrated sulfuric acid, and 95% ethanol are used, it is necessary to calculate the weight of the solute present.

$$\text{Weight of solute (g)} = \text{weight of solution} \times \frac{\text{g of solute}}{\text{g of solution}}$$

$$= \text{weight of solution} \times \frac{\%\text{ of solute by weight}}{100}$$

Another general statement of the equation is

$$\text{Weight of solute} = \text{volume (mL) at } T°$$
$$\times \text{ density of solution (g/mL) at } T°$$
$$\text{concentration by weight}$$

For convenience, the physical constants of solutions of the common acids and of ethanol are included in the Appendix. More complete tables may be found in chemical handbooks and in reference works.[4]

Example. Let us suppose that we wish to find the weight of pure ethanol present in 30 mL of ordinary ethanol (95% by volume or 92.5% by weight). By reference to an "alcohol table" the density of 92.5% (by weight) ethanol at 20° is found to be 0.8112 (d_4^{20}). Using the preceding equations, we have

$$\text{Weight of 30 mL of 92.5\% ethanol} = 30 \text{ mL} \times 0.8112 \text{ g/mL} = 24.34 \text{ g}$$

$$\text{Weight of pure ethanol (100\%)} = 24.34 \text{ g} \times 92.5/100 = 22.5 \text{ g}$$

1.4 Heat Sources

Many different sources of heat are used in the organic laboratory. Unfortunately, each has its limitations, and no single device can be recommended for all situations. Sooner or later you are likely to encounter a need for each of them, so it is useful to have some idea about their characteristics.

The gas burner is the traditional means of heating and has the unique advantage of providing a wide range of heating that can be altered quickly if the situation requires it. Gas burners have the obvious disadvantage of an open

[4] Such as the Chemical Tables in the *Handbook of Chemistry and Physics*, CRC Press, Inc, Boca Raton, FL.

flame, which requires constant awareness of any flammable vapors nearby. If you do use a gas burner in an organic laboratory, remember the cardinal rule: *flames and organic vapors cause fires and explosions!*.

Steam baths are excellent for heating or distilling when the required temperature is below 100°. Vessels placed on top of the bath will usually reach only a temperature of about 90° because of heat losses from the sides of the vessel. A general practice to be observed in using a steam bath is to avoid excess steam. Not only are clouds of steam unnecessary and annoying, but condensation of escaping steam may introduce moisture into the reaction vessel, which could ruin some reactions.

For temperatures above 90°, an oil bath works well. Because of the high heat capacity of the oil, an oil bath provides a very steady source of heat. For the same reason the temperature cannot be lowered quickly if an emergency requires it, the bath temperature should be raised cautiously. A solution to this problem is to mount the bath on a "Lab-Jack," which allows rapid adjustment of height. A less expensive but more hazardous solution is to use wooden blocks cut from "2 × 4's" or plywood. A serious problem with oil baths is that they are easily spilled and the hot oil can cause severe burns. A related problem is that if water is spilled into a hot oil bath, the agitation caused by the boiling of the water can easily spatter the oil. If a flame is used to heat the bath, *particular care* must be taken not to ignite the hot oil vapors. Because of these major safety concerns, it is not advisable for beginning students to use oil baths, in spite of their other advantages. In academic research laboratories, where oil baths are widely used, they have frequently been heated by a coil of resistance wire placed directly in the bath. This method, because of the electrical shock hazard, has been banned by OSHA.[5] Use of an electrically insulated immersion heater is legal and practical. However, with any electrically heated (or driven) device there is the problem of the cords to the voltage controller and the wall socket. If not arranged neatly, they can be snagged and cause accidents.

For very high temperatures, a sand bath has many desirable features. In addition to being inexpensive and a steady source of heat, sand, unlike oil, is not subject to fire hazards and does not decompose on prolonged heating. Its main limitations are poor heat transfer and slow response to altered temperature settings.

Electrical hot plates, particularly those with built-in magnetic stirring motors, are widely used in research laboratories. Their use calls for caution because solvent vapors can ignite if they come in contact with the hot surface. A hidden danger for the unwary is that many hot plates are built with exposed "on-off" temperature regulators that produce sparks as the contacts open and close. Unless the contacts are properly sealed, the sparks can ignite combustible vapors.

---

[5] The Occupational Safety and Health Administration is a federal body charged with the responsibility of setting safety and health standards in the workplace and has the authority to enforce its regulations.

Another widely used heating device is the Glas-Col heating mantle, constructed by weaving nichrome wire heating elements into fiberglass cloth to form a hemispherical jacket that fits snugly against the lower part of the flask. Loose-fitting mantles tend to overheat, which is dangerous, so that a different mantle *must* be used for each size of flask. Other shapes are available for special types of apparatus. Current OSHA regulations require that the outside case be grounded.

Closely related to the Glas-Col mantle is the Thermowell[6] heating mantle, in which the heating element is imbedded in a rigid hemispherical shell fabricated from a refractory ceramic material. In concept, the Thermowell unit is intermediate between a hot plate and a Glas-Col mantle. Since a snug fit is not required, one unit can be used with several sizes and shapes of flasks. The Thermowell units are excellent heating devices for beginning students.

Some electrically heated devices contain their own current controllers, but more frequently a separate external controller is required. Usually, a Variac or Powerstat is used; these are autotransformers capable of converting 110-V alternating current into any voltage in the range of 0–130 V by the twist of a dial on the top of the case. Autotransformers, when kept clean, do not spark and do not pose a fire hazard.

Another device that is widely used is a proportional controller in which the full 110 V is applied but only for short time intervals. A setting knob on the device controls the time ratio of "off" to "on." Inside the box this ratio can be controlled either by a thermal on-off switch or by a solid-state device. If the former is used, the box *must* be sealed against organic vapors.

1.5 Laboratory Notes[7]

It is essential to have a suitable notebook in which to record directly the observations made during, and to assemble information that will aid in the performance of, experiments. For this purpose, obtain a stiff-covered bound notebook, about 8″ × 10″, preferably with cross-ruled paper (to simplify the preparation of tables of physical constants required in the later experiments). It is convenient to use a notebook with numbered pages and tearout carbon-copy duplicate pages. The use of spiral or loose-leaf notebooks for laboratory records is not satisfactory, and the recording of observations on loose sheets or scraps of paper is not permitted. Notebook entries should be made in ink, and any necessary corrections should be made by adding notes rather than erasing.

---

[6] Thermowell units are available only from Laboratory Craftsmen, P.O. Box 148, Beloit, WI 53511.

[7] The following specific directions for the preparation of notebooks and the general laboratory procedure are based on those used in the organic laboratory courses at Cornell University. For the particular conditions that pertain in other laboratories, your instructor may alter these directions or substitute others.

Before you come to the laboratory, the following steps should be carried out for the exercises on separation and purfication. Other directions are given in Chapter 12 for the preparative experiments. The schedule of exercises will be announced beforehand, so that you will have an opportunity to prepare your notebook and be ready to start work at the beginning of the laboratory period.

1. Read the descriptive pages concerning the laboratory operations to be carried out (these are found immediately preceding each experiment). In your notebook, write a title and general statement of the process to be studied.
2. Read the laboratory directions for the entire experiment, note particularly the ▮▮▶ *CAUTIONS* and other warnings about handling materials, and think about the reasons for the procedure to be followed. In your notebook jot down any points that require special observation or reminders of specific details.
3. For each chemical to be used look up and record its properties[8] and note any chemical or biological hazards it presents.
4. Write the names and formulas of the compounds to be used and, where chemical tests are to be made, write equations for the reactions.

Careful planning of laboratory work is essential. Effective use of laboratory time requires that you know in advance just what you are going to do in the laboratory. Instead of watching idly while a liquid is being heated for an hour or more, you can use periods when full attention is not required to conduct another experiment, to clean apparatus, or to prepare for subsequent work.

When you come to the laboratory, proceed as follows.

1. Arrange the apparatus for the experiment and secure the approval of the laboratory instructor for the setup.
2. Perform the experiment according to the laboratory directions and record observations directly in your notebook. When the exercise has been completed, dismantle the setup and immediately clean the glassware and apparatus, as described in Section 1.2.

After you have completed the laboratory and gone home,

1. Spend a few minutes to review what you have accomplished in the laboratory and reflect on how you could prepare better for the next laboratory period.

---

[8] An excellent single-volume source of such information is M. Windholz, Ed., *The Merck Index*, 10th ed. (Rahway, NJ: Merck & Co., Inc., 1983). The material data safety sheets described earlier in Section 1.1 under "Toxic Chemicals and Mutagens" are another excellent source of information.

2. Fill out the report forms provided by the instructor.
3. Write answers to the questions assigned by the instructor. Make complete statements in answering the questions.

1.6 Exercises on Basic Laboratory Operations

The first order of business is to check the equipment in your desk. After you have done so, the remainder of the period will be spent carrying out several very basic laboratory operations. These may seem trivial, but the point is to ensure that in the subsequent experiments you can focus on the new material.

The directions given here assume that you will be using an electrically heated Thermowell mantle and controller. If you are not, small adjustments in the procedure will be required.

The main difficulty in using a heating mantle is its temperature lag. These devices require time to heat and cool; there is a natural (impatient) tendency to set the controller at too high a setting only to find that the flask is eventually overheated. It is a good practice to arrange the apparatus so that the heating mantle can be easily lowered or even removed if the situation calls for it.

(A) Heating Rate

Support the heating mantle on a ring about 3–5″ above the base of a ring stand. Place a 100-mL round-bottomed flask in the mantle and clamp the flask in place with a buret clamp. In the flask place 50 mL of water. Next, support a thermometer with a three-finger clamp in such a position that the bulb is immersed in the water but the bottom of the bulb is about 5 mm (1/4″) above the bottom of the flask.

Turn the controller on and set it to a midrange setting (remember to record the time and setting). Every 5 min for the next half hour record the time and temperature. If the water boils, note the time and stop the heating. From the data draw a graph of the temperature versus time.

If time permits, repeat this procedure for another setting of the controller.

These temperature-vs.-time curves will be useful later in other experiments when you have to make an initial guess at the controller setting needed to maintain a particular temperature.

(B) Calibration of Pasteur Pipet Drop Size

Determine the weight of an empty 10-mL graduated cylinder. Attach a rubber bulb to a Pasteur pipet and draw up water into the pipet (but *not* into the bulb) from a beaker half filled with water. Slowly squeeze the bulb so that the water is expelled a drop at a time. In this fashion add exactly 25 drops of water to the graduated cylinder and determine the weight of cylinder with the added water. From the difference in the two weights calculate the average weight of one drop, and from the density of water calculate the volume of one drop.

(C) Identification of an Unknown from Its Density

Your instructor will give you a liquid that is one of the following three compounds.

Compound	Density at 25°
Water	1.00
Acetone	0.786
Dichloromethane	1.325

Weigh an empty, dry 10-mL graduated cylinder, fill it with your unknown as closely as possible to the 10-mL line (recall that the measuring point is the bottom of the meniscus), and reweigh the filled cylinder. From the difference in weight calculate the density of the liquid and identify your unknown.

I

Separation and Purification of Organic Compounds

2 Simple and Fractional Distillation

Since organic compounds usually do not occur in a pure condition in nature and are accompanied by impurities when synthesized, the purification of materials forms an important part of organic laboratory work. There are four frequently used separation procedures: distillation, chromatography, crystallization, and extraction. Sublimation is used occasionally, and various special techniques such as electrophoresis and zone-refining are available for advanced work. The process used in any particular case depends upon the characteristics of the substance to be purified and the impurities to be removed.

The basic techniques are so easily learned that it is tempting to use them in a purely mechanical, rote fashion. However, you should aim for a higher standard. In any new situation, to select the most appropriate process and to employ it effectively, you must understand the principles involved as well as the correct methods of manipulation. This book starts with distillation because the principle of vapor pressure on which it depends is familiar to you from your freshman chemistry course. The immediate goal here is to understand how vapor pressures of mixtures depend on the structures of the components and how, in turn, the vapor pressure controls the distillation behavior and separation efficiency.

2.1 Principles of Distillation

Boiling. In a liquid the molecules are in constant motion and have a tendency to escape from the surface and become gaseous molecules, even at temperatures far below the boiling point. When a liquid is placed in an enclosed space, the pressure exerted by the gaseous molecules increases until it reaches the

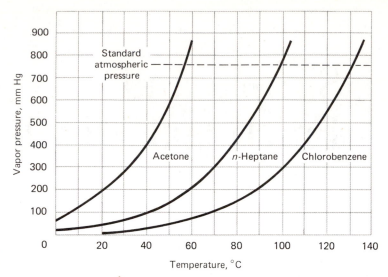

FIGURE 2.1
Temperature Variation
of Vapor Pressure

equilibrium value for that particular temperature. The equilibrium pressure is known as the vapor pressure and is a constant characteristic of the material at a specific temperature. Although vapor pressures vary widely with different materials, vapor pressure always increases as the temperature increases (Figure 2.1). The vapor pressure is commonly expressed as the height, in millimeters, of a mercury column that produces an equivalent pressure (mm Hg—also now called torr after Torricelli, the famous Italian scientist who discovered the effects of vacuum). The addition of soluble substances to a pure compound alters the measured vapor pressure.

In ordinary glass flasks, there are microscopic pockets of air trapped in the pores and crevices of the walls. With a liquid in the container the pockets are filled with vapor of the liquid at its equilibrium vapor pressure. When the temperature of the liquid is raised, the vapor remains compressed until the vapor pressure exceeds the applied pressure (the pressure at the liquid surface plus the hydrostatic pressure), whereupon the trapped vapor rapidly expands to form bubbles that rise to the surface and expel their vapor. The resulting agitation (boiling) churns more air bubbles into the liquid where they continue the process after receiving new charges of vapor. Liquids heated in containers that have been degassed do not boil, although they vaporize explosively if heated to a sufficiently high temperature. To avoid the hazards associated with sudden, irregular boiling (bumping), a dependable source of bubbles should always be introduced into a flask before its contents are heated to boiling. When a liquid is boiled at atmospheric pressure, the bubble source is customarily a boiling chip (see Section 2.3); with vacuum distillations boiling chips do not work as reliably and other sources are usually used (see Chapter 3).

Boiling Point and Boiling Temperature. The boiling point of a liquid is defined as the temperature at which its vapor pressure equals the atmospheric pressure. By convention, boiling points reported in the scientific literature are at a pressure of 1 atm (the so-called "normal" boiling point) unless otherwise specified. The boiling temperature is the actual observed temperature when boiling occurs and is generally a few hundredths to a few tenths of a degree above the true boiling point because of experimental difficulties involved in the measurements.

Distillation of Pure Compounds. Distillation consists of boiling a liquid and condensing the vapor in such a manner that the condensate (distillate) is collected in a separate container. A simple apparatus assembly for this operation is shown in Figure 2.10.

When a pure substance is distilled at constant pressure, the temperature of the distilling vapor will remain constant throughout the distillation provided that sufficient heat is supplied to ensure a uniform rate of distillation and superheating is avoided. In actual practice these ideal conditions are not obtained; drafts in the laboratory can cause momentary condensation of vapor before they reach the thermometer, which lowers the temperature sensed by the thermometer. On the other hand, the distilling vapors after they leave the surface of the liquid may be heated above the liquid's boiling point (superheating), which increases the temperature sensed by the thermometer. Because of these two contrary effects, a distillation range of 1–2° actually represents an essentially constant boiling point. With somewhat more refined apparatus and technique a distillation range of 0.1° can be observed for a pure compound.

The temperature reading of a thermometer *in the distilling vapor* represents the boiling point of that particular portion of the distillate. This temperature will be the same as the boiling point of the liquid in the distilling flask only if the distilling vapor and the boiling liquid are identical in composition. Since a pure liquid fulfills this condition, a constant thermometer reading is sometimes used as a criterion of purity of a liquid. It should be noted, however, that certain mixtures (such as azeotropes—page 22) also give constant thermometer readings. Occasionally two liquids have such similar boiling points that no appreciable change in thermometer reading will be observed when a mixture of them is distilled.

Ideal Solutions. The pressure and composition of vapor above an ideal mixture of liquids at a given temperature can be calculated if the composition of the mixture and the vapor pressures of the pure components are known. The total pressure is the sum of the partial vapor pressures of all components. The partial pressure of each component is given by Raoult's law.

$$p_A = p_A^\circ X_A \qquad (2.1)$$

where p_A (the partial pressure of A) is the vapor pressure of A above the mixture, p_A° is the vapor pressure of pure A, and X_A is the mole fraction

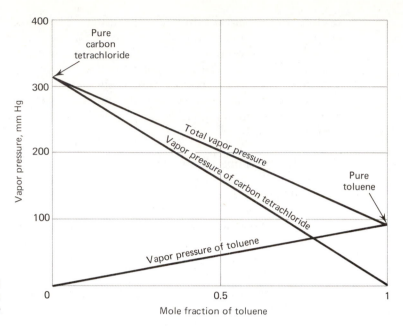

FIGURE 2.2
Graphical Application
of Raoult's Law

of component A in the mixture. Because there is a fixed number of molecules in a mole, Raoult's law states in molecular terms that the vapor pressure of A above a solution is proportional to the mole fraction of the molecules of A in the liquid. Application of Raoult's law to the two-component mixture of carbon tetrachloride and toluene is illustrated graphically in Figure 2.2.

The composition of the vapor, with respect to each component, can be calculated from Dalton's law.

$$Y_A = \frac{p_A}{\text{total vapor pressure}} = \frac{p_A}{p_A + p_B + \cdots} \tag{2.2}$$

where Y_A is the mole fraction of component A in the vapor. Dalton's law and Raoult's law together show that for an ideal mixture at any temperature the most volatile component has a greater mole fraction in the vapor than in the solution. In terms of the previously defined symbols, if A is the most volatile component of the mixture, Y_A is greater than X_A.

The boiling point of a mixture is defined as that temperature at which the *total* vapor pressure equals the pressure above (or on) the solution. From Raoult's law (also see Figure 2.1) it is apparent that the total vapor pressure of an ideal mixture is intermediate between the vapor pressures of the pure components. This means that the boiling point also will be intermediate between the boiling points of the pure substances. The general dependence of boiling point on composition of ideal binary mixtures resembles that depicted in Figure 2.3 of the specific system of carbon tetrachloride and toluene. The boiling point of any particular mixture is

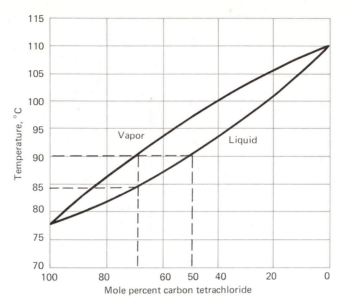

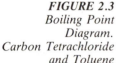

FIGURE 2.3
Boiling Point
Diagram.
Carbon Tetrachloride
and Toluene

obtained by erecting a vertical line from the horizontal composition axis until it intersects the *liquid* curve. For example, from Figure 2.3 it will be found that a 50-mole-percent (0.5 mole fraction carbon tetrachloride) mixture of carbon tetrachloride in toluene boils at 90°. A thermometer placed *in the boiling* liquid would read 90°, but a thermometer placed in the distillation head, as shown in Figure 2.10, would read lower, since it is in contact with the condensing vapor, which is richer in the lower-boiling component.

The composition of the vapor in equilibrium with any particular liquid composition is obtained from Figure 2.3 by projecting a horizontal line from the vertical intersection of the *liquid* curve over to the *vapor* curve and from that intersection back vertically to the composition axis. The vapor above a 50-mole-percent solution of carbon tetrachloride in toluene contains 71 mole percent of carbon tetrachloride. Figure 2.3 demonstrates graphically the previous conclusion that the vapor is richer in the lower-boiling, more volatile component. The reading of a thermometer placed in the distillation head would be the boiling point of the 71-mole-percent mixture, which can be seen from Figure 2.3 to be close to 84°.

The boiling point and vapor composition calculated in this manner apply only to the initial state of distillation. Because of the higher concentration of carbon tetrachloride in the vapor compared to the liquid remaining in the boiler, the composition of the liquid *gradually* shifts toward pure toluene as the distillation proceeds; the boiling point, reflecting this composition change, climbs gradually also. The actual rate of change of boiling point depends on how rapidly the mixture is distilled, but a typical set of observations would resemble those of Figure 2.4, obtained with a 50:50 (by volume) carbon tetrachloride–toluene mixture. A 50:50 mixture by volume corresponds to 0.52 mole fraction carbon

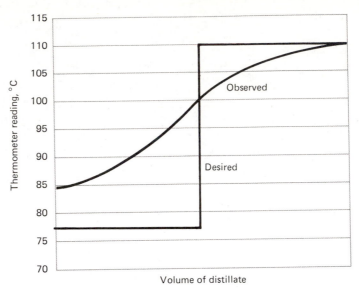

FIGURE 2.4
Distillation Curve for
a 50:50 Carbon
Tetrachloride–Toluene
Mixture

tetrachloride and can be found from Figure 2.3 to have a boiling point of about 88°
and an initial distillation temperature of about 84°. The technique of fractional
distillation (discussed in Section 2.2) is a method for more nearly approaching this
perfect separation.

Nonideal Solutions and Azeotropes. Many actual solutions depart widely
from Raoult's ideal law. An expression for vapor pressure that encompasses
both ideal and nonideal behavior is

$$p_A = f_A X_A \tag{2.3}$$

where, as before, p_A is the vapor pressure of A above the mixture, X_A is the
mole fraction of component A in the mixture, and f_A is the effective vapor
pressure of A. If the solution is ideal, $f_A = p_A$. If the solution shows negative
deviations from Raoult's law, that is, if $f_A < p_A$, component A behaves in
solution as though the vapor pressure of pure A were less than it actually is. The
analogous description applies for positive deviation from Raoult's law for which
$f_A > p_A$.

The methanol–water system is typical of those that show positive deviations
from Raoult's law (Figure 2.5). The boiling point composition curve for
methanol–water mixtures shown in Figure 2.6 also reflects the nonideal behavior
by its distorted shape (compare with Figure 2.3). The origin of the positive
deviations for methanol–water is the disruption of hydrogen bonding between
the hydroxyl groups, which leads to enhanced escaping tendencies for the two
components. In contrast, there are some binary mixtures in which the two

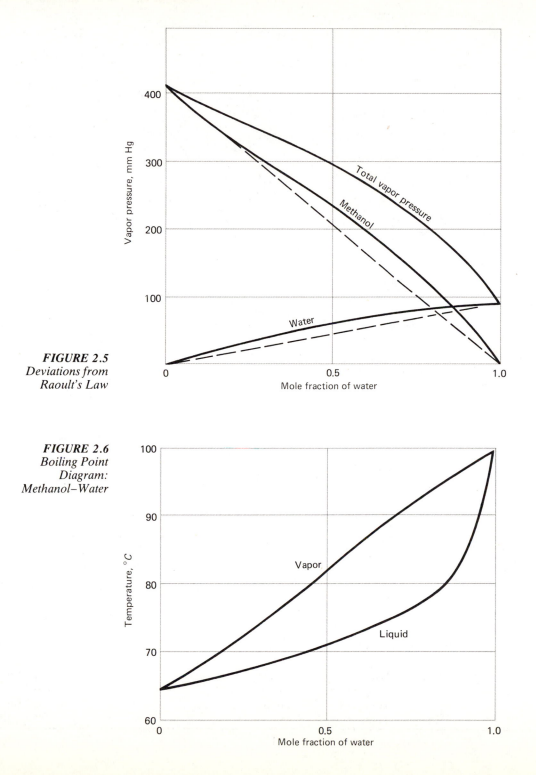

FIGURE 2.5
Deviations from Raoult's Law

FIGURE 2.6
Boiling Point Diagram: Methanol–Water

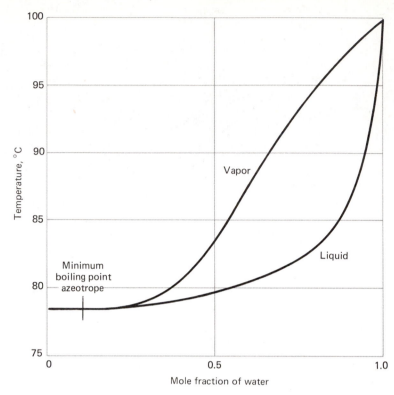

FIGURE 2.7
Boiling Point
Diagram:
Ethyl Alcohol–Water

components attract each other particularly strongly and cause the vapor pressure to be lower than ideal (negative deviations from Raoult's law). Other mixtures show positive deviations because the molecules of one component are attracted to each other more strongly than they are to molecules of the other component.

TABLE 2.1 *Binary Azeotropic Mixtures*

Component A		Component B		Azeotropic mixture		
Substance	Boiling point, °C	Substance	Boiling point, °C	Percent of A (weight)	Percent of B (weight)	Boiling point, °C
Acetone	56.4	Chloroform	61.2	20	80	64.7 (max.)
Nitric acid	86.0	Water	100.0	68	32	120.5 (max.)
Formic acid	100.7	Water	100.0	77.5	22.5	107.3 (max.)
n-Propyl alcohol	97.2	Water	100.0	71.7	28.3	87.7 (min.)
t-Butyl alcohol	82.5	Water	100.0	88.2	11.8	79.9 (min.)
Ethanol	78.3	Water	100.0	95.6	4.4	78.1 (min.)
Ethanol	78.3	Chloroform	61.2	7	93	59.0 (min.)
Ethanol	78.3	Toluene	110.6	68	32	76.7 (min.)
Acetic acid	118.5	Toluene	110.6	28	72	105.4 (min.)

Frequently, the deviations from ideality are so extreme that boiling point-composition diagrams have a maximum or a minimum (Figure 2.7). With such mixtures the vapor curve and the liquid curve coincide at the maximum (or minimum). If the mixture has a composition corresponding to this coincidence point (an *azeotropic mixture*), it will behave like a pure liquid and show a constant boiling point. The components of an azeotropic mixture cannot be separated by ordinary distillation processes because the vapor in equilibrium with the liquid has the same composition as the liquid itself. Table 2.1 gives the composition and boiling point of several examples of binary azeotropic mixtures. Azeotropic mixtures containing three components (ternary systems) are also encountered; for example, benzene–water–ethanol or ethanol–water–ethyl acetate give minimum-boiling-point azeotropic mixtures.

The effect of even more extreme deviations from Raoult's law will be considered in connection with steam distillation (Section 4.1).

2.2 Fractional Distillation

The common use of the term *fractional distillation* refers to a distillation operation where a *fractionating column* has been inserted between the boiler and the vapor takeoff to the condenser (see Figure 2.13). The effect of this column, when properly operated, is to give in a single distillation a separation equivalent to several successive simple distillations. This saves considerable time and makes the selection and proper operation of fractionating columns an important subject for chemists.

Fractionating Columns. The easiest way to understand the principles by which fractionating columns give their superior separations is to consider first a rather special type of column known as a *bubble plate column*. The essential features of a bubble plate column, illustrated in Figure 2.8, consist of (1) a series of horizontal plates, A, which support a layer of distillate; (2) capped risers, B, through which the distilling vapors ascend; and (3) overflow pipes, C, which return any excess distillate to the next lower plate. At the beginning of a distillation, the vapors coming up from the boiler pass through the first riser and are deflected downward by the cap onto the first plate, where they are condensed. As simple vaporization and condensation continue, the rising vapors are forced to bubble through the liquid on the plate. The liquid level rises to the top of the overflow tube and then flows downward to the boiler. The liquid on the first plate corresponds to the first fraction in a simple distillation—it is enriched in the lower-boiling component. It follows that the temperature of the vapor bubbling through the liquid is above the boiling point of the liquid on the plate; through heat exchange the liquid is brought to its boiling point and its vapor rises to the second plate where the same processes are repeated. As the distillation continues, each plate becomes filled with a layer of liquid whose composition is that of the vapor rising from the next

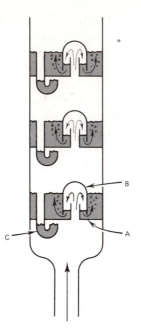

FIGURE 2.8
Bubble Plate Column

lower plate. Under ideal circumstances, each plate achieves an increment of separation equivalent to one simple distillation.

The overflow tubes serve a more important function than just acting as returns for excess condensate. Since the vapor leaving any plate is richer in the lower-boiling component than the vapor entering the plate, the higher-boiling materials tend to accumulate on the plate. The overflow returns this higher-boiling material to the lower plate, so that an equilibrium balance of low-boiling to high-boiling components is maintained. In effect, vapor and condensate are passing in opposite directions through the column; the more volatile component ascends the column in the vapor stream, while the less volatile components descend. The counterflow is essential for effective separation in a fractionating column.

The separation process can be understood more clearly by reference to a liquid–vapor composition diagram such as that shown in Figure 2.9 for carbon tetrachloride (bp 77°) and toluene (bp 111°). A liquid mixture containing 50 mole percent carbon tetrachloride (point A on the liquid line) is in equilibrium with vapor containing 71 mole percent carbon tetrachloride (point A' on the vapor curve). If liquid with composition A is partially vaporized and the vapor with composition A' condensed completely on the first bubble plate, the condensate is represented by B (on the liquid line). Repetition of the vaporization–condensation process with liquid B yields a new distillate, C, containing 85% carbon tetrachloride, which condenses on the second bubble plate. Each successive bubble plate achieves an additional increment of separation.

Bubble plate columns have the drawback of requiring large samples for

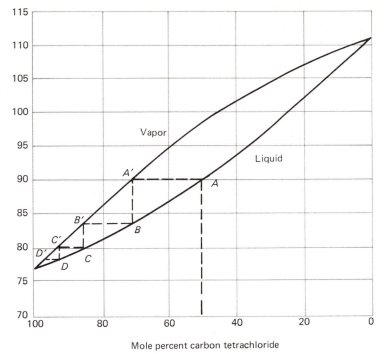

FIGURE 2.9
*Boiling Point
Diagram:
Carbon Tetrachloride–
Toluene*

effective operation, and a substantial portion of material is withheld on the plates (*holdup*). To overcome these disadvantages, small-scale laboratory fractionations are usually done with cylindrical columns packed with materials having large surface area (glass beads or helices, small sections of twisted metal, Carborundum chips, and the like). The principles of operation of *packed columns* are quite similar to those of the bubble plate column. The layers of packing material, like the bubble plates, serve as support for films of condensate; vapor passing through the layers is enriched in the lower-boiling component, and the higher-boiling components move downward to lower layers. The scrubbing action of the packing material effects the counterflow of vapor and condensate that is essential for fractionating efficiency.

Relative Efficiency of Fractionating Columns. Since column packings differ widely in efficiency, it is desirable to have a means of comparing their effectiveness for separating mixtures. The enrichment factor, α, which relates the volatilities of two components of a mixture, is the ratio of their effective volatilities, which, from equations (2.2) and (2.3), can be expressed as the quotient of the ratio of the mole fractions of the components in the vapor to the ratio of their mole fractions in the liquid.

$$\alpha = \frac{f_1}{f_2} = \frac{Y_1/Y_2 \text{ (vapor)}}{X_1/X_2 \text{ (liquid)}} \tag{2.4}$$

A theoretical plate is the unit of separation corresponding to the composition ratio, α, that exists at equilibrium between a liquid mixture and its vapor. This concept may be illustrated by considering a 50:50-mole-percent mixture of carbon tetrachloride and toluene. The vapor in equilibrium with the liquid (bp 90°) contains 71 mole percent carbon tetrachloride and 29 mole percent toluene. This amount of enrichment corresponds to one theoretical plate.

$$\alpha = \frac{71/29}{50/50} = \sim 2.5$$

The length of packed column required to obtain this degree of separation in the mixture is known as the height equivalent to a theoretical plate (usually abbreviated HETP). The smaller the value of the HETP, the more efficient the column is. Although the exact HETP of any given packing depends on operating factors (diameter of the column, density of packing, rate of distillation, etc.), it is useful to have rough estimates of relative values. Table 2.2 records representative values of HETP for several packings as measured under normal working conditions using student apparatus to separate a benzene–toluene mixture. Also shown in Table 2.2 are representative values of the column holdup per plate. Both the HETP and the holdup values will vary with the manner of packing and subsequent treatment of the column.

In addition to packed columns, special columns are available that achieve mixing of the ascending vapor and the descending condensate by their special construction. One of the simplest, least expensive, and most widely used is the Vigreux column illustrated in Figure 2.14. It is essentially an empty tube with many finger-like indentations that point downward at a 45° angle. The rising vapors condense on the fingers, and any excess liquid drips down to lower parts of the column. The film of condensate on each finger equilibrates with the rising vapor. Under normal working conditions the Vigreux column has a relatively low efficiency (high HETP of 10 cm), but its low resistance to vapor flow permits a large throughput (volume of distillate per unit of time) that makes the column well suited to distillation of bulk solvents. Because of its small surface area, the column has a low holdup and is sometimes used for preliminary purification of small samples.

TABLE 2.2
HETP for Common
Packing Materials[a]

Packing	HETP, cm	Hold/plate, g
Carborundum chips	6	1.2
Glass beads	8–9	0.9
Glass helices	4–5	0.6
Metal helices	8–9	0.9
Metal sponge	30	1.6

[a] Obtained with 25-cm packed column using a benzene–toluene mixture.

The spiral wire column is also widely used. It consists of a wire wound spirally on a glass rod that is held concentrically within an outer glass tube. Spiral wire columns are slightly more efficient than columns packed with glass beads (HETP of 2 cm) and have about half the holdup of a packed column capable of the same throughput. Their limitation is throughput, which is essentially fixed (0.5 mL/min maximum), whereas packed columns can be scaled up as needed. Because of their simple construction, spiral wire columns are generally built in the laboratory rather than purchased.

A unique column deserving special mention is the "Auto Annular Still."[1] This unit, resembling a spiral wire column, contains an annular Teflon helix wrapped around a Teflon rod. A motor spins the Teflon helix at high speed, so that the ascending vapors are effectively mixed with the descending condensate. It is claimed that the 60-cm column (115 cm with accessories attached) produces more than 150 theoretical plates with a holdup of less than 0.5 mL and a throughput of 15–60 mL/hr. This remarkable unit unfortunately costs several thousand dollars.

Separation Efficiency. The total number of theoretical plates, n, present in a column is equal to the height of the packed portion of the column divided by the HETP of the packing material. The composition of the vapor at the top of the column, (Y_1/Y_2), is related to the composition in the boiler, (X_1/X_2), in the following way.

$$\frac{Y_1}{Y_2} = \alpha^{n+1} \left(\frac{X_1}{X_2}\right) \qquad (2.5)$$

The exponent of α is $n + 1$, rather than n, because in vaporizing the mixture in the boiler an additional enrichment factor is introduced. Although this equation has theoretical significance, it is more practical to have an expression for the number of theoretical plates required to separate a given mixture. An approximate expression (equation 2.6) has been derived for fractional distillation of 50 : 50 mixtures, which gives the number of theoretical plates required to achieve a separation such that the first 40 percent of the material distilled will have an average purity of 95% in the lower-boiling component. Equation (2.6) shows that as the relative volatility, α, approaches unity, the number of theoretical plates required to achieve 95% purity increases sharply.

$$\begin{array}{l} \text{Number of theoretical plates required} \\ \text{to achieve a standard separation} \end{array} = \frac{1.53}{\log_{10} \alpha} \qquad (2.6)$$

This relationship becomes still more useful (but also more approximate) if one substitutes for $\log_{10} \alpha$ an expression involving the difference in boiling points of

[1] This column, introduced by the Nester Faust Mfg. Corp., is now manufactured and distributed by the Perkin-Elmer Corp., Norwalk, CT 06856.

the two components. Equation (2.7) gives results for ideal mixtures being fractionally distilled under perfect conditions.[2] Under practical conditions, discussed in the next section, the number of plates required can double. It is clear that tall, high-platage fractionating columns are required for clean separation of materials boiling a few degrees apart. When the required number of theoretical plates is unavailable for practical reasons, it is necessary to collect a smaller portion of the low-boiling distillate.

$$\text{Number of theoretical plates required for standard separation} = \frac{120}{T_2 - T_1} \qquad (2.7)$$

Reflux Ratio and Holdup. Equations (2.5) and (2.6) were derived for an ideal fractional distillation where there is equilibrium between the rising and descending counterflowing streams of materials. For this equilibrium to be attained, it is essential that vapor reaching the top of the column be condensed and the liquid returned to the column (*reflux*). If a large portion of the vapor reaching the top of the column is removed as a distillate (*takeoff*), the equilibrium is seriously disturbed and much lower separation efficiency results. The extreme modes of operation are known as *total reflux* and *total takeoff*. Since the first mode yields no distillate and the second gives distillate of much lower purity than is possible with the column, in practice some intermediate ratio of takeoff to reflux is employed. The best practical compromise is to adjust the reflux ratio so that it equals the number of theoretical plates of the column. Higher rates of collecting distillate (lower reflux ratios) give poorer separations; slower rates are overly time-consuming and do not provide significantly better separations.

Another factor that seriously affects separation efficiency is the total amount of liquid and vapor in the column at any instant (holdup). A great drop in separation efficiency occurs if the holdup is more than about 10% of the amount of sample to be distilled.

Under practical conditions of partial takeoff and some holdup, the number of plates required to achieve a given separation can be twice as many as predicted by equation (2.7). If the number of plates available is insufficient, the distillation rate must be slowed to more nearly approach ideal conditions.

2.3 Laboratory Practice

Apparatus for Simple Distillation. A simple distillation apparatus suitable for distillation of samples in the range of about 1–50 mL in volume is shown in Figure 2.10. The key part of this assembly is the distillation "head" shown in

---

[2] Equations (2.6) and (2.7) differ from those given in the literature by not including the effect of partial equilibration when the reflux ratio is small. The present equations were derived by assuming a constant and an average boiling point of 100°. They give the theoretical minimum required number of plates without regard to practical factors. Liquid mixtures showing deviations from ideality may require fewer plates (for positive deviations) or more plates (for negative deviations).

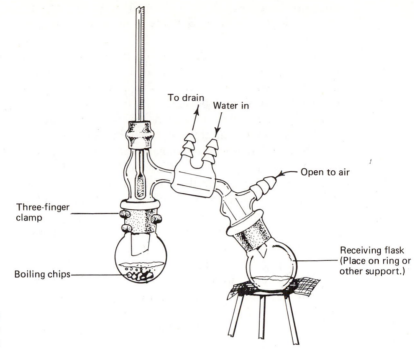

FIGURE 2.10
Apparatus for Simple Distillation

the center of the figure. This "head" is a single glass unit that contains a water-cooled condenser, an opening for a thermometer, and ground joints at both ends for attaching a boiling flask and receiver. A thermometer is held in place in the vertical arm of the distillation "head" by a special rubber connector[3] at a height adjusted so that the top of the mercury bulb is *even* with the bottom of the opening of the side arm.

The distilled liquid is collected in a clean, dry receiver, commonly a round-bottomed flask with its ground-glass joint mated to the lower joint of the distillation "head." It is permissible to use an Erlenmeyer flask or a graduated cylinder as a receiver if vapor losses and fire hazards are minimized by inserting the lower end of the outlet tube well into the mouth of the receiver. *A distilling assembly must have an opening to the atmosphere* to avoid development of a dangerously high pressure within the system when heat is applied. When a mated round-bottomed flask is used as the receiver, the side arm on the distillation "head" becomes the opening, and this arm *must not be sealed*.

If you are using a heating mantle as a heat source, it should be supported on an iron ring to avoid overheating the bench top. This arrangement also allows you to interrupt the distillation easily without having to handle hot glassware. With the Thermowell ceramic heating mantle, the size of the distillation flask

---

[3] The thermometer can be attached to the vertical arm by means of a short length of soft rubber tubing, but this procedure is quite hazardous because of the large difference between the diameters of the thermometer and the arm.

does not have to match the size of the mantle. However, soft mantles must fit closely; otherwise they are likely to burn out. An alternative, best reserved for advanced workers, is to heat the flask in an oil bath. Oil baths have the advantage of being very even heat sources, but they can cause severe burns if the hot oil spills on the skin. If an electrically heated bath is used, the cords for the voltage controller should be arranged neatly at the back of the bench and the controller should be placed so it can be adjusted conveniently.

The distilling flask should be of such size that the material to be distilled occupies between one third and two thirds of the bulb. If the bulb is more than two-thirds filled, there is danger that some of the liquid may splash into the distillate. If the bulb is less than one-third filled, there will be an unnecessarily large loss resulting from the relatively large volume of vapor required to fill the flask. This loss is particularly serious with compounds of high molecular weight. The only exception to the one-third to two-thirds rule is with liquids, such as cyclohexanol, that foam badly on distillation. These liquids require a much larger distillation flask to contain the foam.

Whether a pure compound or a mixture is distilled, a small portion of liquid will always be left in the flask upon cooling. The flask containing the material to be distilled should never be heated to dryness because the flask might crack.

Apparatus for Micro Distillations. With less than 1 mL of a liquid to distill, the apparatus described in the previous section gives large losses because of holdup on the glass surfaces. An alternative for small volumes is the Hickman still shown in Figure 2.11. The liquid is placed in the bottom bulb and on *slow* heating gradually distills and collects in the well part way up the still, where it is removed with a Pasteur pipet. The main disadvantage of the Hickman still is that its small size does not allow inclusion of a thermometer and therefore the distillation temperature is unknown. The boiling points of the distillate fractions have to be determined subsequently by a micro-boiling-point procedure such as that described in Chapter 9. Nevertheless, the Hickman still is very useful where only one fraction is to be collected.

A substitute for a Hickman still can be assembled from a boiling flask and a Claisen adapter as shown in Figure 2.12. The apparatus is tilted during the distillation so that the bend in the arm of the Claisen adapter forms a well to collect the distillate. A thermometer can be inserted in the vapor stream to measure the condensation temperature of the distillate.

Apparatus for Fractional Distillation. Figure 2.13 shows a fractional distillation apparatus that uses for the distillation column a cooling condenser packed with glass beads or other materials. The column packing can be held in place with a wad of steel wool inserted loosely in the bottom. Alternatively, one can use a Vigreux column (Figure 2.14) in place of the packed column.

Successful fractional distillation demands a column with an adequate number of plates. As simplified as it is, equation (2.7) is a useful guide to the minimum required number. Estimation of the desired number of plates requires knowledge of the composition and boiling behavior of the mixture to be separated.

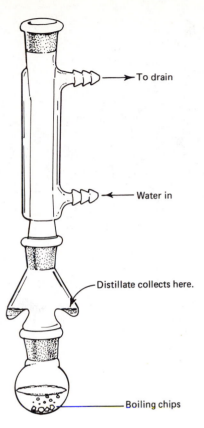

FIGURE 2.11
Hickman Still for
Small Samples

→ To drain

← Water in

Distillate collects here.

Boiling chips

FIGURE 2.12
Substitute for
Hickman Still

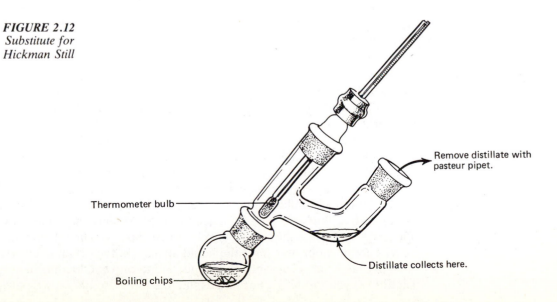

Thermometer bulb

Boiling chips

Remove distillate with
pasteur pipet.

Distillate collects here.

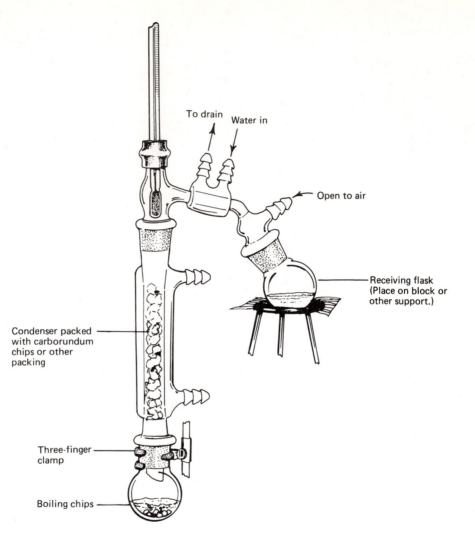

FIGURE 2.13
Apparatus for
Fractional Distillation
(with condenser for
a column)

When this information is lacking, it is desirable to run a preliminary simple distillation and to plot a graph showing the actual relation of distillation temperature to volume of distillate.

Perhaps no available fractionating column has an adequate number of plates. In this case it will be necessary to separate the mixture into a number of fractions of progressively higher boiling point and to refractionate each one separately in a systematic way until an acceptable separation of the components is achieved.

Increasing the reflux ratio or the length of a column improves the efficiency of separation, but care must be taken to avoid flooding the column with liquid. Flooding diminishes the contact area between vapor and liquid, and the pressure of ascending vapor may force the liquid upward in the column. To obtain good heat exchange between vapor and liquid, and to prevent flooding, a long column should be well insulated. For liquids that distill below 100°, a wrapping of glass

FIGURE 2.14
Vigreux Column

wool is usually sufficient; for higher-boiling liquids or very long columns, an evacuated or electrically heated jacket may be used.

In a packed column it is essential to leave sufficient free space for the countercurrent flows of liquid and vapor. With packing materials like Carborundum chips, glass beads, or short lengths of glass tubing the column may be filled simply by pouring in the packing, but with glass or metal spirals the best results are obtained by dropping the spirals into the column one at a time.

Distillation Procedure. The proper method of carrying out a distillation is to supply just enough heat at the distilling flask so that the liquid distills regularly at a uniform rate. Insufficient supply of heat will stop the distillation temporarily, and the bulb of the thermometer will cool below the distilling temperature, resulting in erratic temperature readings. Overheating and unsteady application of heat increase the opportunity for superheating the liquid and cause bumping. Even with proper heating, it is necessary to introduce one or two tiny boiling chips of a porous substance,[4] or some other antibumping agent,[5] into the liquid before heat is applied.

---

[4] Carborundum chips, No. 12 mesh, are suitable. They are available from Carborundum Electro Metals Co., Niagara Falls, NY, 14302.

[5] A convenient substitute is a Peerless wood applicator, which may be purchased in any drugstore. The effective surface of these wooden splints is greatly increased by breaking them and inserting the broken end into the liquid.

Superheating occurs because the transformation of a liquid into the vapor phase will not take place immediately, even at the boiling point, unless the liquid is in contact with a gaseous phase. Consequently, in a distilling flask, the liquid can vaporize only at the surface unless gas bubbles are introduced into the body of the liquid. Boiling chips are chemically inert porous materials containing a large amount of air. The trapped air expands on heating and furnishes bubbles that initiate vaporization throughout the liquid. Boiling chips lose their effectiveness after a single use and must be discarded; indeed, fresh boiling chips should be added before resuming a distillation that has been interrupted. It is dangerous to introduce boiling chips into a liquid that is at or near its boiling point, as this will induce sudden and violent bumping.

When the distillation assembly has been completed, it is checked for tightness of all connections and for physical stability. The liquid to be distilled is introduced through the neck of the distilling flask with the aid of a funnel to prevent it from contaminating the ground-glass joint. When a condenser is being used, the flow of water through the jacket is started before heat is applied. The water should enter the lower end of the jacket and flow in a direction opposite to that of the organic vapor (countercurrent cooling). The rate of flow through the condenser should not be excessive but adequate to keep the jacket cool; this may be tested from time to time by carefully touching the underside of the adapter through which the distillate is running to see if the distillate is too warm.

With less than about 10 mL of liquid to be distilled there frequently will be insufficient heat flowing into the system to compensate for the heat loss due to radiation. In this circumstance, it is good practice to wrap the top of the flask and the vertical section of the "head" with an insulating layer of glass wool. Do *not* place the glass wool between the flask and the heat source.

The rate of heating must be adjusted (see Section 1.4) so that the liquid boils gently and distills slowly at a uniform rate, generally between 30 and 60 drops (1–2 mL)/min for simple distillation. A slower rate is used for fractional distillation or for vacuum distillation. Heating should be stopped before the last few drops of liquid have been vaporized to avoid decomposition and charring in the flask.

The thermometer reading is recorded when the first drops of distillate appear at the end of the side arm or on the walls of the condenser; this is called the "initial boiling point."[6] Thereafter the temperature and the volume of the distillate are recorded at frequent intervals. If the purpose of the distillation is to determine the composition of the liquid, many temperature and volume readings are required, and it is convenient to collect the distillate directly in a graduated cylinder. The results should be recorded in a tabular form.

It is useful to plot a temperature-vs.-volume curve, from which the presence and amount of low-boiling impurities, the approximate distilling range of

[6] An incorrect value will be obtained if the temperature is read when the first drops of liquid appear on the thermometer bulb. A short time lag is necessary to permit the thermometer to warm up to the temperature of the distilling vapor. During this lag the thermometer is coming into equilibrium with the distilling vapor, and the mercury column is rising rapidly.

constant-boiling components of a mixture, and so on, can be determined. When a substance containing small amounts of impurities is distilled, the first portion of distillate (called the forerun, or low-boiling fraction) will contain the more volatile impurity and a certain amount of the main liquid that is carried with it. As the temperature continues to rise, the bulk of the principal liquid will distill over a short temperature range, usually 2–3° (called the principal fraction, or main fraction). After this fraction has distilled, the boiling point will rise, owing to the presence of the less volatile impurity. The next fraction (called the afterrun or high-boiling fraction) will consist of a mixture of the principal liquid and the less volatile impurity. The residual liquid in the distilling flask will contain the less volatile impurity along with some of the principal liquid, which it holds back from distilling. However, even a pure substance will always leave a small amount of residual liquid because the vapor in the boiler flask will condense when the heat is removed.

When the distillation is being carried out to purify a liquid, it is better to use tared[7] flasks to collect the diferent fractions. If the distillation behavior is known or can be estimated (as when a liquid of known boiling point is being purified), it is a simple matter to use three receivers and to collect the forerun, the main fraction, and the afterrun over the proper temperature ranges. When a liquid with unknown properties is being purified and a sufficient sample is available, it is a good strategy to determine first the temperature-vs.-volume distillation curve. If the losses of two distillations cannot be tolerated, it is necessary to deduce the boiling behavior of the sample as the distillation proceeds. This requires close attention to the thermometer readings; it is desirable that several extra tared flasks be available in case the collection of the main fraction is begun or ended prematurely.

Correction of Boiling Temperatures. The temperature registered directly on an ordinary thermometer in the course of laboratory distillations (or determinations of melting points) are subject to several sources of error. Probably the most common for beginning students is the placement of the thermometer bulb too high in the distilling head so that the vapors do not condense on the thermometer bulb and low temperature readings result. Two other less serious sources of errors are exposure of a portion of the mercury column to atmospheric cooling and inaccurate or incorrect graduations of the thermometer scale.

A typical thermometer with a long scale (250–300 mm) does not register the true boiling point (or melting point) because the mercury column is not entirely at the temperature of the mercury in the bulb of the thermometer. The portion of the mercury column that extends above the stopper of the distillation adapter (or the surface of a melting-point bath) is cooled by the surrounding atmosphere, and the registered temperature is therefore below the true temperature

---

[7] A tared flask is a vessel that is weighed when it is clean and dry. The amount of liquid distilled is then easily calculated by subtracting the weight of the tared flask from the total weight of liquid plus flask.

TABLE 2.3
Reference
Temperatures for
Calibration

Liquid	Boiling point, °C (at 760 mm)[a]	Solid	Melting point, °C
Acetone	56.1	Water–ice	0.0
Water	100.0	Benzoic acid	121.7
Aniline	184.4	Benzilic acid	150
Nitrobenzene	210.9	Hippuric acid	187.5
2-Bromonaphthalene	281.1	3,5-Dinitrobenzoic acid	204
Benzophenone	305.9	p-Nitrobenzoic acid	241

[a] The boiling points of azeotropes also may be used for reference temperatures.

of the vapor in the distilling flask. For temperatures below 100° this cooling effect does not cause any considerable error, but for high temperatures the observed reading may be several degrees below the true temperature. This error can be corrected by adding a stem correction calculated by the formula

$$\text{Stem correction (deg)} = 0.000154(T - T')N$$

where 0.000154 is the coefficient of apparent expansion of mercury in glass, N is the number of degrees on the stem of the thermometer from the lower exposed level to the temperature read, T is the temperature read, and T' is the average temperature of the exposed mercury column. In practice, this correction is subject to an error, since T' is not accurately known, but it may be taken roughly to be one half of the difference between room temperature and the observed temperature.

Some thermometers have graduated scales that already include a correction for an assumed 3″ (76-mm) immersion of the stem, and temperature readings taken with them should *not* be corrected. Such partial-immersion thermometers are designated by an engraved line circling the stem 76 mm above the bottom of the mercury bulb.

Many important errors of temperature readings in ordinary laboratory work may be due to incorrect graduation and calibration of the thermometer scale. To determine whether or not a thermometer registers correctly you may test it by verification at several temperatures against the boiling points of pure liquids or the melting points of pure solids (Table 2.3) or by comparison with previously standardized thermometers.

In a simple distillation, the pressure on a liquid is the atmospheric pressure. For ordinary work, the variation in boiling point due to small deviations in pressure from 1 atm (760 mm) may be neglected, but for accurate work or when you are working at higher elevations, it is necessary to record to barometric pressure during distillation. Examples of the effect of pressure changes are shown in Table 2.4.

The boiling point of a reference liquid must be corrected if the atmospheric pressure during standardization is other than 760 mm. For water and several other liquids, the changes of boiling point at pressures near 760 mm are given in Table 2.4. The boiling point at pressures in the region of 760 mm can be

| *TABLE 2.4* | Boiling point, °C | | | |
Effect of Pressure on Boiling Points **Pressure, mm**	**Ethanol**	**Benzene**	**Water**	**Aniline**
780	79.0	81.1	100.73	185.6
770	78.6	80.6	100.37	184.9
760	78.32	80.2	100.00	184.4
750	78.0	79.8	99.63	183.8
740	77.6	79.4	99.26	183.3
730	77.3	79.0	98.88	182.5
100	34.3	25.8	51.58	121.0
20	7.1	−5.6	22.14	81.9

calculated with sufficient accuracy for most purposes by the rule of Crafts, in the following convenient form.

$$\text{Bp at } P \text{ mm} = \text{bp at } 760 \text{ mm} - \frac{(273 + \text{bp at } 760 \text{ mm})(760 - P)}{10,000}$$

No correction for variations from 760 mm is needed when standardizations are made by means of melting points, since the effect of small pressure changes on melting points is negligible.

It is best to calibrate the thermometer using the same conditions under which the thermometer is to be employed. Thus, a thermometer normally should be calibrated for a fixed partial immersion of the stem; for example, a thermometer to be used for distillation should be calibrated for 3″ (76-mm) immersion of the stem, and one to be used for melting-point determinations should be calibrated for 1″ (25-mm) immersion of the stem.

2.4 Representative Simple and Fractional Distillations

The purpose of this section is to provide sufficient practice in purification of liquids by distillation so that this operation can subsequently be carried out skillfully and without reference to detailed directions. Usually only one or two of these procedures will be assigned.

Simple Distillation.   Arrange a distillation assembly similar to the one shown in Figure 2.10. Use a 25-mL boiling flask and follow the correct methods for supporting the apparatus and lubricating the joints as described in Section 2.3.

(A) Distillation of a Pure Compound

Into a 25-mL boiling flask introduce 10 mL of pure, dry methanol (caution—*flammable liquid*) by means of a clean, dry funnel. Add one or two tiny boiling chips, attach the boiling flask, and make certain that all connections are tight.

Place a graduated cylinder beneath the drip tip of the condenser. Heat the flask gently until the liquid begins to boil. Adjust the heating rate until the ring of vapor condensation moves up the wall of the flask and past the thermometer into the condenser. Record the temperature when the first drops of distillate collect in the condenser. Continue to distill the liquid slowly (not over 2 mL/min) and record the distilling temperature at regular intervals during the distillation— when the total distillate amounts to 1, 2, 3, etc., mL. Discontinue the distillation (and turn off the heat source) when all but 1 mL of the liquid has distilled. Record the temperature range from the beginning to the end of the distillation; this is the observed boiling point. If the boiling point differs from the literature value, record the correction in your laboratory notebook for future reference.

Transfer the used methanol to a bottle provided for this purpose. From your data, draw a rough distillation graph for pure methanol, plotting distilling temperatures on the vertical axis against total volume of distillate on the horizontal axis.

(B) Distillation of a Mixture

By means of a funnel,, introduce 12 mL of a mixture of methanol and water into the distilling flask, add a few tiny boiling chips, and distill the mixture slowly. Follow the same procedure used for distilling pure methanol. Draw a rough distillation graph and compare it with the observed boiling point for pure methanol. From the graph, estimate the composition of the liquid and record the analysis in your notebook.

Transfer the distillate to a bottle for this purpose (labeled "Recovered Methanol–Water Mixture").

(C) Purification of an Unknown Liquid

From your instructor obtain a 10-mL sample of an impure unknown. Carry out a preliminary distillation to determine the distillation behavior of the mixture and its approximate composition. Redistill the liquid and collect the fraction boiling over a 4–5° range. Record the boiling range and weight of the main fraction.

Fractional Distillation. Arrange an assembly for fractional distillation as shown in Figure 2.13. A short condenser makes an effective thermally insulated distillation column. A small wad of stainless steel sponge or a coil of copper wire poked into the lower end of the column will hold the packing without interfering with the flow of vapor or liquid reflux. If a condenser is used, one of the side arms should be stoppered with a dropper bulb to prevent loss of heat by convection.

Your laboratory instructor will indicate what kind of packing is to be used and issue any special instructions for placing it in the column. The column should have about five plates if a "standard separation" is to be achieved. Prepare five dry receivers (10–50-mL Erlenmeyer flasks), provide each with a tight-fitting cork, and label them A, B, C, D, and E (residue). The boiling flask should be selected so that it is approximately 60% full at the beginning of the distillation.

Before starting the distillation, check your apparatus carefully and have it approved by the laboratory instructor. Be careful to carry out a fractional distillation methodically, since haste will lead to sharply lowered separation efficiency.

(D) Cyclohexane and *n*-Heptane[8]

Into a 50-mL distilling flask place 25 mL of a mixture of cyclohexane and *n*-heptane, 1 : 1 by volume, add two boiling chips, and fit the flask securely to the column.

◄ *CAUTION* If you are using a burner as a heat source, remember that cyclohexane and *n*-heptane are highly flammable. Extinguish or remove any flame when transferring fresh fractions into the round-bottomed flask.

Heat the flask until the mixture starts to boil, and then regulate the heat source with particular care so that the liquid distills slowly and regularly at a rate of about 1 mL (30 drops)/min. In the first distillation, collect in flask A the fraction (if any) that distills between 81 and 84 (the 81–84° fraction); in flask B, the 84–88° fraction; in flask C, the 88–92° fraction; and in flask D, the 92–96° fraction.[9] After fraction D has distilled, remove the heat source, cool the flask, allow the column to drain, disconnect the flask, and pour the residue into E. Measure the volume of each fraction and record the results in tabular form (Table 2.5).[10]

If the separation efficiency of the column was not adequate, it will be necessary to redistill the different fractions. In the subsequent distillations proceed in the following way: Pour the contents of flask A into the round-bottomed flask, add one or two tiny boiling chips,[11] and redistill, collecting the 81–84° distillate in the same flask A. Distill until the thermometer reaches 84°, then stop

---

[8] The composition of the distillate may be determined by gas chromatography using a 5-ft 5% SE 30/Chromosorb W column at 25°. With these data, the number of theoretical plates can be calculated from equation (2.5), and an approximate average value for α of 1.9.

[9] In the first distillation there may be no distillate in range 76–81°, so that flask A will remain empty and flask B will be used. In the subsequent distillations, likewise, there may be little or no distillate in the intermediate ranges B, C, and D. The results vary widely depending on the type of column and the care used in operation.

[10] If you desire to record the weight of each fraction instead of the volume, it is convenient to weigh each receiver empty, with its cork, and record this weight (called the tare) on the label of the receiver. It is good practice also to record the tare of each receiver in your notebook; when the receivers with distillate are weighed after a distillation, the gross weight is recorded, the tare subtracted, and the net weight of the fraction entered in the tabular form. If all of the weights are recorded in the notebook, you can check the figures at a later date for arithmetic errors if a discrepancy shows up.

[11] It is desirable to add a tiny fresh boiling chip each time the distillation is stopped and a new fraction introduced. At the end of the distillation series, the residual liquid in the still is poured off and the accumulated used chips are discarded.

TABLE 2.5
Fractional Distillation
of Cyclohexane and
n-Heptane Mixture

Fraction	Temperature range, °C	Volume, mL		
		1st dist'n	2nd dist'n	3rd dist'n
A	81–84			
B	84–88			
C	88–92			
D	92–96			
E	Residue			
Total volume of fractions				

the distillation and add the contents of flask B. Resume the distillation, and collect the 81–84° fraction in flask A, and the 84–88° fraction in flask B. When the thermometer reaches 88°, stop the distillation and add the contents of flask C. Resume the distillation and collect the 81–84° fraction in flask A, the 84–88° fraction in flask B, and the 88–92° fraction in flask C. When the thermometer reaches 92°, stop the distillation and add the contents of flask D. Resume the distillation and collect the fractions A, B, C, and D. When the thermometer reaches 96°, stop the distillation and add the contents of flask E. Resume the distillation and collect the fractions A, B, C, and D. After fraction D has distilled, extinguish the flame, cool the flask, allow the column to drain, disconnect the flask, and pour the residue into E. Measure the volume of each fraction, and record the results in tabular form.

If flasks B, C, and D at this stage contain a total of more than 3–4 mL of liquid, carry out a third distillation in the same manner. If necessary, carry out a fourth or fifth distillation, so that the fraction A will contain almost all of the cyclohexane and the residue E almost all of the *n*-heptane. To obtain almost pure *n*-heptane, E may be redistilled from a small distilling flask without a column.

Draw rough distillation graphs for each successive distillation, plotting the midpoint of the temperature range of each fraction against total volume of distillate.

In a research laboratory, it is customary to follow the progress of a distillation by some convenient analytical procedure. Gas chromatography (Section 7.4) is commonly used.

(E) Acetic Acid and Water

In this experiment you will fractionally distill a mixture of glacial acetic acid and water (100 : 31.5 by volume, 1 : 1 mole ratio) and follow the progress of separation by titrating 0.5-mL portions of several fractions against standardized aqueous sodium hydroxide with phenolphthalein indicator to determine the acetic acid content. The acetic acid content of the original mixture should be determined in the same way before the material is fractionated. If a column having a large number of plates is used, it will be desirable to use larger portions of the early fractions.

Obtain a 35-mL supply of a 1 : 1 molar solution of acetic acid and water. Fill a 50-mL buret with 1.0 N sodium hydroxide solution. With the aid of a 0.5-mL or 1.0-mL pipet, place 0.5 mL of the 1 : 1 molar solution of acetic acid and water in a 50-mL Erlenmeyer flask and add 10 mL of water and a few drops of phenolphthalein indicator. Titrate to a slightly pink end point and record the volume of titrant. Repeat the titration on two more 0.5-mL samples of the 1 : 1 molar solution of acetic acid and water and compute the average titer.

Assemble a fractional distillation apparatus using a 50-mL round-bottomed flask for the boiler and a 25-mL graduated cylinder for the receiver. Place 30 mL of the 1 : 1 mixture in the flask and add a boiling chip. You will need a small test tube that has been marked to show the liquid level when it contains exactly 0.5 mL of liquid.

Heat the mixture until it boils, and then adjust the heating rate so that the mixture distills at a *maximum* rate of 1 drop/sec. Note the temperature at which the first drop distills. Collect the first 0.5 mL of distillate in your marked test tube and the next 4.5 mL in the graduated cylinder. Record the distillation temperatures at each 1-mL interval. Transfer the 0.5-mL sample to a 50-mL Erlenmeyer flask (rinse the tube with a total of 10 mL of distilled water and add the rinse to the Erlenmeyer flask). Mark the flask to indicate the sample it contains.

When the volume of distillate reaches 5 mL, collect another 0.5-mL sample in the test tube and transfer it in the same manner to another Erlenmeyer flask. Collect the next 4.5 mL of distillate in the graduated cylinder, recording the distillation temperatures at each 1-mL interval. Repeat this process at 10, 15, 20, and 25 mL of distillate.

Titrate the six samples with the sodium hydroxide solution (the early samples will require very little titrant) and calculate the mole fraction of acetic acid present. In the calculations assume that the volumes of acetic acid and water are additive so that the mole fraction in any sample is simply proportional to the titer value obtained for the initial 0.5-mole-fraction mixture.

Prepare a plot of boiling point (ordinate) vs. the total volume of distillate (abscissa) and a second plot of the mole fraction of acetic acid vs. the total volume of distillate.

From the measured mole fraction of the first 0.5 mL of distillate, using equation (2.5) with a value of 1.8 for α, calculate the number of theoretical plates provided by the column.

(F) Methanol and Water

For the separation of a mixture of methanol and water, the following temperature ranges are satisfactory for the fractions: A, 64–70°; B, 70–80°; C, 80–90°; D, 90–95°; and E, residue. A simple way to verify the quality of separation is to determine the specific gravity of each fraction.

Questions

1. (a) Define accurately the term boiling point.
   (b) What effect does a reduction of external pressure have upon the boiling point?

2. What effect on the temperature of a boiling liquid and on its distillation temperature is produced by each of the following?
   (a) a soluble, nonvolatile impurity
   (b) an insoluble, admixed substance such as sand or fragments of wood or cork

3. Why should a distilling flask at the beginning of a distillation be filled to not more than two thirds of its capacity but not less than one third of its capacity?

4. Calculate the weight of vapor of each substance required to fill a 50-mL flask at the boiling point under normal atmospheric pressure
   (a) cyclohexane, C_6H_{12}
   (b) toluene, C_7H_8
   (c) carbon tetrachloride, CCl_4

5. Why is the apparatus shown in Figure 2.10 not suitable for distillation of samples with a volume of 5 mL or less?

6. Why is it dangerous to heat an organic compound in a distilling assembly that is closed tightly at every joint and has no vent or opening to the atmosphere?

7. Calculate the stem correction for observed temperature readings of 125°, 175°, and 250°; assume that the thermometer scale is exposed above the 25° mark and the average temperature of the exposed portion is half the difference between the observed temperature and room temperature (20°).

8. The composition of the vapor above methanol–water mixtures has been measured and is given by the following formula in which X is the mole fraction of methanol in the liquid and Y is the mole fraction in the vapor.

$$Y(CH_3OH) = 6.05X - 21.37X^2 + 39.05X^3 - 34.02X^4 + 11.29X^5$$

What are the mole fractions of methanol in the vapors above mixtures with X equal to 0.25, 0.50, and 0.75?

9. Why is it necessary to have liquid flowing back through the fractionating column to obtain efficient fractionation?

10. (a) What is an azeotropic mixture?
    (b) Why cannot its components be separated by fractional distillation?

11. What physical constants may be used to test the purity of the samples of purified material obtained after a fractional distillation?

12. (a) What is meant by the temperature gradient of a column?
    (b) Why is it desirable to maintain a uniform temperature gradient, and how is this achieved?

3 Vacuum Distillation

Since the boiling temperature of a liquid is decreased by diminishing the pressure on its surface, one can distill at a lower temperature by using an apparatus inside which the pressure is maintained at a low value by an attached vacuum pump. This procedure is useful for purifying liquids (or low-melting solids) that decompose at elevated temperatures. For example, glycerol boils with some decomposition at 290° under 760-mm pressure but may be distilled without decomposition under 12-mm pressure, where its boiling point is 180°. A possible disadvantage of fractional distillation under reduced pressure is the reduction in separation efficiency of most fractionating columns.

In planning a vacuum distillation, three aspects must be considered: the pressure needed to achieve the desired boiling point, the type of vacuum pump needed to lower the pressure to the required level, and finally, the associated glassware, pressure measuring devices, and heat sources.

3.1 Effect of Pressure on Boiling Point

Estimation of Boiling Point. One useful relationship between pressure and boiling point is given in equation (3.1), where P is the pressure over the liquid and T is the boiling point at this pressure.[1] In this equation, both boiling temperatures are expressed in kelvins (K = °C + 273). Equation (3.1) is fairly precise for most organic liquids, but is in error for substances possessing unusually large attractions between molecules (water, alcohols, acids). More precise relationships have been

[1] In equation (3.1) the value of the constant has been changed from 4.81, its usual theoretical value, to 5.46 to correlate a wider range of data.

developed, but the extra work required to use them is not justified for preparative organic chemistry.

$$\log_{10}\left(\frac{760}{P}\right) = 5.46\left(\frac{\text{normal boiling point}}{T} - 1\right) \tag{3.1}$$

The use of equation (3.1) is illustrated here for nitrobenzene.

1. Normal boiling point of nitrobenzene = 211°C = 484 K.
2. If the desired boiling point is 100°C = 373 K, the equation becomes

$$\log_{10}\left(\frac{760}{P}\right) = 5.46\left(\frac{484}{373} - 1\right) = 1.62$$

3. The expression is solved for P.

$$P_{\text{predicted}} = 18.0 \text{ mm}$$

4. At a pressure of 18.0 mm, it is observed that nitrobenzene boils at 98°C instead of the desired 100°C. That is about as good agreement as you could hope for.

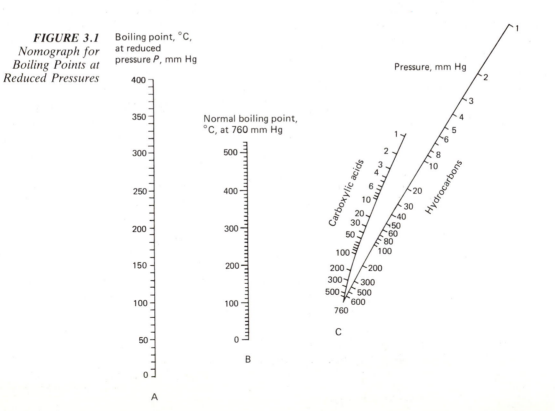

FIGURE 3.1
Nomograph for
Boiling Points at
Reduced Pressures

Another approach to estimating the boiling points at reduced pressure is to use a nomograph, such as the one shown in Figure 3.1. To estimate the boiling point at some reduced pressure of a hydrocarbon for which boiling point at 760 mm is known, place a straightedge (preferably transparent) on the nomograph connecting the known boiling point on scale *B* and the reduced pressure on line *C* for hydrocarbons. The boiling point of the compound at the reduced pressure is read from the intersection of the straightedge and the left-hand *A* scale. If the compound contains a carboxylic acid functional group, line *C* for carboxylic acids is used instead. The boiling points of polar molecules and less strongly hydrogen-bonded molecules can be estimated as the average of the boiling points by using both the hydrocarbon and carboxylic acid lines. Note that for compounds boiling near 100° both lines give about the same result; it is only for high-boiling liquids that different predicted boiling points are obtained.

3.2 Vacuum Pumps

The pump used to reduce the pressure in the system is selected according to the range of pressure required. An aspirator (water pump) is used for pressure above about 25 mm, a rotary oil pump is usually used for the range of 0.01–25 mm, and a diffusion pump is used for pressure below about 0.01 mm. Portable rotary vane pumps are available that are suitable for the pressure range of 75–760 mm. You should be aware of the operating characteristics of each type of pump.

An aspirator in good condition can produce a vacuum almost down to the vapor pressure of the water flowing through it. With room-temperature water, a pressure of about 25 mm can be produced, but during the winter, if the water is quite chilled, pressures of 20 mm or less may be obtained. It is not feasible to vary the water flow through the aspirator to regulate the pressure. Instead, turn the water flow fully on and control the pressure by adjusting a valve that leaks air into the system. An inexpensive valve can be obtained by using the valve base of an adjustable Bunsen burner. A hose is connected from the apparatus to the gas inlet tube of the burner, and the reverse flow of air into the apparatus is regulated at the burner base.

A troublesome characteristic of aspirators is their tendency to allow water to flow back into the system if the water pressure drops momentarily. Modern aspirators come with internal check valves to prevent this backflow, but they should not be trusted; it is wise to include a safety trap between the aspirator and the rest of the system. The safety trap also provides a convenient means of interconnecting the pump, the air leak, and the manometer used to measure the internal pressure. A practical arrangement is shown in Figure 3.4.

The pump oil in a rotary oil pump has such a low vapor pressure that contamination of the distillate is improbable; instead, one must guard against contamination of the pump oil by any uncondensed materials from the distillation. Oil pumps work by compressing small samples drawn from the vapor inside

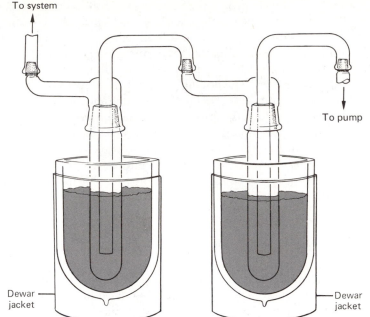

FIGURE 3.2
Trap Arrangement for
Rotary Oil Pumps

the distillation apparatus to a pressure above the atmospheric pressure and then releasing them from the pump. This pumping operation requires good seals between two reciprocating vanes and a rotating eccentric piston. If the oil becomes contaminated with acidic materials, the movable vanes will corrode and the pumping capacity of the pump will be sharply diminished. Some nonacidic organic materials may polymerize in the pump and form sludges that also will damage the movable vanes, with a corresponding reduction in pump effectiveness. It is important that the pump be protected by at least one vapor trap that is maintained at the dry ice sublimation temperature ($-78°$) or lower. If the mixture being distilled evolves gases that would sweep vapors into the pump, much more efficient trapping devices are required. Figure 3.2 displays one widely used trap arrangement for rotary oil pumps.

An alternative to the rotary oil pump, suitable for the higher pressure range of 75–760 mm, is the much less expensive pressure-vacuum pump, such as the Gast pump. This unit also uses rotating vanes to develop a vacuum, but it is built to lower tolerances and requires less care in its operation. The vacuum (or pressure) is adjusted easily with a needle valve built into the pump. Because the motor is connected directly to the pump, the hazard of moving belts is eliminated. The most significant drawback of this pump, beside its limited range, is its noisiness.

Like the water aspirator, diffusion pumps work on the Bernoulli principle,[2]

---

[2] When liquids or gases flow through a pipe of variable cross section, the pressure is smallest where the cross section is least and the velocity is greatest.

except that in place of a rapidly flowing stream of water they use a jet of mercury or oil vapor. Diffusion pumps produce pressures in the range from 10^{-2} to 10^{-6} mm. They require heaters to vaporize the mercury or oil and cooling devices for recondensing the vapor after it has passed through the jet. Unlike aspirators and mechanical pumps, diffusion pumps require an additional fore-pump ("backing pump") that reduces the pressure in the pump to a critical level (usually about 0.1–0.01 mm). Using an oil or mercury diffusion pump is much more involved than using the other pumps described here.

Manometers. There are many types of manometers for measuring the pressure in the system, each designed for maximum precision over a small range of pressures. Two general-purpose manometers that together cover a sufficient range with adequate precision for preparative work are the tilting McLeod gauge (0.01–10 mm) and closed-end U-tube mercury manometer (5–200 mm), which are shown in Figure 3.3.

The McLeod gauge is operated by tilting the movable section of the gauge toward the upright position until the higher of the two mercury columns is level with the top of the bore of the lower capillary column. If the gauge has been calibrated properly, the pressure can be read directly from the height of the lower capillary column against the gauge markings.

With the open-tube manometer shown in Figure 3.4, the pressure in the system is obtained by subtracting the difference in height of the two mercury columns from the barometric pressure.

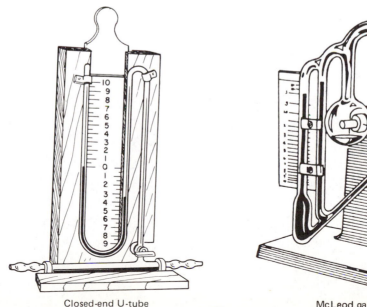
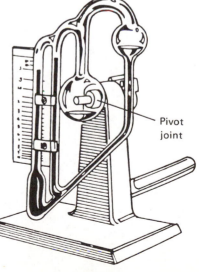

FIGURE 3.3
Manometers

Closed-end U-tube
manometer

McLeod gauge

3.3 Laboratory Practice

An apparatus for moderate-scale vacuum distillation is shown in Figure 3.4. The purpose of the Claisen adapter is to reduce the chance of contamination of the distillate from frothing or violent bumping, both of which are more troublesome in vacuum distillation than in ordinary distillation. For very small-scale distillations, because one must minimize mechanical losses from holdup, the Claisen adapter is left out and the sample is distilled *slowly* and *carefully*. The size of the boiler flask should be such that it is not quite half filled at the start of the distillation. It is preferable to use an oil bath or sand bath to ensure regular heating, but a heating mantle, used with care, will do the job. The bath temperature is usually 15–25° higher than that of the distilling liquid.

Boiling chips are frequently ineffective in vacuum distillation because of their limited air supply. Somewhat better are long wooden splints of the kind that can

FIGURE 3.4 *Assembly for Distillation Under Diminished Pressure*

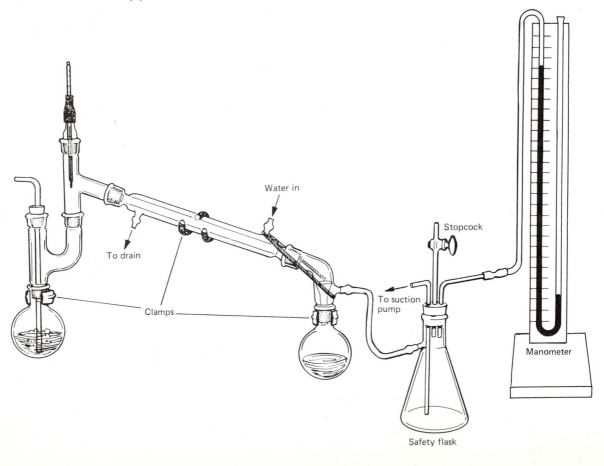

be purchased in any drugstore. These can be broken off to a size that will fit in the flask comfortably. The best procedure, although it is somewhat inconvenient, is to introduce a fine stream of air bubbles through a thin, flexible capillary tube.[3] If the substance to be distilled is easily oxidized, argon or nitrogen should be substituted. A different approach that works quite well is to stir the boiling mixture rapidly with a magnetic stirrer.

With vacuum distillations using ground-jointed apparatus, it is imperative that the joints be properly lubricated to prevent leaks during the distillation and simplify separation of the joints afterward. Remember that if a distillation is carried out at a pressure even as high as 100 mm, there is still almost 1 atm ($660/760 = 0.9$ atm) pressing in on the walls of the flasks and squeezing the joints together.

In advanced work the single receiver is replaced by a device consisting of a flask with two or more arms (known picturesquely as a "cow"), to each of which is attached a receiving flask. By rotation of the "cow" the different receivers in turn can be brought in line with the drops of condensate without disruption of any of the vacuum seals.

The following points should be observed in carrying out a vacuum distillation.

1. Never use Erlenmeyer flasks as receivers. Even with the small 50-mL Erlenmeyer flask, the force acting on the flat bottom is about 50 lb. Remember that the force at any point is proportional to the difference between the internal and external pressure. A distillation at 100 mm places nearly as much stress on the apparatus as one at 10^{-6} mm.

2. Test the completely assembled apparatus before placing the liquid in the boiler flask, to detect leaks and to make certain that all of the parts of the apparatus will withstand the external atmospheric pressure. Because apparatus containing a vacuum might *implode* without warning, it is important to wear safety goggles at all times to protect your eyes.

3. When using a water aspirator, turn on the water to the full pressure, otherwise water may be sucked back into the safety flask.

4. To release the vacuum in the distilling flask, open the stopcock on the safety flask and gradually allow the pressure to reach atmospheric pressure in the apparatus before shutting off the aspirator.

5. When using a rotary pump, be certain it is adequately guarded with traps and that they are filled with coolant. A slush of dry ice and isopropanol

---

[3] The capillary is prepared by drawing out the center of a 6″ length of soft-glass tubing, 7–8 mm in diameter, to capillary dimensions and drawing out this first capillary, in a small luminous flame, to an extremely fine and flexible capillary thread. The capillary thread is tested by blowing into the tube while the thread is held under ethanol or methylene chloride; it should emit a fine stream of very minute bubbles. The top of the tube, which bears the capillary, should be bent at right angles to facilitate adjustment of the depth of the capillary tube so that it will reach exactly to the bottom of the distilling flask.

(−78°) is reasonably effective and inexpensive. In some research laboratories liquid nitrogen (−196°) is used because of the greater protection it affords the pump.

6. When changing receivers, remove the heat and allow the distilling flask to cool slightly before releasing the vacuum. After the receiver has been changed, the system should be evacuated again before heating is resumed.

7. If a closed-tube manometer is used, close the stopcock of the manometer before releasing the vacuum. If this is not done, the abrupt surge of the mercury column may break the glass tube.

3.4 Representative Vacuum Distillations

(A) Purification of Benzaldehyde

Purify a 15-g (15-mL) sample of technical benzaldehyde[4] in the following way. Wash with two 5-mL portions of sodium carbonate solution (10%), then with water, and dry over anhydrous magnesium sulfate. It is advantageous to add a few small crystals of hydroquinone or some other antioxidant during the drying operation.

Arrange an assembly as shown in Figure 3.2 with a boiler flask of appropriate size, and decant the benzaldehyde through a fluted filter into the boiler. Distill under diminished pressure (preferably below 30 mm) in the manner previously described. Be certain to wear your *safety goggles*. Place a few small crystals of the antioxidant in the receiving flask in which the purified benzaldehyde is to be collected. To determine the boiling point of benzaldehyde under the particular pressure in your apparatus, use the nomograph in Figure 3.1; the boiling point at 760 mm is 180°.

Autooxidation. In common with many oxidizable substances, benzaldehyde is capable of combining directly with oxygen of the air (autooxidation) and is converted eventually to benzoic acid.

Autooxidation is extremely sensitive to the effect of catalysts, which are considered to act on an unstable intermediate complex of "peroxide" character that is formed in the initial step of oxidation. Catalysts that accelerate autooxidation are called prooxidants; those that retard or inhibit autooxidation are called antioxidants. The latter find important technical application for the preservation of organic materials; for example, the deterioration of rubber is greatly retarded by the incorporation of antioxidants. Likewise, the autooxidation of benzaldehyde can be effectively inhibited by the addition of a trace (less than 0.1% is sufficient) of hydroquinone or some other antioxidant.

---

[4] Technical benzaldehyde usually contains a small amount of benzoic acid.

(B) Purification of Ethyl Acetoacetate

Place a 15-g (15-mL) sample of technical ethyl acetoacetate[5] in a vacuum distillation assembly of suitable size and distill under diminished pressure (preferably below 30 mm) in the manner described earlier. Be certain to wear your safety goggles.

During the distillation of the low-boiling fraction, the high vapor pressure of the ethyl acetate that it contains may raise the pressure in the system above 30 mm. If this occurs, the distillation of the first fraction may be carried out at a higher pressure, but the pressure should be maintained below 30 mm for the distillation of the remaining fractions. To obtain the correct boiling point of ethyl acetoacetate under the particular pressure in your apparatus, use the nomograph of Figure 3.1; the boiling point at 760 mm is 180°.

Collect the purified ethyl acetoacetate over an interval within ±3 of the boiling temperature determined from the nomograph, and calculate the percentage recovery from the crude product.

Questions

1. Estimate from the nomograph in Figure 3.1 the boiling point of each component at 1 mm pressure.
   (a) pentanoic acid (bp 184° at 760 mm)
   (b) ethylbenzene (bp 136° at 760 mm)
   (c) *o*-chlorotoluene (bp 159° at 760 mm)

2. Estimate the pressure under which each compound could be vacuum distilled at 80°.
   (a) pentanoic acid
   (b) ethylbenzene
   (c) *o*-chlorotoluene

3. Why is bumping more troublesome in vacuum distillation than in ordinary distillation?

4. What precautions must be observed in using an aspirator for vacuum distillations?

5. In a vacuum distillation what is likely to happen if
   (a) the boiling flask is filled two-thirds full?
   (b) you failed to introduce a new boiling chip after you had broken the vacuum to collect a new fraction?
   (c) you failed to cool the flask before opening the system to the atmospheric air?

6. Prepare a diagram that shows the interconnections of the apparatus used in a vacuum distillation. Include the vacuum pump and manometer.

---

[5] Technical ethyl acetoacetate may contain small amounts of acetate, acetic acid, and water. Since ethyl acetoacetate decomposes to some extent on heating to its boiling point at atmospheric pressure, with the formation of dehydroacetic acid, it is purified by distillation under diminished pressure. Small quantities may also be purified by rapid distillation at atmospheric pressure.

4 Steam Distillation

Steam distillation consists of distilling a mixture of water and an immiscible or partly immiscible substance.[1] The practical advantage of steam distillation is that the mixture usually distills at a temperature below the boiling point of the lower-boiling component. Consequently, it is possible to effect steam distillation of a high-boiling organic compound at a temperature much below its boiling point (in fact, below 100°) without resorting to vacuum distillation. Steam distillation is useful also in separating mixtures when one component has an appreciable vapor pressure (at least 5 mm) in the vicinity of 100° and the other has a negligible vapor pressure. The process of steam distillation is widely employed in the laboratory and in industry; for example, for the isolation of α-pinene, aniline, nitrobenzene, and many natural essences and flavoring oils.

4.1 Principles of Steam Distillation

Mixtures of two immiscible substances behave quite differently from homogeneous solutions, and the description of their behavior requires a different physical law. The basis of this law can be grasped by considering the consequence of increasingly positive deviations from Raoult's law (see Section 2.1). One symptom of small positive deviations is a skewed boiling point–composition diagram, as is found with methanol–water solutions (see Figure 2.6). Greater positive deviations, as occur in ethanol–water solutions, lead to maxima in the total vapor pressure curve and to low-boiling azeotropes (Figure 2.7). With still

[1] Occasionally the principle of steam distillation is extended to mixtures of two immiscible organic compounds such as ethylene glycol and a hydrocarbon; then it is called codistillation.

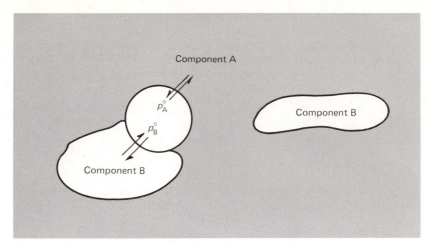

FIGURE 4.1
Vapor Pressure Inside
Bubble During Steam
Distillation

greater positive deviations the two components can separate into two immiscible layers. In the limit of very large positive deviations from Raoult's law, the two components are essentially insoluble and each component vaporizes independently of the other to give a total vapor pressure that is the sum of the individual vapor pressures. The physical basis for this independent behavior of the two components is depicted in Figure 4.1. In this diagram component B is represented as two globules suspended in component A; an incipient bubble in contact with both components is shown in the center of the diagram. The vapor pressure of component A inside the bubble is p_A° (the vapor pressure of pure A). just as it would be if no B were present. In the same fashion the vapor pressure of B inside the bubble is p_B°. Most water-insoluble organic compounds approximate this extreme behaviour so that steam distillation calculations are normally based on the simple law; *the total vapor pressure equals the sum of the vapor pressures of the two components.*

4.2 Distillation Temperature and Composition of Distillate

As with ordinary distillations the boiling point is the temperature at which the total vapor pressure equals the atmospheric pressure. If the vapor pressures of the two components are known at several temperatures, the distillation temperature is found readily by plotting the vapor pressure curves of the individual components and making a third curve showing the sum of the vapor pressure at the various temperatures (Figure 4.2). The steam distillation temperatures will be the point where the sum equals the atmospheric pressure.

Knowing the distillation temperature of the mixture and the vapor pressures of the pure components at that temperature, one can calculate the composition of the distillate by means of Dalton's law of partial pressures.

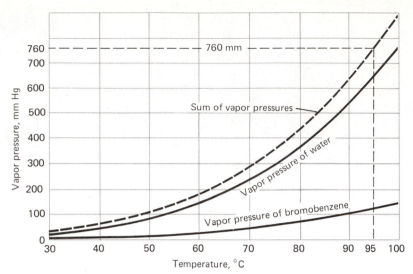

FIGURE 4.2
Vapor Pressure
Curves for Water and
Bromobenzene

According to Dalton's law, the total pressure (P) in any mixture of gases is equal to the sum of the partial pressures of the individual gaseous components (p_A, p_B, etc.). The proportion by *volume* of the two components in the distilling vapor will consequently be equal to the ratio of the partial pressures at that temperature; *the molar proportion* of the two components (n_A and n_B) in steam distillation will be given by the relationship $n_A/n_B = p_A/p_B$, where $p_A + p_B$ equals the atmospheric pressure. The weight proportion of the components is obtained by introducing the molecular weights (M_A and M_B).

$$\frac{\text{Weight of A}}{\text{Weight of B}} = \frac{p_A \times M_A}{p_B \times M_B}$$

Example Consider a specific case, such as the steam distillation of bromobenzene and water. Since the sum of the individual vapor pressures (see Figure 4.2) attains 760 mm at 95.2°, the mixture will distill at this temperature. At 95.2° the vapor pressures are bromobenzene, 120 mm; water, 640 mm. According to Dalton's law, the vapor at 95.2° will be composed of molecules of bromobenzene and of water in the proportion of 120:640. The proportion by weight of the components can be obtained by introducing their molecular weights.

$$\frac{\text{Weight of bromobenzene}}{\text{Weight of water}} = \frac{120 \times 157}{640 \times 18} = \frac{1.63}{1.00}$$

The weight composition of the distillate will therefore be 62% bromobenzene and 38% water.

$$\text{Bromobenzene} = \frac{1.63}{1.00 + 1.63} \times 100 = 62\%$$

$$\text{Water} = \frac{1.00}{1.00 + 1.63} \times 100 = 38\%$$

This calculation gives the minimum amount of water in the distillate. In practice, an excess of water or steam is used in the distilling flask to sweep out the vapor mixture and to compensate for imperfect mixing.

It can be seen from calculations of the type illustrated above with bromobenzene that there are several requirements for the practical use of steam distillation in the laboratory: the substance to be steam distilled must be insoluble, or only sparingly soluble, in water; it must not be decomposed by prolonged contact with boiling water or steam; and it must have an appreciable vapor pressure (preferably, at least 5 mm) in the neighborhood of 100°. That water has a very low molecular weight (18) compared with those of typical organic compounds is a favourable circumstance for steam distillation, since this permits a substance to be steam distilled at a practical rate even though its vapor pressure is relatively small near 100°.

4.3 Laboratory Practice

A simple assembly for small-scale distillations is shown in Figure 4.3. A large round-bottomed flask, which serves as the boiler, is fitted with a Claisen adapter. The central arm of the adapter is stoppered; the side arm is attached to a distillation "head" equipped with a thermometer.

The receiver does not have to fit tightly and can be any flask of convenient size. The mixture to be steam distilled and about twice the volume of water anticipated to be needed for the distillation is placed in the boiler flask.[2] The size of the flask is chosen so that it is less than half filled. If a large enough flask is not available, the stopper in the Claisen adapter can be replaced by a dropping funnel, so that additional water can be added as the distillation proceeds. In carrying out a steam distillation; it is essential that the mixture be boiled vigorously, since success depends on thorough mixing of the water-insoluble compound with the boiling water. The purpose of the Claisen adapter is to prevent fine spray from being carried over mechanically into the distillate. If the mixture tends to froth, the Claisen adapter will help trap it, or at least give some warning that unvolatilized material is about to be carried over and allow the operator to lower the heat.

Many substances that are solids at room temperature can be steam distilled. With such materials, which may solidify in the condenser and form a mass that

---

[2] If the boiling points of the components are known, their vapor pressures at 100° can be estimated by the nomograph in Figure 3.7; the volume of water can be estimated from these data using the method of the preceding section.

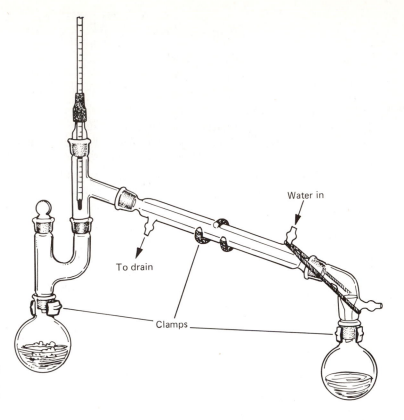

FIGURE 4.3
Small-Scale Steam Distillation Assembly

obstructs the tube, it is essential to watch the condenser tube carefully. If a mass of crystals forms, the flow of water through the jacket is stopped temporarily, and the water is allowed to drain out of the jacket. The heat from the hot distillate will cause the crystalline mass to melt and flow out into the receiver. When the tube is clear, the flow of water through the jacket is started again.

4.4 Representative Steam Distillations

(A) Steam Distillation of Turpentine[3]

Arrange an apparatus for steam distillation as shown in Figure 4.3 using a 50-mL distilling flask and a 10-mL graduated cylinder as the receiver. In the flask, place 5 mL (4.3 g) of turpentine(bp 156–165° at 760 mm) and 15 mL of water. Add two boiling chips and adjust the heating mantle to give vigorous boiling. It is

---

[3] This experiment illustrates the level of agreement to be expected between predictions from the ideal steam distillation law and an actual distillation. Commercial turpentine is largely α-pinene and can be treated as such for the purposes of this experiment. The vapor pressures of water and α-pinene ($C_{10}H_{16}$) are given in Appendix A. The density of α-pinene at 25° is 0.86.

essential for the success of this experiment that the mixture boil rapidly with good mixing of the two phases. Because the point of this experiment is to measure an equilibrium composition and the initial distillate may not have time to equilibrate, discard the first 1.5 mL of distillate, and collect the next 5 mL. Record the volumes of the water and turpentine layers in this distillate. Compare the ratio of the volumes actually found with the ratio calculated from the ideal steam distillation law using the tabulated vapor pressures and densities. Compare the observed distillation temperature with the calculated value.

(B) Separation of a Mixture by Steam Distillation

Arrange an apparatus for steam distillation as shown in Figure 4.3 using a 100-mL round-bottomed flask as the boiler and a 50-mL Erlenmeyer flask as a receiver. Make sure that the apparatus is supported firmly and that all stoppers and connections are tight.

Weigh out 1-g samples of *p*-dichlorobenzene and of salicylic acid, and determine the melting point of each sample.[4]

In a porcelain mortar, thoroughly mix the samples of *p*-dichlorobenzene and salicylic acid and determine the melting point of the mixture. Transfer the mixture to the steam distillation flask, add about 40 mL of water, and heat the mixture until it boils vigorously. Continue to distill with steam until a test portion of the distillate shows that no more water-insoluble material is passing over. When the distillation is finished, save the distillate and the residue in the round-bottomed flask for further examination.

▐▶ *CAUTION*   Since the material that distills with steam may solidify in the condenser, watch carefully to avoid the formation of a crystalline mass that could completely obstruct the condenser tube. If a large crystalline mass collects in the tube, stop the flow of water through the condenser momentarily, and drain the water from the condenser jacket. The heat from the vapors will then melt the crystals, and the obstruction will be removed. As soon as this occurs, start the water again through the condenser jacket.

Before the residue in the flask cools, filter the solution through a fluted filter (see Figure 5.5) and collect the filtrate in a clean beaker. Cool the filtrate in an ice bath and collect the crystals with suction in a Büchner funnel (see Figure 5.7). Allow the crystals to dry and determine their melting point. What is the substance that did not distill with steam?

Separate the solid material in the distillate by filtering with suction, and press it as dry as possible with a cork or spatula. Allow the crystals to dry completely, and determine their melting point. What is the substance that distilled with steam?

---

[4] The determination of melting points is described in Section 5.1.

Questions

1. What properties must be a substance have for steam distillation to be practical?

2. What are the advantages and disadvantages of steam distillation as a method of purification?

3. At 100° the vapor pressure of limonene, $C_{10}H_{16}$, is 70 mm. Estimate the amount of steam required to steam distill 13.6 g (0.1 mole) of limonene.

4. If a mixture of bromobenzene and water were subjected to steam distillation at 100 mm pressure (see Appendix A for vapor pressure data), what would be the temperature of distillation and the weight composition of the distillate? Compare the results with the composition at 760 mm, in Figure 4.2.

5. (a) Calculate the distillation temperature and the theoretical weight composition of the distillate for the steam distillation of *p*-dibromobenzene at 760 mm.

   (b) How would a mixture of bromobenzene and *p*-dibromobenzene behave when subject to steam distillation ?

Melting Points, Crystallization, and Sublimation

This chapter is concerned with solids—their melting behavior as a criterion of purity—and two common methods of purification: crystallization and sublimation.

5.1 Melting Points

The Theory of Melting-Point Depression. The molecules of crystals are aligned in regular patterns. As the temperature of a crystal is raised, the increasing vibrational motions of the molecules makes it more difficult for the regularity to be preserved. Eventually a temperature is reached (the *melting temperature* or *melting point*) at which the pattern is broken and the solid *melts* and turns into a disordered liquid. A pure crystalline substance usually possesses a sharp melting point; that is, it melts completely over a very small temperature range (in practice, not more than 0.5–1.0°, provided good technique is used). The presence of even small amounts of impurities soluble in the molten compound will usually produce a marked depression of the temperature at which melting begins as well as a smaller depression of the temperature at which the last crystal disappears, resulting in a large increase in the melting-point range. The amount of lowering of the final temperature at which the last crystal disappears is called the melting-point depression.

The melting-point depression by impurities is a consequence of the different effects of impurities on the vapor pressures of solutions and of solid mixtures. As illustrated in Figure 5.1, the melting point of a pure substance is the temperature at which the solid and liquid have the same vapor pressure. Addition of a soluble

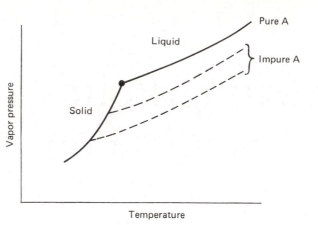

FIGURE 5.1
*Vapor Pressure of
Substance A in Solid
and Liquid Phases*

impurity to a molten substance will reduce the mole fraction of the substance (below 1.0) and lower the vapor pressure above the solution. The dashed lines in Figure 5.1 represent vapor pressures at different temperatures for two concentrations of impurities.

The situation is entirely different in the solid phase. Here, the original substance and the impurities usually form a *heterogeneous* mixture of crystals of each substance. The crystals are so intimately mixed that it is impractical to try to separate them; yet at a molecular level, they behave as though they were independent of each other. As a consequence, the vapor pressure of the solid substance is essentially unaltered by the presence of impurities, which may be thousands of angstroms (Å) away. This behavior is shown in Figure 5.1 as a vapor pressure curve for the solid that is independent of the presence of impurities. Since by definition the temperature at which melting ends is the temperature at which the solid and melt have the same vapor pressure (the point of intersection of the solid and liquid curves), that temperature will be depressed by the presence of impurity. The greater the amount of impurity (at least up to a point, to be described later), the larger the melting-point depression is.

Molecular-Weight Determination. An instructive application of the phenomenon of melting-point depression is the Rast method for determining molecular weights. Many substances, frequently those that have approximately spherical shapes, show unusually large depressions of melting points when impurities are added. For example, the melting point of camphor is depressed by 38.5° when the concentration of impurity is 1 mole/1000 g of camphor. In the Rast method a weighed sample of the unknown is intimately mixed with a weighed sample of camphor, and the final melting point of the mixture determined. From the observed lowering of the camphor's melting point, the molecular weight of the unknown is calculated from the following expression.

$$\text{Molecular weight} = 38.5 \times 1000 \times \frac{\text{weight of unknown}}{\text{weight of camphor} \times \Delta T}$$

In this expression, the figure 38.5 is called the melting-point-depression constant and is characteristic of camphor.

A useful variation of the Rast method is to determine the freezing point of cyclohexanol solutions of the unknown. Cyclohexanol (mp 24.7°) has a melting-point-depression constant of 42.5.

Substance Identification.

The melting points of mixtures (colloquially referred to as *mixed melting points*) are used frequently to establish the identity of two samples.

A typical situation is when a substance to be identified, compound X, is suspected of being identical with one or the other of two known substances, A or B, and where the three pure compounds, A and B and X, have approximately the same melting point. Mixtures of about equal amounts of A and X, and B and X, are prepared and the melting points of these mixtures are determined.[1] If A and X are identical, the melting point of a mixture of A and X in any proportion will always be the same as that of A or X alone, apart from any slight differences resulting from different purities of the samples. If A and X are different from one another, the melting point of the mixture of A and X will usually (but not invariably) be lower than that of either A or X. Similar reasoning is applied to the mixture of B and X. If the melting points of the mixtures of A and X, and of B and X, are both below the original melting points of the pure substances, one can safely conclude that X is not identical with either A or B. There are some mixtures that are exceptions to the general rule that the melting point of a pure compound will be lowered by the presence of a soluble impurity or by admixture with a different substance (see the next paragraph). Consequently, if the melting point of X is not depressed by mixing with A (or with B), one can conclude that X is probably identical with A (or B), but one cannot assert this with absolute certainty. Comparisons of other physical constants of the two materials are necessary to establish an identity beyond any reasonable doubt.

If one is making a thorough comparison of substances A and X, it is better to prepare a number of mixtures that contain varying proportions of the substances to be identified and the known substance; for example, a mixture containing 10% of A and 90% of X, another containing 50% of each, and a third mixture containing 90% of A and 10% of X. The results of the melting-point determinations of these mixtures may be then plotted and a rough melting-point curve may be drawn. For this purpose, the temperature plotted is that at which the mixture liquefies completely. The nature of the melting-point curves thus obtained is shown in Figure 5.2.

---

[1] In practice, it is highly desirable to determine the melting points of the mixtures and of a control sample of pure X in a single operation, so that the rate of heating will be the same for the three samples. To avoid confusion, the individual tubes must be marked in some way (by making one, two, and three small file scratches at the upper end, or cutting the tubes to three different lengths, etc.) and arranged in a definite manner on the thermometer. The symbols and arrangement should be jotted down in the notebook. Differences in the behavior of the samples are more significant than the actual temperature readings.

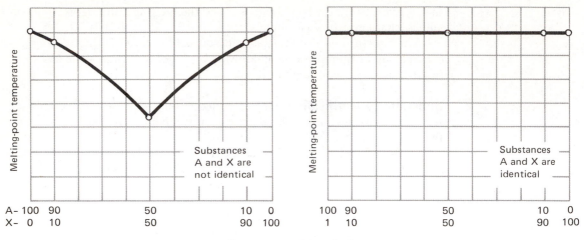

FIGURE 5.2 *Melting-Point Behavior of Mixtures*

Apparatus for Melting-Point Determination.

One common method for determining melting points is to use the Thiele apparatus illustrated in Figure 5.3. The sample is contained in a capillary tube. The thermometer is inserted into a drilled cork or rubber stopper and supported by means of a buret clamp. The thermometer position is adjusted so that it is centered vertically in the Thiele tube, with the upper end of the mercury bulb 1 cm below the side arm of the Thiele tube. The capillary tubes usually come with one end sealed. If yours do not, a tube can be sealed by touching one end to a small, hot flame.

A small amount of the material to be examined (0.05 g is ample) is pulverized finely by crushing with a spatula or knife blade on a piece of smooth hard paper or a watch glass. The crushed material is collected into a small mound, and the open end of the capillary tube is thrust into it. The tube is then inverted, and the solid is shaken down into the tube by drawing a triangular file gently along the upper portion of the tube and then tapping the lower end on the desk top; alternatively, the solid may be forced down by dropping the tube (sealed end downward) through a 2′ length of ordinary glass tubing onto the desk top. Further increments of the sample are introduced in the same way, until the material forms a compact column 3–5 mm high at the bottom of the tube after repeated tapping. It is essential that the material be pulverized finely and packed tightly to ensure transfer of heat throughout the sample.

Although the capillary tube will usually adhere to the thermometer by capillary action of a thin film of bath liquid,[2] it should be attached more firmly by means of a thin slice of rubber tubing or a small rubber band. The tube should be

---

[2] Liquids suitable for bath temperatures up to about 250° are medicinal paraffin oil (Nujol, etc.), anhydrous glycerol, di-*n*-butyl phthalate, and Dow-Corning Silicon fluid #500 (mixtures of organosilicon compounds). Concentrated sulfuric acid is unsafe and should be avoided.

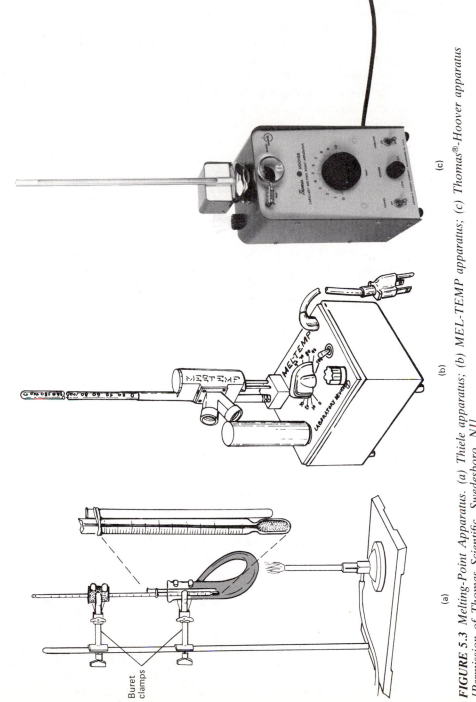

FIGURE 5.3 *Melting-Point Apparatus. (a) Thiele apparatus; (b) MEL-TEMP apparatus; (c) Thomas®-Hoover apparatus [Permission of Thomas Scientific, Swedesboro, NJ].*

Buret clamps

(a)

(b)

(c)

adjusted so that the sample is just alongside the mercury bulb of the thermometer, and the rubber fastening should be above the level of the bath liquid (to avoid softening of the rubber and discoloration of the bath).

The Thiele tube is heated at its lowest point. The resulting convection currents will circulate the oil in a counterclockwise direction if the apparatus is set up as shown in Figure 5.3. The tube may be heated at a fairly rapid rate until the bath temperature approaches within about 15–20° of the melting point (roughly determined in a preliminary trial, if necessary). Heating is then continued with a very small flame, adjusted so that the temperature rises slowly and regularly, at a rate of about 2°/min. The observed melting point is reported as the temperature range beginning with the thermometer reading when the substance starts to liquefy and ending with the reading when the melt becomes clear. The temperature readings and any other observations are recorded at once in the notebook.

The thermometer used for melting points should be one that has been standardized by one of the methods described in Section 2.3. Ideally, the observed readings should also be corrected for the exposure of the mercury column to atmospheric cooling, but usually this stem correction is omitted, even in research. Whether or not this correction has been made should be indicated by a notation such as, "mp 172–173° (uncor)" or "mp 172–173° (cor)." Melting points tabulated in books are frequently given as a single value, and generally refer to the upper, liquefaction temperature. After a single use the melting point tube is discarded in a "waste glass" container.

A more refined version of the simple Thiele apparatus includes a small electrical stirrer and is heated electrically by a coil of resistance wire. Another form uses a beaker, in place of the Thiele tube, with an electrical stirrer and heater. Usually these devices are equipped with some optical magnifying device for easier viewing of the sample.

One of the problems with the Thiele apparatus and its more refined variations is that at temperatures above 220–250° the usual bath liquids begin to smoke and decompose. For melting points in the higher temperature ranges, a metal block with vertical holes for the thermometer and capillary tube and a small window for observation of the sample may be used. A very convenient commercial version of such a device is the MEL-TEMP unit[3] (Figure 5.3). It is electrically heated, allows up to three samples to be melted simultaneously, and contains a built-in light and optical viewer. Heated blocks depend on thermal conduction rather than convection for heat transfer from the heat source. For this reason, a heating block is less responsive to changes in the heat level, and care must be taken not to raise the temperature too quickly and shoot past the melting-point range.

Formation of Eutectic Mixtures. From the previous discussion one might assume that as the amount of impurity increased, the magnitude of the melting-

[3] The MEL-TEMP unit is available from Laboratory Devices, P.O. Box 68, Cambridge, MA 02139.

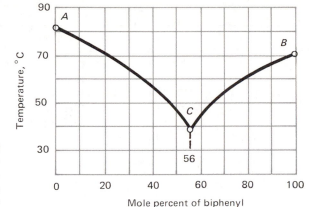

FIGURE 5.4
*Typical Melting
Point–Composition
Diagram for a
Mixture*

point depression would increase indefinitely. Actually, if the impurity concentration becomes very large, it changes roles and becomes the main substance. If one examines the equilibrium melting temperatures of the entire range of concentrations of two substances, the behavior illustrated in Figure 5.4 is typically observed. This is the curve for mixtures of naphthalene and biphenyl. The curve *A–C* shows the lowering of the final melting point of napthalene (*A*, mp 80°) by addition of biphenyl, and *B–C* the lowering of the melting point of biphenyl (*B*, mp 69°) by naphthalene. The intersection *C* is called the eutectic point and is the temperature (39.4°) at which both crystalline solids can exist in equilibrium with a melt of fixed composition (44 mole percent naphthalene + 56 mole percent biphenyl).

5.2 Crystallization

In a typical laboratory preparation, a crystalline solid separating from a reaction mixture is contaminated with some impurities. Purification is accomplished by crystallization from an appropriate solvent. The procedure consists of the following steps.

1. Dissolving the substance in the solvent at an elevated temperature.
2. Filtering the hot solution to remove insoluble impurities.
3. Allowing the hot solution to cool and deposit crystals of the substance.
4. Filtering the cold solution to separate the crystals from the supernatant solution (quaintly still known as the mother liquor).
5. Washing the crystals to remove adhering mother liquor.
6. Drying the crystals to remove the last traces of solvent.

There are several factors to consider in selecting a suitable solvent for crystallization. A good solvent for crystallization is one that will dissolve a

TABLE 5.1
Common Solvents for
Crystallization

Solvent	Bp, °C	Solvent	Bp, °C
Water	100	Methylene chloride[b]	40
Methanol[a]	64	Carbon tetrachloride[b]	77
Ethanol (95%)[a]	78	Cyclohexane[a]	81
2-Propanol[a]	82	Benzene[a,b]	80
Diethyl ether[a]	34	Petroleum ether[a]	60–90
Acetone[a]	56	Ligroin[a]	90–150
Ethyl acetate[a]	78	Toluene[a]	111
Acetic acid (glacial)[a]	118	Xylene[a]	139

[a] Flammable liquid—requires caution against fire hazards.
[b] Prolonged exposure may induce cancer.

moderate quantity of the substance at an elevated temperature but only a small quantity at low temperatures. The solvent should dissolve the impurities readily (except for mechanical impurities), even at low temperatures, and should be easy to remove from the crystals of the purified substance. It is essential that the solvent not react in any way with the substance to be purified. Other factors such as flammability and cost should also be taken into consideration. Table 5.1 lists some of the more common solvents used for crystallization.

In selecting a solvent for the purification of a given substance, one should consider its effectiveness for removal of the particular impurities that are likely to be present. The following general categories of impurities may be encountered.

Mechanical impurities (dust, grit, particles of paper, etc.) are readily removed by filtering the hot solution, since they are insoluble in all of the common solvents. Inorganic salts may often be separated in this way by using an organic solvent in which they are insoluble; an alternative method is to wash the crystals before recrystallization with a solvent such as water, in which the inorganic salts are soluble and the organic compound is insoluble.

Traces of coloring matter and resinous impurities may often be removed by warming the solution with a small amount of decolorizing carbon (about 0.2 g/100 mL of solution) or other adsorbent (Norit, Darco, Nuchar, etc.) before filtering the hot solution. The action of decolorizing agents varies widely, and effectiveness in removing a particular impurity may differ markedly from one solvent to another. An excessive amount of decolorizing agent should be avoided, since it will also adsorb the compound that is being purified, thus reducing the amount of pure compound isolated.

Impurities more soluble in the solvent are readily removed by crystallization, since they will be retained in the mother liquor. Likewise, impurities having about the same solubility as the substance being purified, when present in moderately small amounts, are readily eliminated in the mother liquor.

Impurities less soluble in the solvent are very difficult to remove if they are present in considerable amount, since the hot solvent will dissolve an appreciable amount of the impurity and, on cooling, the impurity will crystallize out and contaminate the product. It is for this reason that one tries to select a solvent that will readily dissolve the impurities, even at room temperature. Unfortunately, one usually does not know the structures of the impurities and must resort to trial and error.

If information is lacking about the solvents suitable for crystallizing the substance to be purified, the common solvents must be tested experimentally, on a small scale, to select a satisfactory one.[4] This is done conveniently by using a series of small test tubes and placing in each tube a small quantity of the substance to be purified and a small quantity of the solvent to be tested. The solubility is tested in room-temperature solvent and at its boiling point, and one notes whether well-formed crystals are produced abundantly on allowing the hot solution to cool. If two or more solvents appear to be promising, each may be tested more thoroughly with larger, weighed samples of material, to determine the loss of weight in the crystallization process and to compare the purities of the recrystallized samples.

To secure a satisfactory recovery of the purified material it is essential to avoid using an unnecessarily large amount of solvent. The quantity of the substance lost through retention in the mother liquor will be minimized if the sample is dissolved in the smallest possible amount of the hot solvent. In practice it is desirable to employ slightly more than the minimum quantity of hot solvent required to dissolve the sample (2–5% more), so that the hot solution will be not quite saturated with the solute. This aids in avoiding separation of crystals as a result of slight cooling during filtration of the hot solution, which may clog the filter and funnel. With a substance that melts at a temperature below the boiling point of the solvent, enough solvent should be used so that the dissolved substance does not begin to separate at a temperature above its melting point; otherwise, the material will separate at first in oily droplets instead of well-formed crystals. If this happens, the only practical solutions are to try cooling the solution more slowly and to use more solvent. Using more solvent will increase the loss of product.

During the preparation of the solution in the hot solvent, the liquid should be stirred or shaken to aid the solution process, as many organic substances dissolve quite slowly. It is advantageous to crush any large crystals or lumps of the sample beforehand. One should avoid using an excessive amount of solvent through haste or attempting to bring insoluble foreign matter into solution. It is better to err on the side of an insufficient amount of solvent; the undissolved residue remaining after filtration or decantation of the hot solution may then be tested with a fresh portion of solvent to see if some of it will dissolve.

Frequently a mixture of two solvents (a solvent pair) is more satisfactory than a single solvent. Such solvent pairs are made up of two mutually soluble liquids, one of which dissolves the substance readily and another that dissolves it very sparingly. Examples of solvent pairs are ethanol and water, glacial acetic acid

---

[4] For hints on general solubility relationships, see Section 6.4; the experimental procedure is illustrated in Section 5.4F.

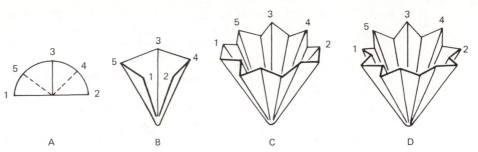

FIGURE 5.5
*Preparation of a
Fluted Filter Paper*

A   B   C   D

Select a filter paper and glass funnel of such dimensions that the top of the folded paper is 5–10 mm *below* the upper rim of the funnel (120-mm-diam. paper for a 65-mm funnel, 185-mm paper for a 100- to 120-mm funnel, etc.). *A*: Fold the paper in half, and again into quarters (fold *3*). *B*: Bring each edge *in to* the center fold and crease again, producing new folds at *4* and *5*. Do not crease the folds tightly at the center—this would weaken the central portion so that it might break during filtration. Grasp the folded filter cone (*B*) in the left hand, and make a new fold in each segment—between *2* and *4*, between *4* and *3*, and so on—*in a direction opposite* to the first series. The result is a bellows or fan arrangement. *C*: Open the filter paper completely and note the two places, next to folds *1* and *2*, where the paper would lie flat against the side of the funnel. *D*: Fold each of these sections in half, forming two smaller flutings, only half as deep as the others. Strengthen all of the flutings by creasing lightly a second time, and the paper is ready for use.

and water, 2-propanol and petroleum ether, and cyclohexane and ethyl acetate. A typical procedure in using a solvent pair, such as ethanol and water, is to dissolve the substance in the better solvent at the boiling point and add the poorer solvent dropwise, with shaking or stirring, until a faint turbidity persists in the hot solution. A few drops of the better solvent are then added, slightly more than the minimum amount required to form a clear solution, so that the hot solvent mixture will not be quite saturated with the solute. The hot solution is subsequently treated in the usual way (clarified with carbon, filtered, etc.). The main difficulty with using solvent pairs is the tendency of one of the components to evaporate preferentially during the heating process so that the solvent composition and sample solubility change with time. Care must be taken to minimize the evaporation.

Filtration of the hot solution to remove insoluble impurities, decolorizing carbon, and so on, must be carried out rapidly and efficiently to avoid crystallization of the dissolved substance in the filter and funnel. There are two ways to do this. The traditional method is to use a 60° funnel fitted with a fluted filter paper (Figure 5.5), which allows rapid filtration. Another method, better suited to small samples, employs a filter tube of the type shown in Figure 5.6. The constricted neck of the filter tube holds a small wad of glass wool that substitutes for the filter paper of the traditional method. With either setup a very short wide stem is employed to avoid clogging due to separation of crystals in the stem. Just before the filtration is started the funnel (or tube) should be pre-

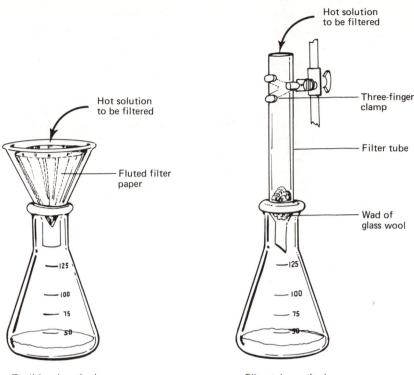

FIGURE 5.6
Filtration of Hot Solution by Two Methods

Hot solution to be filtered

Fluted filter paper

Traditional method

Hot solution to be filtered

Three-finger clamp

Filter tube

Wad of glass wool

Filter-tube method

heated by placing it in the mouth of the flask in which the hot solution is being prepared and allowing the hot vapors to warm the funnel.

It is usually desirable to produce small, uniform crystals of the purified material. To accomplish this, reheat the hot filtered solution to redissolve any crystals that have deposited during filtration; crystals that have formed in the funnel may be recovered by placing the funnel in the mouth of the flask containing the heated solution, where condensed vapor of the solvent will redissolve the crystals. The clear solution is covered and set aside to cool undisturbed. With a substance that contains only traces of impurity and crystallizes readily, it may be advantageous to cool the hot solution rapidly with vigorous stirring, so that small, uniform crystals are obtained. Since organic substances differ greatly in their rates of crystallization and materials of varying degrees of impurity will be encountered, it it necessary to adapt the crystallization procedure to the specific material at hand.

Not infrequently, one encounters a crystalline organic compound that exhibits a marked tendency to separate from solution, even at temperatures well below the melting point of the pure substance, in the form of an oily liquid that cannot easily be induced to crystallize. This situation arises especially with low-melting solids and with mixtures of closely related compounds. In such cases satisfactory crystallization may be obtained by allowing the solution to cool slowly and to remain undisturbed for a considerable period of time. Often it is advantageous to inoculate the solution with a few tiny crystals of the desired product (seed crystals), which will serve as crystal nuclei, and then to allow the solution to remain undisturbed. The necessary seed crystals may be secured by inducing crystallization in a test portion by vigorous scratching with a glass rod or by reserving a small portion of the original sample of material. If a solution is already supersaturated, cooling to a very low temperature may not hasten crystallization, since the rate of crystallization is apt to be reduced by lowering the temperature.

It may be necessary to prepare a dilute solution and allow the solvent to evaporate slowly to secure crystals, but this practice should be used only as a last resort, since a pure product is difficult to obtain in such a manner.

The purified crystals can be collected and freed from the mother liquor by filtration using either of the assemblies shown in Figure 5.7. On the left is shown a Büchner funnel connected to a heavy-walled suction filtering flask with a side tube for attachment to a water pump to furnish suction. This assembly is suitable for collecting amounts of solid down to about half a gram. For smaller amounts the Hirsch funnel, shown at the right of Figure 5.7, is better. Whichever funnel is used, it is important that the inner portion, beneath the perforated plate, should be perfectly clean and all of the holes in the plate should be open (clean with a pin if necessary). The filter paper should be selected (or trimmed if necessary) so that it lies flat on the perforated plate and does not fold up against the side. It is desirable to moisten the filter paper with the solvent used in the crystallization process and apply suction before the filtration is started. The suspension of crystals is then poured onto the filter in such a way that a layer of uniform thickness is collected. If crystals adhere to the walls of the flask in which crystallization was carried out, they may be washed out with some of the filtrate.

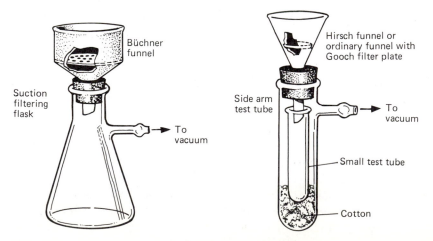

FIGURE 5.7
Apparatus for Suction Filtration

Büchner funnel

Suction filtering flask

To vacuum

Hirsch funnel or ordinary funnel with Gooch filter plate

Side arm test tube

To vacuum

Small test tube

Cotton

An important objective in collecting the recrystallized product is the *complete separation of the crystals from the mother liquor containing the dissolved impurities*. The proper procedure is as follows. The crystals are first pressed down firmly on the filter with a spatula and sucked as dry as possible. Before the crystals are washed, the suction tube is disconnected and the crystalline cake is loosened *carefully* (avoid tearing the filter paper) with a spatula. The fresh cold solvent for washing is added in small portions and the material is stirred into a smooth paste. When this has been accomplished, suction is applied again and the crystals are pressed down firmly as before to remove the wash liquid as completely as possible. Washing the crystals with two or three small portions of solvent is more effective than a single washing with the same total amount of solvent. It is particularly important in the washing operations to stop the suction and break up the crystalline cake, so that the whole mass comes into contact with the fresh solvent.

Another method, suitable for separating small amounts of crystals from the mother liquor, is to centrifuge the mixture in a centrifuge cone and draw off the liquid with a Pasteur pipet. This method fails if the densities of the solid and the solvent are close.

After the washed crystals have been pressed firmly and sucked as dry as possible, the crystalline cake is removed and spread out on a watch glass to permit evaporation of remaining traces of solvent. The final stages of drying may be done in a desiccator over an appropriate solid drying agent or in a drying oven regulated to a temperature well below the melting point of the substance. All traces of the solvent should be removed before the melting point of the purified substance is taken because the presence of solvent will appreciably lower the melting point.

5.3 Sublimation

The sublimation of a solid substance is an unusual type of purification process in which the solid undergoes direct vaporization and the vapor is condensed back to the solid state, *without the intermediate formation of a liquid state*. For sublimation to occur, the solid must have a relatively high vapor pressure at a temperature below its melting point. As few organic compounds fulfill this condition, the sublimation method is not used frequently in ordinary laboratory work. However, when it is applicable, it yields a product of high purity; it is particularly valuable for the isolation of a volatile solid from colored gums and tars and is best adapted to use with small amounts of material.

A simple apparatus suitable for sublimation of 0.2–1.0 g of sample at ordinary or reduced pressures is diagrammed in Figure 5.8. The sample is contained in a large side-arm test tube. The test tube is fitted with a rubber stopper that has been drilled to accept a 16×50-mm test tube, which, when filled with ice, serves as a "cold finger" to condense the rising vapors. If the sample has a high enough vapor pressure to be sublimed at atmospheric

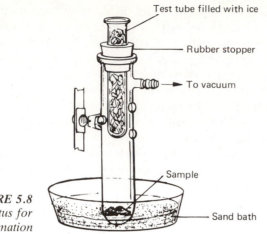

Test tube filled with ice

Rubber stopper

To vacuum

Sample

Sand bath

FIGURE 5.8
Apparatus for
Sublimation

pressure, the side arm of the test tube is connected to a drying tube (Section 8.1) to prevent atmospheric moisture from condensing on the test tube. If the sample has too low a vapor pressure, the drying tube is omitted and the side arm is connected directly to the vacuum system.

Examples of substances that can be purified by sublimation at atmospheric pressure are camphor, hexachloroethane, anthraquinone, and many other quinones.

5.4 Representative Procedures

(A) Melting Point of Mixtures

Prepare three different mixtures of benzoic acid and 2-naphthol having roughly the following proportions of the two components: (1) about 90% benzoic acid and 10% 2-naphthol; (2) about 50% of each; (3) about 10% benzoic acid and 90% 2-naphthol. For the purpose of this experiment it is permissible to judge the proportions very roughly by the relative lengths of long thin piles of the powdered crystals (9 parts benzoic acid and 1 part 2-naphthol for mixture No. 1, etc.). The components must be powdered finely, and each sample mixed *very thoroughly*, on a piece of smooth hard paper or a watch glass, by means of a clean spatula or knife blade.

Introduce one of the mixtures into a melting point tube marked appropriately for identification (see footnote 1, Section 5.1), and determine the melting point of the mixture by the following procedure.

Apply heat at a moderate rate until the bath liquid is within 15–20° of the melting point (for this pair of compounds to about 100°; when necessary, make a rough preliminary determination). Continue the heating so that the temperature rises slowly and at a uniform rate (about 2°/min). Observe carefully the sample in the melting point tube and the thermometer reading. Record as the observed melting point the range between the thermometer reading when the sample starts to liquefy and that when the melt is clear. Note also whether the sample undergoes preliminary fusing together (sintering) or discoloration, melts sharply or slowly over a wide range, and so on. Repeat this process for each of the other samples, being careful to allow the bath to cool below 100° before inserting the next sample. If care is taken, it is possible to observe the melting point behavior of all three samples at once.

After the samples have melted, allow the bath to cool. Melting point tubes are discarded into the waste crock (not into the sink) after a single use. Record the observed melting points directly in your notebook. From your results, draw a rough melting-point curve for mixtures of benzoic acid and 2-naphthol, plotting compositions on the x-axis and melting points on the y-axis. For this purpose, the temperature used is that at which the mixture liquefies completely.

(B) Crystallization of Benzoic Acid from Water

Weigh out 0.5 g of impure benzoic acid for recrystallization,[5] and transfer it to a 50-mL Erlenmeyer flask. Add about 10 mL of water, and heat the mixture to the boiling point. Add successive small portions of hot water, while stirring the mixture and boiling gently, until the benzoic acid has dissolved completely; then add an additional 2–4 mL of the hot solvent. The objective is to dissolve the solid in slightly more than the minimum amount of hot solvent. Do not use too much solvent in an attempt to dissolve resins, mechanical impurities, and the like.

Remove the hot solution from the heat source and allow it to cool slightly (for a few minutes). To the hot solution add gradually, with care to avoid excessive foaming, about 0.05 g (1/10 teaspoon) of decolorizing carbon (Norit, Darco, etc.); boil again for a short while to aid in removing small amounts of colored impurities. Meanwhile prepare a fluted filter paper[6] (Figure 5.5) and place it in a warmed funnel supported in a clean flask that will receive the filtrate.

The funnel is warmed so that it does not cool the solution prematurely. One method is to heat the funnel, inverted, on a steam bath, and then dry it with a towel. Alternatively, one can place the funnel in the neck of the receiving flask, to which a little fresh solvent has been added, and heat the flask so the hot vapors condense on the funnel. Place the fluted filter in the funnel and, without allowing the funnel or the solution to cool, pour the solution into the filter. If the

---

[5] The impurities may include soluble and insoluble materials as well as colored substances.

[6] Filter tubes prepared with glass wool plugs do not work well for removing charcoal because it is so fine. Even with filter paper a few fine particles of the decolorizing agent may pass through. Often this may be remedied by heating the filtrate to boiling and filtering again through the same filter; or a small amount of filter aid (Filter-Cel, Super-Cel, etc.) may be added. If the filter has been damaged in folding or is defective, a fresh one must be used.

solution cannot be filtered in a single portion, reheat the remainder briefly and then filter it. As soon as all of the solution has been filtered, cover the mouth of the flask containing the hot filtrate with a watch glass and allow it to stand and cool undisturbed. Do not close the mouth of the flask tightly with a stopper.

If crystals have separated in the hot filtrate during filtration, the filtrate should be heated to redissolve them. Crystals that have formed in the funnel may be recovered by placing the funnel in the mouth of the receiver and allowing condensing vapor from the hot solution to redissolve them.

When the filtrate has cooled and the product has separated completely, filter the crystals with suction through a Hirsch funnel. Wash the crystals twice with a little cold water, then press them as dry as possible on the funnel with a spatula. Spread the crystals on a watch glass, *allow them to dry completely*, and weigh them. Determine the melting point (see part (A)); if the product is not sufficiently pure, as shown by a melting range greater than 1–2°, repeat the recrystallization procedure. Calculate the percentage recovery of pure benzoic acid.

(C) Crystallization of Acetanilide from Water[7]

Recrystallize a l-g sample of impure acetanilide from about 20 mL of water using a 50-mL Erlenmeyer flask. Follow the general directions given in part (A). About 0.05 g of decolorizing charcoal should be sufficient. In washing the crystals, be careful to use ice-cold water and not too much of it, since acetanilide is somewhat soluble even in cold water (about 0.5 g/100 mL).

(D) Crystallization from a Flammable Solvent

Weigh out 0.3 g of impure dimethyl terephthalate for recrystallization and transfer it to a 50-mL Erlenmeyer flask. Add 5 mL of 95% ethanol and warm the mixture on a steam bath until the solvent boils. Add successive small portions of ethanol (not more than 1–2 mL total) and boil gently after each addition, until the dimethyl terephthalate has dissolved; then add 0.2–0.4 mL more of the solvent. Do not attempt to dissolve admixed particles of sand, grit, and so on. Meanwhile, prepare a fluted filter (Figure 5.5) and arrange a funnel (65-mm diameter, 10-mm stem) and a flask to receive the filtrate.

⚠ *CAUTION*   Flammable solvents such as ether, alcohols, and hydrocarbons must never be heated in an open flask over a burner, or manipulated near a flame. These solvents should be heated on a water or steam bath (or an electric hot plate having no exposed hot filament or switching element). If a burner must be used, it is essential that the flask be fitted with an upright reflux condenser. Take particular care to ensure that no lighted burner is nearby during the filtration of the hot solution!

---

[7] Acetanilide is a challenge to recrystallize because in concentrated aqueous solutions it tends to separate as an oil rather than a crystalline solid. At the same time, excess solvent must be avoided because acetanilide is somewhat water soluble at room temperature.

Remove the boiling solution from the steam bath, add gradually 0.05 g of decolorizing carbon (**Caution**—*frothing!*),[8] and swirl the solution gently. Reheat to boiling and pour the hot solution into the fluted filter (see footnote 6). Cover the mouth of the flask containing the hot filtrate with a watch glass and allow it to stand and cool undisturbed. When the product has separated completely, collect the crystals with suction on a Hirsch funnel; wash all the crystals into the funnel by rinsing the Erlenmeyer flask with part of the filtrate. Discontinue the suction and wash the crystals with two 1-mL portions of fresh ethanol (cold). Apply suction again and press the crystals firmly with a spatula. Spread the crystals on a watch glass, allow them to dry thoroughly, and record the weight of purified dimethyl terephthalate. Calculate the percentage recovery and determine the melting point of the purified product.

(E) Crystallization of Naphthalene from Ethanol

Recrystallize a l-g sample of naphthalene from about 5 mL of ethanol in a 50-mL Erlenmeyer flask. Follow the general directions given in part (C). About 0.05 g of decolorizing charcoal should be sufficient. Care is required in drying the crystals, since naphthalene is volatile and will sublime slowly at room temperature.

(F) Selection of a Recrystallization Solvent

The purpose of this experiment is to illustrate the process for selecting a recrystallizing solvent. Obtain from your instructor 1 g of a solid unknown and a list of solvents you are to try.

In a 1-mL test tube place 0.1 g of the solid and add 1 mL of the first solvent. Stir the sample with a rounded stirring rod and gently crush the crystals to increase their surface area. If the solid dissolves at room temperature, the solvent is not suitable for recrystallizing the unknown. If the solid does not dissolve, warm the test tube cautiously until the solvent begins to boil. Stir the mixture for a few minutes near the solvent's boiling point to give the solid a chance to dissolve. If the solid dissolves, cool the solution to room temperature and note if the solid crystallizes. If the amount of crystals is approximately the same as the amount of crude solid you started with, then you have found a suitable recrystallizing solvent.

It may happen that the solid only partially dissolves at the boiling point of the solvent. In that case, you can add more cold solvent and repeat the heating and cooling process. You should not use more than a total of about 5 mL of solvent since it is difficult to separate a small amount of crystals from a large volume of solvent.

Repeat the process described above using the other solvents listed by your instructor.

---

[8] It should be emphasized that decolorizing charcoal is added only if colored impurities are present. The use of charcoal always causes some loss of sample.

If indicated by your instructor, recrystallize the remainder of the solid unknown using the best recrystallization solvent that you have identified; adjust the volume of solvent to match the weight of solid.

(G) Sublimation

Assemble the apparatus pictured in Figure 5.8. Place 0.3 g of impure *p*-dichlorobenzene in the side-arm test tube. Insert the inner test tube, attach a drying tube to the side arm of the flask, and fill the test tube with ice. Warm the flask gently with a heating mantle. When no more material sublimes, remove the condenser and scrape the condensate into a tared bottle. Record the weight of purified material and determine its melting range.

Another apparatus that can be used for sublimation of small samples at atmospheric pressure is a covered Petri dish. The sample is placed in the bottom dish, the bottom is covered with the top dish, and the pair heated gently on a hot plate. The sublimed material will collect on the top cover from which it can be scraped off.

Questions

1. Define the term *melting point*.

2. What general conclusions can be drawn from the determination of melting points of mixtures?

3. Suppose that two different organic compounds, M and N, have about the same melting point and that you are given an unknown organic compound, X, which also has the same melting point and is suspected of being identical with either M or N. Describe a procedure for identifying X and state the results you would obtain in each of the following situations.
   (a) X is identical with M.
   (b) X is identical with N.
   (c) X is not identical with either M or N.

4. (a) What physical constants other than melting point and melting points of mixtures may be used to aid in establishing the identity of an organic solid?
   (b) What constant other than boiling point can be used for identifying an organic liquid?

5. Explain why it is essential
   (a) to pack the sample densely and tightly in the melting point tube and
   (b) to heat the bath slowly and steadily in the vicinity of the melting point.

6. Outline the successive steps in the crystallization of an organic solid from a solvent and state the purpose of each operation.

7. What properties are necessary and what properties are desirable for a solvent to be well suited for recrystallizing a particular organic compound?

8. Not counting solubility relations and effectiveness for removal of impurities, in what respects would *n*-hexyl alcohol be a less desirable crystallization solvent than methanol or ethanol?

9. **(a)** Mention at least two reasons why suction filtration is preferable to ordinary gravity filtration for separating the purified crystals from the mother liquor.

   (b) Why is it desirable to release the suction before washing the crystals with small portions of the fresh solvent?

10. What means other than crystallization from a solvent may be used to purify an organic solid or to effect preliminary separation of a solid mixture?

6

Extraction with Solvents

There are two main applications of extraction in the organic laboratory: (a) the separation and isolation of substances from mixtures of solids, typically those that occur in nature and (b) the selective isolation of substances from solutions of mixtures that arise in synthetic chemistry.

6.1 Extraction of Solids

Examples of extractions of solid mixtures are the extraction of alkaloids from leaves and bark, flavoring extracts from seeds, perfume essence from flowers, and sugar from sugar cane. Solvents commonly used for this purpose are ether, methylene chloride, chloroform, acetone, various alcohols, and water. A common form of apparatus for continuous extraction of solids by means of volatile solvents is called a Soxhlet extractor. A typical laboratory setup employing this extraction device is shown, mounted and ready for use, in Figure 6.1.

Vapors from the solvent boiling in the pot rise through the vertical tube at the left into the condenser at the top. The liquid condensate drips into the filter paper thimble in the center, which contains the solid sample to be extracted. The extract seeps through the pores of the thimble and eventually fills the siphon tube at the left, where it can flow back down into the pot. If the sample being extracted is not volatile, it gradually accumulates in the pot. With the apparatus shown the siphoning action is intermittent. No liquid will flow through the siphon until the liquid level in the thimble reaches the top of the siphon tube. At that point almost all of the liquid in the siphon and the thimble drain out and the cycle of filling and draining starts again.

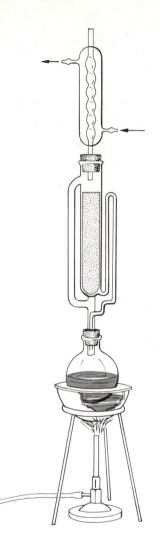

FIGURE 6.1
*Soxhlet
Extractor
Assembly*

6.2 Extraction of Solutions

A more common application of extraction is in "liquid-liquid" extraction, which is used to isolate a substance dissolved in one solvent by shaking the solution with another solvent immiscible with first, in a separatory funnel (see Figure 6.2). In the ideal situation, the substance is extracted into the second solvent, the impurities are left behind, and after the two layers are separated, the substance is isolated by removal of the solvent.

The general principle underlying this process is known as the distribution law. In dilute solutions a substance distributes itself between two immiscible solvents so that the ratio of the concentration in one solvent to the concentration in the second solvent always remains constant (at constant temperature). This constant ratio of concentrations for the distribution of a solute between two particular solvents is called the distribution coefficient or the partition coefficient for the substance between the two solvents.

$$\frac{\text{Distribution coefficient}^{1}\text{ of S}}{\text{between solvents A and B}} = \frac{\text{concn of S in A}}{\text{concn of S in B}}$$

$$= K \text{ (at constant temperature)}$$

where the concentrations are expressed in grams per milliliter of the solution.[2] It is important to observe the manner in which the ratio is expressed; for example, at 20° the distribution coefficient of butanoic acid between ether and water is approximately 5 (concn in ether/concn in water = 5), but if the ratio is expressed as the distribution between water and ether, the value is 0.2 (concn in water/concn in ether = 1/5).

In actuality, no two solvents are totally immiscible (insoluble in each other); there is always some solubility, small though it may be. In practice the most common application of the distribution law is to the extraction of dissolved substances from aqueous solutions by *almost water-insoluble* solvents such as ether, methylene chloride, and hexane. Ether, which is used frequently as an extraction solvent because so many neutral organic compounds are soluble in it, is somewhat more soluble in water than other common solvents; for example, at 30° diethyl ether is soluble to the extent of 1 g in 18.8 g of water (water is soluble to the extent of 1 g in 73 g of ether). In addition, ether presents a serious fire hazard. Methylene chloride (bp 40°) is less soluble in water (about 2 g/100 mL at 20°) and has the important advantage over ether that it is not flammable under ordinary conditions. Hexane is for all practical purposes insoluble in water, but it is highly flammable, although not so dangerously as diethyl ether.

The actual distribution coefficient is determined by bringing the solvents and solute into equilibrium at a given temperature and measuring the concentration of solute per unit volume of each separate phase. A rough approximation to the distribution coefficient can be made by determining the solubility of the solute in each pure solvent independently, since the distribution coefficient is approximately the ratio of the solubilities in the two solvents. The values obtained in this way are subject to several errors, but are usually good enough for simple laboratory calculations.

For application of the distribution law to laboratory extractions a convenient formula is shown in equation (6.1), which gives the fraction of compound

---

[1] In the case of dissolved substances that are ionized or associated in one of the solvents, the distribution law holds true for the ratio of the concentrations of the simple molecules only. To obtain an expression that holds for the total concentrations, it is necessary to introduce an expression for the ionization or association equilibrium.

[2] The values of the distribution coefficient found in the chemical literature are usually based on concentrations per volume of solution. However, for rough calculations of the type illustrated here, concentration per unit of solvent is much simpler to use. For dilute solutions, the error introduced in this way is negligible.

S remaining in volume V_A of solvent A after extraction with volume V_B of solvent B.

$$\text{Fraction remaining in A} = \frac{C_{final}}{C_{initial}} = \frac{1}{1 + \dfrac{V_B}{V_A K}} \qquad (6.1)$$

where C_{final} is the final concentration of S in A, $C_{initial}$ is the initial concentration of S in A, and K is the distribution coefficient (concentration of S in A divided by the concentration of S in B).

Example. Let us apply equation (6.1) to the situation where the distribution coefficient of the compound S between hexane and water (K hexane/water) is 1/3 at 25°. If a hexane solution containing 8 g of S in 100 mL of hexane is extracted at 25° with 100 mL of water, the fraction of S remaining in the hexane is

$$\frac{1}{1 + \dfrac{100 \text{ mL}}{100 \text{ mL} \times 1/3}} = 0.25$$

from which it follows that the weight of S in the hexane layer is 2.0 g (i.e., a reduction from 1.0 to 0.25, or three fourths) and the amount in the aqueous layer is 6.0 g.

    If the hexane layer is separated and extracted with a fresh 100-mL portion of water the amount of S will again be reduced by three fourths to give only 0.5 g remaining in the hexane layer. The second aqueous extract contains 1.5 g of S, which, when added to the 6.0 g of S in the first aqueous extract gives a total of 7.5 g of S in the combined aqueous layers.

6.3 Multiple Extractions

One can ask whether, with a specified quantity of solvent, it is preferable to make a single extraction with the total quantity of solvent, or to make several successive extractions (multiple extraction) with portions of the solvent. The answer, practical considerations aside, is that the amount of material extracted is greater with the second method.

    The general multiple extraction formula shown in equation (6.2) gives the fraction of compound S remaining in volume V_A of solvent A after n extractions with solvent B, each of volume V_B/n.

$$\text{Fraction remaining in A} = \frac{C_{final}}{C_{initial}} = \left(\frac{1}{1 + \dfrac{V_B}{V_A n K}} \right)^n \qquad (6.2)$$

where the terms have the same meaning as in equation (6.1).

In the limit of very large n it can be shown that equation (6.2) reduces to equation (6.3), which gives the theoretical maximum amount of S that can be extracted from a volume, V_A, of solvent A with a given volume, V_B, of solvent B.

$$\text{Fraction remaining in A} = e^{-V_B/V_A K} \qquad (6.3)$$

for large n, where $e = 2.718$, the base for natural logarithms.

With the example used in the previous section and assuming that both V_A and V_B equal 100 mL, the greatest possible reduction in the amount of S in the hexane is

$$e^{-100\,\text{mL}/(100\,\text{mL} \times 1/3)} = e^{-3} = 0.05$$

Thus, with just 100 mL of water it is theoretically possible to extract 95% (7.6 g) of S if the 100-mL portion is divided into a large number of smaller portions, which are used consecutively to extract the hexane layer. Note that multiple extractions are more effective than the previous example, which after extraction with a single 100-mL portion of solvent had removed only 75% (6.0 g). The comparison demonstrates the potential advantage of carrying out many extractions with small volumes of solvent instead of a single extraction with the same total volume of solvent.

Whether the increased efficiency of extraction is worth the trouble depends very much on the circumstances. It is worth noting that essentially the same amount of S can be extracted by successive extractions with two 100-mL portions of water. In practice, most chemists prefer to use a larger total volume of solvent and do only a few successive extractions.

6.4 Solubility Relationships

The distribution coefficient of an organic compound between two solvents cannot be predicted accurately, but it usually can be estimated with enough precision to make practical laboratory decisions. Since the distribution coefficient is very nearly equal to the ratio of the solubilities of the compound in the two solvents, its estimation reduces to estimating the relative solubilities. At a rough, practical level there are only three factors to consider: the ability of the compound to form hydrogen bonds (acidity and basicity), molecular charge, and molecular dipole moment. Saturated hydrocarbons have no charge, no significant dipole moment, and neither acidic or basic properties. As their older Latin-based name, paraffins, suggests, they do not interact significantly with any solvent. Their only interaction with themselves and other molecules is through weak van der Waals forces, and for this reason they represent an extreme molecular type. By contract, water is polar and both accepts and donates hydrogen bonds. Water interacts strongly with polar molecules and with ionic salts and represents another extreme molecular type. Hydrocarbons and water are quite insoluble in each other; hydrocarbon molecules that find their way into water interrupt the water's hydrogen-bonding network, an energy-costly process.

Alcohols contain a combination of hydrocarbon and hydroxyl groups and have properties intermediate between the two extremes. If the alcohol's alkyl group is small (e.g., methanol), the hydroxyl character dominates and the alcohol readily dissolves in water; if the alkyl group is large (e.g., octanol) the hydrocarbon character dominates and alcohol is only slightly water soluble but readily soluble in hydrocarbons.

Liquid ammonia and amines represent the analogous spectrum of solubility behavior among basic solvents for which the hydrogen-bond-accepting quality dominates.[3] Methylamine readily dissolves in water; octylamine is only slightly water soluble but readily soluble in hydrocarbons.

Quantitative Estimation of K. As it happens, there are many data for the distribution of organic compounds between water and octanol. If K_X is defined as the distribution coefficient of substance X between water and octanol (defined so that large values of K_X indicate a high octanol solubility and a low water solubility), then it is found empirically that

$$\log_{10} K_X = 0.50 + 0.54 n_{\text{alkyl}} + 2.63 n_{\text{phenyl}} - 1.98 n_{\text{OH}} - 1.92 n_{\text{\textbackslash C=O}}$$

$$- 1.38 n_{\underset{\text{OR}}{\text{\textbackslash C=O}}} - 1.92 n_{\text{\textbackslash N-}} - 4.24 n_{\underset{\text{ONa}}{\text{\textbackslash C=O}}}$$

where n_{alkyl} is the number of CH_3, CH_2, CH, and C alkyl groups present in compound X,

n_{phenyl} is the number of benzene rings
n_{OH} is the number of hydroxyl groups
$n_{\text{\textbackslash C=O}}$ is the number of aldehyde and ketone groups
$n_{\underset{\text{OR}}{\text{\textbackslash C=O}}}$ is the number of ester groups

$n_{\text{\textbackslash N-}}$ is the number of amine nitrogens
$n_{\underset{\text{ONa}}{\text{\textbackslash C=O}}}$ is the number of carboxylate salts

In applying this equation you do not count n_{alkyl} carbon atoms that are part of any of the designated functional groups.

---

[3] Another extreme solvent type is observed with nonhydroxylic solvents such a dimethylformamide, dimethyl sulfoxide, and acetonitrile for which polarity is the dominant quality. Such solvents are particularly effective at dissolving salts and in recent years have played an important role in synthetic chemistry.

Example. Cyclohexanone has *five* CH$_2$ carbons and *one* carbonyl group.

$$\log_{10} K_{\text{cyclohexanone}} = 0.50 + 0.54 \times 5 - 1.92$$
$$= 1.28$$
$$K_{\text{cyclohexanone}} = 10^{1.28} = 19$$

The observed value of $K_{\text{cyclohexanone}}$ is 6.5.

From the estimated coefficient you would conclude, correctly, that cyclohexanone could be extracted from water by octanol. Sometimes the estimates give poorer agreement than this, but usually they are slightly better. Any positive value for $\log_{10} K_X$ indicates that the compound could be extracted into octanol, and the larger the value, the more readily it will be extracted. A negative value for $\log_{10} K_X$ indicates that the compound would be extracted from octanol into water.

There are fewer data available for distribution coefficients for compounds between water and other solvents such as diethyl ether and methylene chloride, but the coefficients are approximately the same as for octanol. Because diethyl ether and methylene chloride do not have hydroxyl groups, basic substances will be less soluble in them than they would be in octanol. This is particularly true of carboxylate salts, which will almost invariably be more soluble in water than in diethyl ether or methylene chloride.

Reactive Solvents. Solubility tests, which are described in great detail in Chapter 9, are often used as a preliminary step in the identification of organic compounds. The basis for these tests is precisely the same as discussed above for extraction. Of particular interest for the extraction of mixtures of products are the so-called reactive solvents. In these solvents water-insoluble compounds are converted into water-soluble salts. Salt formation reduces distribution coefficient of the compound by a factor of thousands, and the compound thus is readily extracted into the aqueous phase. Any unreactive components of the mixture will stay behind in the organic solvent. Commonly used reactive solvents are dilute aqueous alkalis (5% sodium hydroxide or potassium hydroxide solutions), dilute aqueous mineral acids (5% hydrochloric acid, etc.), and cold concentrated sulfuric acid.

Dilute sodium hydroxide solution (also sodium carbonate or bicarbonate) can be used to extract an organic acid from its solution in an organic solvent, or to remove traces of acid that are present as an impurity in an organic preparation. The aqueous alkali converts the free acid to the corresponding sodium salt, which is soluble in water or dilute alkali. Thus, butanoic acid may be extracted quantitatively from a benzene solution by dilute sodium hydroxide because this reagent converts the acid to sodium butanoate, which is very soluble in water or dilute alkali but insoluble in benzene. Likewise, an organic acid or mineral acid present as an impurity in a water-insoluble neutral liquid or solid can be removed by washing with dilute alkali.

Dilute hydrochloric acid can be used in a similar way to extract basic substances from mixtures or to remove impurities. Dilute acids convert a base (amines, ammonia, etc.) into a water-soluble salt. Thus in the preparation of benzanilide, unreacted aniline can be removed by washing with dilute hydrochloric acid, and the benzoic acid side product can be removed by subsequent washing with sodium carbonate solution; the anilide does not react with either the acid or the base, but the amine is converted into the water-insoluble aniline hydrochloride, and the benzoic acid is converted into the water-soluble salt sodium benzoate.

Reaction mixture: benzanilide, aniline, benzoic acid

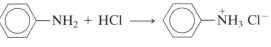

$$\text{Aniline} \quad -NH_2 + HCl \longrightarrow \quad -\overset{+}{N}H_3 \; Cl^- \quad \text{(water-soluble salt)}$$

$$\text{Benzoic acid} \quad -CO_2H + Na_2CO_3 \longrightarrow 2 \quad -CO_2^- \; Na^+ + CO_2 + H_2O \quad \text{(water-soluble salt)}$$

Cold concentrated sulfuric acid can be used to remove unsaturated hydrocarbons, alcohols, ethers, esters, and so forth, present as impurities in inert compounds such as saturated hydrocarbons and alkyl halides. Alkenes and alcohols are dissolved by chemical reaction to form alkyl hydrogen sulfates; ethers, esters, and so on, form addition complexes that are soluble in concentrated sulfuric acid.

Effect of Salts on Solubility. The solubility of organic substances in water is markedly affected by the presence of dissolved inorganic salts. For example, ethanol, which is perfectly miscible with pure water, is only slightly soluble in strong aqueous solutions of sodium chloride, potassium carbonate, and certain other inorganic salts. The same is true of acetone, pyridine, methanol, and many other water-soluble organic compounds. This phenomenon (salting-out effect) occurs commonly with salts that have ions of small radius and concentrated charge. The opposite effect of enhanced solubility in salt solutions (salting-in) occurs frequently when the salt has ions of large radius and diffuse charge. Benzene, for example, is about 40% more soluble in 1 M aqueous tetramethylammonium bromide than in pure water.

In the case of solutions of ionized organic substances, such as metallic salts of organic acids and salts of organic bases with mineral acids, it is possible that the common-ion effect may also decrease solubility of the salt. Thus, sodium benzenesulfonate is quite soluble in water, but is precipitated from an aqueous solution by adding sodium chloride; aniline hydrochloride is readily soluble in water, but is only slightly soluble in strong hydrochloric acid solutions.

6.5 Laboratory Practice

The selection of a solvent for extraction involves considerations analogous to those for crystallization. Properties desired for a suitable solvent are as follows.

1. It should readily dissolve the substance to be extracted (favorable distribution coefficient).
2. It should be sparingly soluble in the liquid from which the solute is to be extracted.
3. It should extract little or none of the impurities or other substances present.
4. It should be capable of being easily separated from the solute after extraction (usually by distillation).
5. It should not react chemically with the solute in an undesired way.

Relative cost, ease of manipulation, flammability, and similar factors also are significant to the choice among various possible solvents.

Large Sample Technique. If one has more than about 5 mL of solution to be extracted, the extraction is usually accomplished by shaking the solution to be extracted, together with the extraction solvent, in a glass separatory funnel. A long tapered funnel of the Squibb type, with a short stem cut off obliquely (Figure 6.2), is particularly convenient for this purpose. A cylindrical or globe-shaped funnel with a short stem may also be used. A long stem is disadvantageous because it holds a long column of the liquid that is being drawn off and makes for difficulties in manipulation.

The separatory funnel should be of such size that it is more than three-fourths filled by the solution and solvent. The funnel is shaken to obtain good physical mixing of the insoluble liquids and is then allowed to stand undisturbed

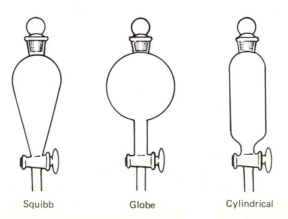

FIGURE 6.2
Extraction Funnels   Squibb   Globe   Cylindrical

FIGURE 6.3
*Correct Manner
of Holding and
Shaking the
Separatory Funnel*

Glass stopcock

until the layers have separated completely. Vigorous shaking is desirable unless the liquids will form an emulsion[4] that interferes with subsequent separation of the layers, in which case the layers are mixed by repeated gentle swirling.

During the shaking operation, it is best to grasp the funnel with both hands, one hand at the top and the other on the stopcock, in such a way that the stopper and stopcock are held firmly in place and the funnel is at an angle with the stem pointing upward (see Figure 6.3). The internal pressure should be released from time to time by opening the stopcock momentarily (with the stem pointing directly up) until the pressure is reduced. This is particularly important when a very volatile solvent, such as ether, is used.

⇒ *CAUTION*   Do not point the stem of the separatory funnel toward anyone when releasing the pressure. Drops of liquid caught in the stem may be ejected forcefully.

When the liquids have separated, the funnel is supported in an iron ring held on a ring stand. The stopper (the glass plug at the top of the funnel) is loosened or removed, and the lower layer is carefully drawn off into an Erlenmeyer flask. The stopcock (the glass plug at the bottom of the funnel) should be rotated with both hands to avoid withdrawing it and losing material. If the liquid remaining in the funnel is corrosive or valuable, place a beaker or pan beneath the funnel

---

[4] Emulsions of water solutions with solvents such as ether and benzene may often be broken by adding a little ethanol or ethyl acetate and swirling the contents of the funnel gently.

when it is to stand for any lenght of time. If only one layer is to be retained, *it is a safe rule to save both layers* until you are certain which one contains the desired material.[5]

As the interface of the two liquids approaches the stopcock, the liquid should be drawn off slowly. After the separation has been made, the funnel is rotated by a twisting motion with the stopcock closed, to assist in draining droplets of the more dense liquid from the walls. This small additional quantity is drawn off into the receiving flask. The upper layer is poured through the mouth of the funnel into a clean flask; it is not allowed to flow through the stopcock, as this would lead to contamination with the liquid adhering to the stem. The organic-solvent layer is usually dried by means of an appropriate solid drying agent (see Section 8.1), and the organic solvent is removed by distillation.

In any operation where a separatory funnel is used, the stopcock must be kept lubricated to avoid sticking or leakage during manipulations. The only exception is with Teflon or Teflon-coated stopcocks, which are self-lubricating. After it is used, the separatory funnel should be cleaned thoroughly and the stopcock freshly lubricated to inhibit "freezing" in a fixed position. If a silicone-based lubricant is used, it should be removed before cleaning the funnel with an oxidizing mixture or a siliceous film will be formed on the glass surface. The problem of "frozen" stopcocks is sufficiently severe that many workers prefer to store funnels disassembled. Stopcocks that have become "frozen" sometimes may be freed by warming the outer barrel with steam and applying gentle pressure to the stopcock plug (using a towel to protect the fingers). A vise-like stopcock plug remover is available from scientific supply firms.

Small Sample Technique (Squirting Technique). If one has less than about 5 mL of solution to be extracted, the separatory funnel technique described in the previous section does not work well because the solution tends to disperse as a film or droplets on the inside of the funnel. A good alternative is the "squirting" technique. The solution to be extracted is placed in a small test tube along with the extracting solvent (the tube should not be more than half full). Then, one of the two layers is drawn into a long-stemmed Pasteur pipet, the tip of the pipet moved into the other layer, and the liquid in the pipet "squirted" into the layer. This process of drawing up one layer and expelling it into the other is repeated about a dozen times or until the two layers have been mixed well. The mixture is allowed to stand until it separates into two layers, which then are separated by drawing up one layer into the same pipet and transferring it to a second test tube. It may be difficult at first to achieve a clean separation of the two layers, but with a little practice the technique works well. Even with good technique there will be a little contamination of one layer by the other.

---

[5] An aqueous layer can be distinguished from a water-insoluble layer by adding a small test portion to a few milliliters of water in a test tube; the aqueous layer will form a homogeneous solution, whereas the nonaqueous layer will form a two-layer system.

Some workers advocate stoppering the test tube with a cork and shaking it. While this procedure mixes the two solutions well, it runs the risk of having the contents leak and of contaminating the solutions by impurities in the cork. A rubber stopper is not much better.

Other Extraction Techniques. For special purposes, laboratory methods and apparatus have been developed for continuous extraction of aqueous solutions with immiscible solvents and for multiple countercurrent extraction. In some instances, a mixture of solvents is employed as an extraction solvent.

6.6 Separation of Mixtures Containing Multiple Components

It frequently happens that a mixture contains acidic, neutral, and basic components. Separation of such a complex mixture is broken down into a series of steps. A general scheme, applied to an illustrative example of a mixture dissolved in 50 mL of solvent, is outlined in Figure 6.4. In this scheme it is assumed that the organic components are *not* water soluble, but that their acid- or base-generated salts are. Note that methylene dichloride is heavier than water and will be the bottom layer throughout the scheme, whereas the reverse is true for diethyl ether.

The first step is to extract the mixture with base in order to remove any acidic component as its water-soluble salt. The extraction scheme shows two base extractions, the second being an "insurance" extraction to scavenge traces of acidic components left behind after the first extraction. Potassium hydroxide rather than sodium hydroxide is shown because potassium salts are usually more water soluble; in most circumstances, however, sodium hydroxide will be adequate. The water-soluble salts contained in the aqueous layer are acidified with hydrochloric acid to restore the water-insoluble acidic forms, which precipitate. These acidic components are then isolated and purified by the appropriate methods (distillation, recrystallization, chromatography, etc.).

The organic layer from the base extraction now contains only the neutral and basic components. These can be separated by extraction (twice for the reason given above) with acid. The basic components, almost always amines, are converted into water-soluble hydrochloride salts and transfer to the acidic aqueous layer; the neutral components remain behind in the organic layer. Neutralization of the acidic aqueous layer converts the salts back into water-insoluble bases, which can be isolated and purified by the appropriate methods.

The organic layer, now containing only the neutral components, is distilled to remove the solvent. The residue of neutral components is purified by the standard methods.

If the original mixture contained water-soluble organic components as well as water-insoluble ones, the separation problem is more complicated. The original mixture can be extracted with water (the organic layer can then be treated as in

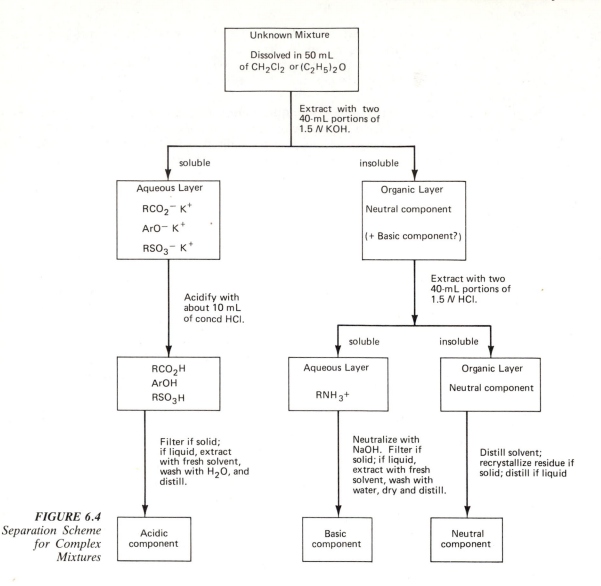

FIGURE 6.4
Separation Scheme
for Complex
Mixtures

Figure 6.4) to remove the water-soluble components, but there is no simple scheme for their further separation. The best general strategy is to remove the water by distillation (but watch for low-boiling organic components) and examine the residue by chromatography (discussed in Chapter 7) to determine the number of components. If you are fortunate enough to have only a single component, it can be purified by the standard methods. If the residue is a complex mixture, you will have to resort to careful fractional distillation, recrystallization, or large-scale chromatography.

6.7 Representative Extractions

(A) Simple and Multiple Extractions

The purpose of this experiment is to demonstrate the effectiveness of a single extraction with a fixed volume of solvent compared to two extractions, each with one half of the fixed volume. An aqueous solution of an intensely purple-colored dye, crystal violet, will be extracted by methylene chloride (dichloromethane) in which the dye is somewhat more soluble. The experiment is written for use of a separatory funnel, but it can be done effectively on one-fifth scale with the "squirting" technique described in Section 6.5.

1. Simple Extractions. Place 10 mL of the stock aqueous solution of crystal violet in a 125-mL separatory funnel and extract with 10 mL of methylene chloride in the following manner. Stopper the separatory funnel, shake gently and turn it upside down. While the separatory funnel is in this position open the stopcock to release the internal pressure, close the stopcock, shake vigorously, and again release the internal pressure.

◖▶ *CAUTION*   Do not point the stem of funnel at anyone when you release the pressure. Any liquid in the stem may be ejected forcefully.

◖▶ *CAUTION*   Prolonged exposure to high concentrations of methylene chloride vapors may induce cancer. As with all organic solvents, work in a well-ventilated area when using it.

Repeat this procedure four or five times, then support the separatory funnel upright in a ring and let it stand undisturbed. When the liquids have separated completely, draw off the lower methylene chloride layer into a Erlenmeyer flask (following the procedure for laboratory extractions given in Section 6.5).

◖▶ *CAUTION*   In the manipulation of organic solvents in a separatory funnel with a glass stopcock, it is important that the stopcock should be kept properly greased to avoid sticking, since the organic solvent removes the stopcock grease. After using an organic solvent in the separatory funnel, it is advisable to clean the separatory funnel thoroughly and to grease the stopcock properly before replacing the apparatus in the laboratory desk. If this is not done, the stopcock is likely to become "frozen" in a fixed position and the separatory funnel will be rendered useless.

Transfer a portion of the remaining aqueous layer to a test tube and set it aside for later comparison.

2. Multiple Extractions. Clean the separatory funnel well with water and place a second 10-mL portion of the stock solution of crystal violet in it. Extract the solution with 5 mL of methylene chloride as described in part 1. Draw off the methylene chloride layer into the Erlenmeyer flask used in part 1 and extract the remaining aqueous layer with a second fresh 10-mL portion of methylene chloride. Draw off the methylene chloride into the Erlenmeyer flask and transfer a portion of the aqueous layer into a test tube of the same size used in part 1. Fill both tubes to the same height.

Compare the effectiveness of extraction by the two different procedures by noting the intensity of color remaining in the aqueous layer. The difference in color will be more noticeable if one looks down the mouth of the test tube.

Pour all of the methylene chloride extracts into a bottle in the hood labeled "Methylene Chloride from Extraction Experiments."

(B) Separation of Benzoic Acid and Acetanilide

Obtain from your instructor 25 mL of a solution containing 2 g of benzoic acid and 2 g of dimethyl terephthalate dissolved in methylene chloride. Calculate the volume of 1.5 N aqueous potassium hydroxide required to react completely with the benzoic acid, and extract the solution twice with the 1.5 N base using the calculated volume each time (follow the procedure for laboratory extractions given in Section 6.5).

⫸ *CAUTION*   Prolonged exposure to high concentrations of methylene chloride vapors may induce cancer. As with all organic solvents, work in a well-ventilated area when using it.

Complete the extraction by washing the methylene chloride solution with 20 mL of water; combine the aqueous layers in a 100-mL beaker. Place the methylene chloride layer in a 125-mL Erlenmeyer flask, add about 0.4 g of anhydrous magnesium sulfate,[6] and set the flask aside to allow the residual water time to be absorbed.

Cool the combined aqueous extracts in an ice batch and add, while swirling, sufficient concentrated (12 N) hydrochloric acid to neutralize the base. Test the acidity of the solution with pH paper, and if it is not distinctly acidic add more acid dropwise until it is. Collect the precipitated benzoic acid and allow it to dry as described in Section 5.2 until the next period. Place the dried sample in a tared bottle and determine and record the weight of recovered benzoic acid. Determine and record the melting-point range of your sample.

Assemble a simple distillation apparatus (see Section 2.3) using a 100-mL round-bottomed flask as the boiler. Transfer the methylene chloride solution into the boiler by passing it through a filter tube containing a wad of glass wool to

---

[6] The anhydrous magnesium sulfate is used as a drying agent to remove droplets of water and traces of dissolved water. The amount is not critical, and only the amount needed to absorb the water should be added. Wetted drying agent is dense and forms clumps at the bottom of the flask, whereas unwetted excess drying agent remains fluffy and floats easily in the liquid when the flask is swirled.

remove the drying agent; distill off all but 3–4 mL of the solvent. Pour the warm concentrated solution of acetanilide into a small beaker and rinse the flask with 2–4 mL of the recovered solvent. Evaporate the combined solutions to dryness on a steam bath *in a hood*. Transfer the solid to a tared bottle and determine the weight of recovered acetanilide. Determine and record the melting-point range of your sample.

If a rotary evaporator (see Section 8.6) is available, it can be used in place of the distillation assembly.

(C) Separation of Benzoic Acid and Dimethyl Terephthalate

Obtain from your instructor 25 mL of a solution containing 3 g of benzoic acid and 2 g of dimethyl terephthalate dissolved in methylene chloride. Calculate the volume of 1.5 *N* aqueous potassium hydroxide required to react completely with the benzoic acid and extract the methylene chloride solution twice with the 1.5 *N* base using the calculated volume each time (follow the procedure for laboratory extractions given in Section 6.5).

➤ *CAUTION*   Prolonged exposure to high concentrations of methylene chloride vapors may induce cancer. As with all organic solvents, work in a well-ventilated area when using it.

Complete the extraction by washing the methylene chloride solution with 20 mL of water; combine the aqueous layers in a 100-mL beaker. Place the methylene chloride layer in a 125-mL Erlenmeyer flask, add about 0.4 g of anhydrous magnesium sulfate,[6] and set the flask aside to allow the residual water time to be absorbed.

Cool the combined aqueous extracts in an ice bath and add, while swirling, sufficient concentrated (12 *N*) hydrochloric acid to neutralize the base. Test the acidity of the solution with pH paper, and if it is not distinctly acidic add more acid dropwise until it is. Collect the precipitated benzoic acid and allow it to dry as described in Section 5.2 until the next period. Place the dried sample in a tared bottle and determine the weight of recovered benzoic acid. Determine and record the melting-point range of your sample.

Assemble a simple distillation apparatus (see Section 2.3) using a 100-mL round-bottomed flask as the boiler. Transfer the methylene chloride solution into the boiler by passing it through a filter tube to remove the drying agent; distill off all but 3–4 mL of the solvent. Pour the warm concentrated solution of dimethyl terephthalate into a small beaker and rinse the flask with 2–4 mL of the recovered solvent. Evaporate the combined solutions to dryness on a steam bath *in a hood*. Tranfer the solid to a tared bottle and determine the weight of recovered dimethyl terephthalate. Determine and record the melting point range of your sample.

If a rotary evaporator (see Section 8.6) is available, it can be used in place of the distillation assembly.

(D) Extraction of Caffeine from Tea

The stimulating effects of aqueous infusions of coffee beans, tea and maté leaves, and cola nuts are due mainly to the presence of caffeine, a nitrogen heterocycle of the molecular formula $C_8H_{10}N_4O_2$. Its structure has been established by study of its degradation products and by synthesis to be 1,3,7-trimethyl-2,6-dioxopurine. Tea leaves contain 3–5% of caffeine and a trace of theophylline, a lower homolog lacking the methyl group of position 7. These compounds are related structurally to the important purines adenine and guanine, which are present in the nucleic acids (RNA and DNA).

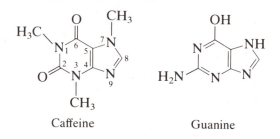

Caffeine   Guanine

In this experiment the caffeine is extracted from tea leaves by hot water, in which it is quite soluble (about 18 g/100 mL at 80°; 2.2 g/100 mL at 20°). Colored impurities such as the tannic acids can be removed as calcium salts by adding calcium carbonate. From the aqueous extract the caffeine is isolated conveniently by multiple extractions with small portions of methylene chloride, in which caffeine is also quite soluble. Caffeine forms a monohydrate that loses water rapidly on warming to give the anhydrous form, mp 238°.

In a 500-mL Erlenmeyer flask, place 30 g of ordinary dry tea, 300 mL of water, and 15 g of powdered calcium carbonate. After boiling the mixture gently for 20 min, occasionally swirling it, add 5 g of Celite or other filter aid, filter the hot mixture on a Büchner funnel, and press the filter cake firmly with a large cork to obtain as much of the liquid as possible.

Cool the aqueous extract to 15–20°, transfer it to a separatory funnel, and extract the caffeine with four successive 25-mL portions of methylene chloride (following the procedure for laboratory extractions given in Section 6.5).[7]

IIII➡ *CAUTION* Prolonged exposure to high concentrations of methylene chloride vapors may induce cancer. As with all organic solvents work in a well-ventilated area when using it.

Transfer the extracts to a simple-distillation assembly (see Figure 2.10), and distill off all but 10 mL of the solvent on a steam bath. Save the recovered

---

[7] Aqueous extracts of plant materials tend to form stubborn emulsions when extracted with organic solvents. An effective means for breaking difficult emulsions is to press, with the aid of a glass rod, a small wad of glass wool into the bottom of the separatory funnel and draw off the lower layer through the glass wool.

solvent. Pour the warm concentrated solution of caffeine into a small beaker and rinse the flask with 5–10 mL of the recovered solvent. Evaporate the combined solutions to dryness on a steam bath or hot-water bath in a hood. To purify the crude product, dissolve it in about 10 mL of hot toluene on a steam bath, add 15–20 mL of petroleum ether (bp 60–90°), and allow the product to crystallize. Collect the green-tinged crystals on a small suction filter and wash them with a little petroleum ether. The melting point of caffeine reported in the literature is 236°. This is too high to be safely determined in an apparatus using a mineral-oil bath.

The green color of the caffeine sample can be removed by sublimation.

(E) Extraction of Trimyristin from Nutmeg

The seeds of plants are frequently rich in triglycerides, the fatty acid esters of glycerol. Many different triglycerides are possible, since glycerol has three hydroxyl groups and there are many naturally occurring fatty acids (differing in chain length and number of double bonds) to combines with it. Nutmeg is remarkable in that the triglycerides in it are mainly the glycerol ester of a single fatty acid, myristic acid. This ester, called trimyristin, represents about 20–25% of the dried weight of ground nutmeg.

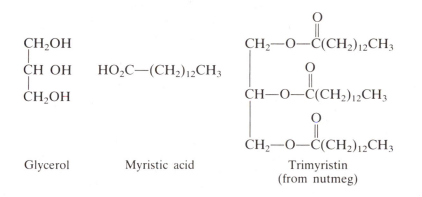

| Glycerol | Myristic acid | Trimyristin (from nutmeg) |

In a 100-mL round-bottomed flask place 5 g of ground nutmeg and 50 mL of methylene chloride.

CAUTION   Prolonged exposure to high concentrations of methylene chloride vapors may induce cancer. As with all organic solvents, work in a well-ventilated area when using it.

Attach a water-cooled condenser and boil the brown mixture for 30 min. Remove the heat, and after the flask has cooled for a short time, filter the warm solution through a fluted filter into a 250-mL beaker. Rinse the solid on the filter paper with 5 mL of fresh solvent and allow it to drain into the beaker.

Heat the beaker on a steam or hot-water bath *in the hood* until all of the solvent has evaporated (about 20 min). Cool the liquid residue in an ice-water bath. If most of the solvent was removed, the residue will solidify. Add 10 mL of acetone to the beaker and stir the liquid and the solid with a spatula in order to wash the crystals; break up any lumps of solid. Prepare a suction filter using a Büchner or Hirsch funnel (Section 5.2); wet the filter paper with a few drops of water so that it will lie flat on the funnel. Collect the fine, white crystals on the filter and then transfer them back to the beaker and repeat the trituration with another 10-mL portion of acetone. Collect the crystals by suction filtration as before. Continue to draw air through the filter for about 5 min in order to evaporate the last of the acetone. The yield of dry trimyristin is about 0.7–1.0 g. The melting point recorded in the literature for pure trimyristin is 56°; your sample will melt somewhat lower because of the impurities in it.

If desired, the crude trimyristin can be recrystallized from 95% ethanol using about 10 mL of solvent per gram of solid.

Questions   1. What conclusions can you draw about the most efficient method of extracting acetic acid from an aqueous solution by means of an immiscible solvent?

2. What is meant by the term *distribution coefficient*?

3. What properties do you look for in a good solvent for extraction?

4. Explain the fact that acetic acid can be extracted quantitatively from an ether solution by dilute aqueous sodium hydroxide solution. Write chemical equations for any reactions involved.

5. In the extraction of an organic compound from a dilute aqueous solution, will the organic solvent form the upper or lower layer when each of the following solvents is used?
   (a) chloroform   (b) cyclohexane
   (c) *n*-heptane   (d) methylene chloride

6. If toluene (density 0.87) were used to extract ethylene bromohydrin (density 2.41) from an aqueous solution,
   (a) Could you be certain that the organic solution would form the upper layer?
   (b) By what test could you identify the nonaqueous layer?

7. Name two organic compounds that cannot be extracted effectively from an aqueous solution by means of an immiscible organic solvent such as diethyl ether or cyclohexane.

7 Chromatography

7.1 Introduction

Chromatography is an exceptionally versatile separation technique that in one or more of its numerous forms is used by just about every chemist. In any chromatographic separation there are two phases (solid, liquid, or gas) that move relative to each other while maintaining intimate contact. The sample is introduced into the moving phase, and the sample components distribute themselves between the stationary phase and the mobile one. The components spend different amounts of time in each of the two phases as determined by the structures of the components and the two phases. If one component spends most of the time in the mobile phase, it will move along quickly; if it spends most of the time in the stationary phase it will move slowly. As with extraction, the degree of separation of a mixture is determined by differences in distribution coefficients, which are related to the same structural factors that control solubility.

A good analogy for chromatography is a moving conveyor belt. A group of packages placed on the start of the belt will move along as rapidly as the belt moves. If a package is taken off for a while at any point along the belt and then put back on, it will lag behind the main group. If this is happening to all the packages, their distribution at the end will depend on the fraction of time spent off the belt.

In this chapter several forms of chromatography will be described. Possible stationary and moving phases will be considered as will the techniques for introducing the sample into the mobile phase and detecting it when it arrives at the end.

7.2 Liquid–Solid Chromatography

The first chromatographic technique to be considered is liquid–solid chromatography. The stationary phase is made up of very small particles of solid packed in a column (hence the common name column chromatography); the mobile phase is a liquid that percolates through the column and past the surfaces of the solid particles.

Solid surfaces adsorb thin layers of foreign molecules as a result of electrostatic and van der Waals forces. Since adsorption strengths differ with the character of the solid surface, a properly chosen solid may adsorb selectively one component of a mixture. An important example of selective adsorption employed in the last chapter is the use of charcoal in crystallization to remove colored impurities. The ideal limiting law governing adsorption from a dilute solution is

$$\frac{\text{Amount of solute A adsorbed per unit surface area}}{\text{Concentration of solute A in solution}} = K_A$$

The structural features that determine the extent of adsorption of a molecule on a solid surface are pretty much the same as those considered when solubility was discussed. A complication in thinking about adsorption is that the solvent and the solute are competing for the same active sites on the surface. For molecules with polar functional groups the value of K_A (the adsorption coefficient) is determined principally by the relative polarities of the group and the solvent. Since highly polar solvents tend to be preferentially adsorbed, a low K_A results for the solute (i.e., the solute is poorly adsorbed and moves along rapidly with the mobile phase). For molecules containing hydroxyl groups, their relative abilities to form hydrogen bonds (proton bonding) to the solid or the solvent are also significant.

Two solutes with different adsorption coefficients for a certain solid can, at least in principle, be separated by liquid–solid chromatography. The traditional method is to prepare a cylindrical column of the solid (with a height about 5–10 times the diameter) and place a concentrated solution of the sample at the top of the column. As the solution penetrates the column, the solutes are adsorbed. As soon as the solution has completely penetrated the column, fresh solvent is added at the top. The solvent flows down the column under the force of gravity and capillary attraction and redissolves the solutes in amounts determined by the adsorption law and carries them to lower clean sections of the column, where they are readsorbed (always in amounts governed by the adsorption law). As more solvent percolates through the column, the cycle of adsorption–solution continues, and the solutes gradually move down the column in concentrated bands (*development*). With solutes having different adsorption coefficients, the least tightly adsorbed material moves ahead. If the adsorption coefficients are sufficiently different or the column is sufficiently long the faster moving component

TABLE 7.1
Adsorbants and
Solvents for
Liquid–Solid
Chromatography

Chromatographic solids in decreasing order of adsorption strength for polar molecules	Solvents in increasing order of eluting ability[b]
Activated alumina,[a] charcoal	Saturated hydrocarbons
Activated magnesium silicate[a]	Aromatic hydrocarbons
Activated silicic acid[a]	Partially halogenated hydrocarbons
Inorganic carbonates	Ethers
Sucrose, starch	Ketones
	Alcohols
	Organic acids

[a] The adsorption strength can be diminished by adding water. Under carefully controlled conditions, this strength is reproducible.

[b] This approximate order only applies to alumina. With nonpolar solids, the order tends to be inverted.

will form a separate band below the slower moving one. At the lower end the solutes are forced off the column (*elution*) and can be collected separately in succesive fractions.

For satisfactory separation by liquid–solid column chromatography, it is essential to choose an appropriate combination of solid adsorbent and eluent that is compatible with the compounds to be separated. Compounds that are adsorbed very tightly require an excessive volume of eluent for development. Compound adsorbed weakly may move too rapidly to give separation before being eluted. Table 7.1 gives some generalizations that are useful as a guide in selecting appropriate solid-solvent combinations.

A common variation of liquid–solid chromatography is the use of a thin film of solid (mixed with a binder such as plaster of paris) on a sheet of glass or plastic. The solution is added as a spot at the bottom of the plate and the plate is dipped vertically into a shallow layer of solvent, which ascends (against gravity) by capillary action and moves the solutes with it. The particular advantage of this technique is that the solutes are exposed and can be isolated readily or treated on the plate at any moment. The method is widely used for qualitative identification of mixture components because of its exceptionally good resolution. For a fixed combination of solid, binder, and solvent, each substance will travel along the thin-layer plate a characteristic fraction of the distance traveled by the solvent. It is customary to report thin-layer chromatography data as R_f values (*retention factor*), define as the distance of the spot from the starting point divided by the distance of the solvent front from the starting point. Thin-layer chromatography is restricted to small samples. Larger samples can be separated by using thick layers of plaster of paris, but there is a practical upper limit of a few tenths of a gram. A method known as "flash" chromatography, which combines the high resolution and speed of thin-layer chromatography with the large sample capacity of regular column chromatography, is described in Section 7.6.

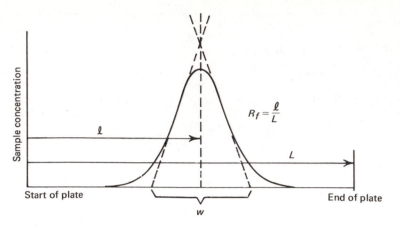

$$R_f = \frac{\ell}{L}$$

FIGURE 7.1
Chromatographic
Sample Distribution

Start of plate   End of plate

More recently, a technique called HPLC (high-pressure liquid chromatography) has been introduced. It is observed that the quality of separation in liquid–solid chromatography improves as the particles are made finer and more uniform in diameter and are packed in narrow diameter tubes. The penalty for using such a combination is that a solvent will not flow through it unless a very high pressure (as high as 1000 psi) is maintained at the entrance to the column. Having to work under a high pressure complicates the introduction of sample and the construction of the detector at the exit. Commercial HPLC instruments, produce truly astounding separations of complex mixtures. These are widely used, particularly in biomedical research, but they cost many thousands of dollars and require considerable care in their operation.

On a chromatographic column or a thin-layer plate each component appears as a more or less uniform spot. However, on careful analysis it is found that the distribution of sample along the length L of the column or plate is as shown in Figure 7.1. The position of maximum concentration when the solvent just reaches the end occurs at length ℓ from the starting point. By definition $R_f = \ell/L$; it can be demonstrated that R_f is also equal to $n/(1 + K)$, where K has the definition given at the beginning of this section and n is the average number of effective exchanges of a molecule of the sample on and off the surface in the time it takes the solvent front to traverse the column or plate.[1] For tightly retained samples, K is very large and n is small. For the ideal case it has been shown that

$$n \approx 16\left(\frac{\ell}{w}\right)^2 \tag{7.1}$$

---

[1] Some workers use the reciprocal of this definition of K, in which case $R_f = nk'/(1 + k')$, where $k' = 1/K$.

where w is the width at the base of the triangle formed by drawing straight lines through the most linear portions of the sample distribution curve, as shown in Figure 7.1.

7.3 Ion-Exchange Chromatography

Ion-exchange chromatography is a special example of liquid–solid chromatography, wherein strong ionic attractions dominate the relatively weak polar adsorption forces.

A column of solid acidic material (such as Amberlite IR-120, a resin of polystyrene beads containing free sulfonic acid groups) can donate protons to any base present in the surrounding liquid phase to form cations that are strongly attracted to the anions bound to the resin. The extent of proton transfer depends on the basicity of the solute and can be described by an equilibrium constant K (analogous to the previously discussed adsorption coefficient).

$$\text{Solid}^- \text{—H}^+ + \text{base} \; \overset{K}{\rightleftharpoons} \; [\text{solid}]^- \cdots [\text{H-base}]^+$$

$$K = \frac{\text{number of donated protons}}{\underset{\text{proton-donating sites}}{\text{number of available}} \times \text{concentration of free base}}$$

Bases with large values of K are present in lower concentration as the free base and descend the column more slowly. A mixture of bases with sufficiently different constants can be separated by this method.

When all components of a mixture are held tightly, as happens frequently, it is necessary to percolate dilute acid through the column to move the components down the column (*displacement development*).

Basic columns also are available (such as Dowex 3, a resin of polystyrene beads containing free amino groups); these accept protons and can be used to separate mixtures of organic acids of different acid strengths. Special column materials that form ionic complexes with various inorganic cations or anions are useful, as are columns containing ions that form complexes with certain organic molecules.

7.4 Liquid–Liquid Chromatography

Superficially, liquid–liquid chromatography (also known as *partition chromatography*) resembles liquid–solid chromatography. As the name indicates, liquid–liquid chromatography employs a liquid moving phase and a second stationary liquid phase. The stationary phase is held immobile by adsorption as a thin film on a solid support. Since the chemical influence of the solid support may largely be ignored, the adhering film behaves essentially like a pure liquid.

A substance added to such a column will be distributed (partitioned) between the two liquid phases. The distribution law is identical with that pertaining to the distribution of a solute between two immiscible solvents.

$$\text{Solid + film of solvent}_1 + \frac{\text{solution of}}{\text{compound in solvent}_2} \underset{}{\overset{K}{\rightleftharpoons}} \text{solid} + \frac{\text{film of solution of}}{\text{compound in solvent}_1} + \text{solvent}_2$$

A convenient form of liquid–liquid chromatography involves the use of a paper strip as the solid support. The sample is placed at a spot near the bottom of a dry paper strip and the strip is dipped into a shallow pool of the mixed solvents. As the solvents ascend by capillary action, the more polar solvent is preferentially adsorbed and becomes the stationary phase. The sample, which rises with the solvent, is partitioned between this stationary liquid phase and what remains of the moving liquid phase.

Some HPLC (see Section 7.2) also use the principle of liquid–liquid chromatography, except that the molecules of the liquid phase are bound by covalent bonds to the solid support to produce an interleaved network of liquid molecules. The bound liquid phase behaves like a thin film of adsorbed liquid. It has the advantage, however, that because it is chemically bound, cleaning solvents can be flushed through the column to remove accumulated slow-moving samples.

7.5 Gas–Liquid Chromatography

In gas–liquid chromatography (GLC or vapor-phase chromatography, VPC) the stationary phase is a film of liquid adsorbed on a solid support and the moving phase is a mixture of vaporized sample and a *carrier gas*, usually helium or nitrogen. The pertinent equilibrium is the distribution of sample between solution in the liquid film and vapor in the moving carrier gas. The rate at which the sample progresses through the column is determined principally by the rate of flow of carrier gas and the equilibrium vapor pressure of the sample in contact with the solution. A diagram of a gas chromatographic instrument is given in Figure 7.4, which shows the essential elements, consisting of the supply of carrier gas, the injection port for introducing the sample, the column for separating the sample, the detector at the end of the column, and the signal recorder attached to the detector.

$$\text{Solid} + \frac{\text{film of}}{\text{nonvolatile solvent}} + \text{vapor of sample + carrier gas} \overset{K}{\rightleftharpoons}$$

$$\text{solid + film of solution + carrier gas}$$

The fundamental data collected in gas chromatography are the periods of

time (*retention time*) required for each of the components of the injected sample
to reach the output detector. Retention time is inversely proportional to the
vapor pressure of the sample in the flowing carrier gas and inversely proportional to the gas flow rate. In discussing gas-chromatographic separations, it
is customary to speak of relative retention times, the ratio of the observed
retention time to the retention time of some standard either already present or
added to the mixture. The relative retention time is independent of the flow rate;
it is determined by the relative equilibrium vapor pressure of the components in
contact with the stationary liquid phase. From the discussion of fractional
distillation in Chapter 2, it can be seen that relative retention time is closely
related to the relative volatility for the sample and reference standard. The
difference is that, in distillation, the liquid phase contains the same components
as the vapor, whereas in gas chromatography the nonvolatile liquid phase must
be considered as well. If the GLC liquid phase is similar in character to the
sample components, the relative retention times will closely reflect the volatility
behavior of the mixture on distillation. For example, if squalane (a fairly
nonvolatile $C_{30}H_{62}$ saturated hydrocarbon) is used as the stationary liquid phase,
pentane (bp 36°), heptane (bp 98°), and octane (bp 125°) have relative retention
times at 100° of 1.00, 2.22, 4.65, and 10.51, respectively. The retention times
parallel the relative vapor pressures of the pure hydrocarbons. By contrast, if
1-propanol (bp 97°) is injected on the squalane column, the relative retention
time is only 1.09; it comes off the column much faster (i.e., it is more volatile)
than would be expected from its boiling point. The difference is that, with pure
liquid 1-propanol, the hydroxyl group of one molecule finds itself immersed as in
sea of hydroxyl groups of other propanol molecules. In the GLC column each
propanol molecule is surrounded mostly by the hydrocarbon molecules of the
liquid phase. In the absence of hydrogen bonding, the propanol becomes quite
volatile. The enhanced volatility is similar to the enhancement of volatility that
occurs in steam distillation.

The situation is completely reversed if the squalane liquid phase is replaced
by Carbowax, a polymer that contains ether groups (hydrogen bond acceptors)
and terminal hydroxyl groups (hydrogen bond donors). The retention data,
given in Table 7.2 along with the squalane retention data for comparison, show
that propanol now has a very long relative retention time; the hydrocarbons
show slightly enhanced volatilities (i.e., lower retention times).

Silicone oil DC-200 contains hindered silcon oxygen bonds (weak hydrogen-
bond acceptors) and gives retention results intermediate between the previous
two liquid phases, as shown in Table 7.2. These data demostrate that, because
relative retention times can be varied by choice of liquid phase, GLC provides a
separation flexibility absent in fractional distillation.

In both GLC and distillation, separation depends on differences in vapor
pressure of the components of a mixture. In fractional distillation the counterflow
of rising vapor and descending liquid establishes (ideally) equilibrium between
the components at each point in the column. In gas–liquid chromatography there

TABLE 7.2
Relative Retention
Times for Different
Liquid Phases at 100°

| | | Relative retention time[a] | | |
Compound	Bp, °C	Squalane	Silicone oil	Carbowax
Pentane	36	1.00	1.00	1.00
Hexane	68	2.22	2.04	2.00
Heptane	98	4.65	4.04	3.62
Octane	125	10.51	7.92	6.75
1-Propanol	97	1.09	1.58	40.5

[a] Relative retention time is the ratio of two times and thus is a pure number without units.

is a unidirectional flow of carrier gas and the components move independently of each other. An expression for the number of plates required in GLC to obtain a 95% pure sample with 80% recovery from a 50:50 mixture is

$$\text{Number of required GLC plates} = \frac{2.0}{(\log \alpha)^2} \tag{7.2}$$

Comparison of this expression with the analogous expression for fractional distillation (equation 2.6) shows that, aside from a small difference in the constant, they differ by the exponent of the log α term. Since α, the relative volatility, is close to unity for any mixture likely to be fractionally distilled or chromatographed, the log α term is near zero. A constant divided by the square of a small number is much larger than the same constant divided by the first power of the small number. It follows that gas–liquid chromatography requires many more theoretical plates to achieve a separation than does fractional distillation. Fortunately, the HEPT of gas–liquid chromatography columns is usually much smaller than those of distillation columns, so that the same length of gas–liquid chromatography column contains many more theoretical plates. Moreover, since separations by gas–liquid chromatography do not depend on gravity return of a counterstream, it is possible to use long lengths of column coiled into a small volume (50-ft columns are common). With narrow-bore tubing (*capillary columns*), columns of 600-ft length and 200,000 plates are common.

Another important factor aiding separations by gas–liquid chromatography is the availability of a wide range of liquid phases, one or more of which may give a greatly enlarged value. This flexibility is inherently absent in fractional distillation. Hundreds of different liquid phases have been employed in gas chromatography, and since these can be supported on over half a dozen different solid phases, it is apparent that the practical art of column selection is complex. Table 7.3 lists several general-purpose liquid phases with a few of their characteristics. A widely used solid support suitable for both polar and nonpolar samples is Chromosorb W, a white, flux-calcined diatomite. The liquid phase is dispersed on the surface of the support by adding the support to a solution of the liquid phase in a volatile solvent and allowing the solvent to evaporate slowly (directions for preparing GLC columns are given in the Appendix). The most

TABLE 7.3 *Liquid Phases for Gas–Liquid Chromatography*

Liquid phase	Temperature limit, °C	Application
Carbowax 20M	250	Separation of high-boiling polar compounds
Silicone oil DC-550	275	For compounds of intermediate polarity
Silicone oil DC-200	250	Separation of nonpolar compounds
Silicone gum rubber GE SE-30	375	Separation of nonpolar compounds
Apiezon (a hydrocarbon-based grease)	300	For low-boiling hydrocarbons

serious deficiencies of gas–liquid chromatography are the related restrictions that the sample be readily vaporized and that it be small (typical sample sizes are 0.1–10 μL). There are elaborate instruments that can handle larger samples (0.1–10 mL), but they are expensive.

7.6 Laboratory Practice

Thin-Layer Chromatography (TLC). Thin-layer plates may either be prepared in the laboratory or purchased. Unless a large number of plates is to be used or some special adsorbent or binder is required, it is not much more expensive (and considerably faster) to purchase the plates rather than prepare them. One convenient type comes as 20×20-cm sheets consisting of a 100-μm layer of adsorbent bound to a 200-μm sheet of plastic. With reasonable care these sheets can be cut with ordinary (sharp) scissors into 2×10-cm strips suitable for analytical separations.

In liquid–solid chromatography, the resolution obtained depends on the ratio of solid to sample. For mixtures that are difficult to separate (R_f values differing by 0.1 or less), the ratio may be reduced proportionately. Because of the small amount of solid on a TLC plate, the sample spot should be applied with a microcapillary tube prepared by drawing out an ordinary capillary tube in a soft flame. In order to simplify the later calculation of R_f values, the plate should be marked lightly with two pencil lines 7 mm from each end. A microdrop of a solution of the sample in a volatile solvent is placed precisely on one of the two lines. When only one sample is being analyzed, the drop should be centered between the edges; when more are to be analyzed on the same plate, the spots should be placed symmetrically along the starting line. But if more than four samples are applied on a 2×10-cm plate, the ones near the edges will develop unreliably. When a low concentration of any component is being sought, it is necessary to superimpose additional drops on the first spot until sufficient sample has accumulated. The solvent should be evaporated between additions.

A convenient developing chamber for TLC plates can be prepared from an ordinary wide-mouth screw-cap bottle. The inside of the bottle is lined with a

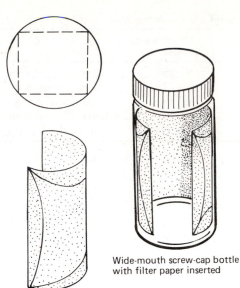

FIGURE 7.2
Developing Chamber
for Thin-Layer
Chromatography

Wide-mouth screw-cap bottle
with filter paper inserted

folded circle of filter paper, which acts as a wick to transfer the developing solvent to the upper portions of the chamber. As shown in Figure 7.2 the circle of filter paper, folded into a rectangle, is inserted in the wide-mouth bottle with the folds against the walls of the bottle. The size of filter paper should be chosen so that the top of the folded paper rises close to the top of the bottle, but there must be a gap between the paper and the top of the bottle so that the approach of the solvent front to the upper line on the plate can be seen without removing the cap. In practice sufficient solvent is added to the bottle to saturate the liner and leave a layer 2–4 mm deep at its shallowest point. The spotted end of the plate is centered in the bottom of the chamber with its upper edge leaning against the wall; the spotted face of the plate should face the gap in the filter paper lining so that the rising spots will be visible. The bottle is capped and gently set aside until the rising solvent front has just reached the upper line. The plate is then removed and the solvent allowed to evaporate from it. Since the solvent vapors may be harmful, it is good practice to do the evaporation in a hood.

If one or more of the components to be identified is colorless, a convenient visualization technique is to place the plate in another screw-cap bottle containing a few crystals of iodine mixed with about a tablespoon of sand, which serves to disperse the iodine. The capped bottle is held horizontally and rotated for a few seconds to bring the plate in contact with the iodine–sand mixture. Iodine vapor is absorbed on the plate wherever there is a concentration of organic material and produces a brown spot (commercial plastic plates do not absorb a significant amount of iodine under these conditions). After the color has developed, the plate is removed and a circle penciled around each spot. On exposure to air, the brown iodine spots gradually evaporate.

Another method for visualization, which works with compounds that absorb ultraviolet light, is to use thin-layer plates that have been impregnated with a fluorescent dye. When the plate is exposed to UV light it will glow everywhere except where the organic compound has absorbed the light and quenched the fluorescence. While the plate is glowing, the dark spots should be circled carefully with a pencil so that their positions can be measured and recorded after the ultraviolet light has been withdrawn. When handling the UV lamp, avoid looking directly at the light source since unfiltered UV light could damage your eyes.

Column Chromatography. A simple apparatus for liquid–solid column chromatography is a glass tube that has been constricted at one end (Figure 7.3). For separation of 0.1 to 0.5-g samples, a convenient tube size is 60 cm of 15-mm diameter tubing. This size will hold about 50 g of solid support and give a 100 : 1 ratio of packing to sample. Other sample sizes may be used with appropriately scaled apparatus.

FIGURE 7.3
Apparatus for Column
Chromatography

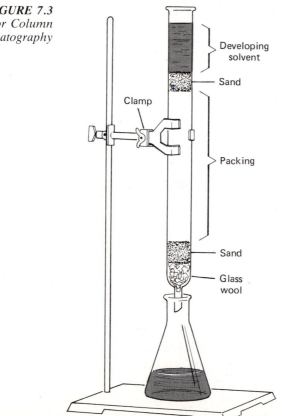

An alternative apparatus widely used in research laboratories are columns assembled from thick-walled glass tubes having molded threads at the ends connected by matching threaded couplers constructed from nylon or other inert plastics. The tubes and couplers come in different lengths and diameters and allow virtually any length of column to be assembled. Special fittings are available that make it easy to attach a stopcock or a porous glass bit to the bottom of the column or to attach a solvent reservoir or a pressure inlet to the top.

A small wad of cotton or (preferably) glass wool is pushed into the tube with a wooden dowel until the wad rests on the constriction. With the tube clamped in an upright position, a l-cm supporting layer of clean coarse sand is poured into the tube. Columns may be packed quickly by pouring in the solid support, but better separations are obtained by using a slurry of the solid in the desired eluting solvent. The slurry is added slowly though a funnel to a column that has been stoppered temporarily at the bottom with a medicine dropper bulb. One advantage of the slurry method is that the solid settles slowly, giving a more uniform packing. Nonuniform packing usually permits channels to develop in the packing, and these seriously diminish the resolution. Another advantage of the slurry method is that any heat of adsorption of the solvent on the support will be given off before the solid is added to the tube. When solid supports containing many active adsorption sites are packed dry and then wetted with solvent, the heat evolved is frequently sufficient to expand the packed column and cause channels to develop.

Alumina frequently produces tightly packed columns that have excessively slow flow rates unless pressure is applied to force the eluent through the column. This problem can be minimized by added about 10% Celite to the alumina to produce a coarser column. When the slurry method is applied to mixtures of adsorbent, they should be added in particularly small portions to prevent segregation of the solid as they sink through the solvent.

After the addition of the solid support has been completed, a second 1-cm layer of clean sand is added, followed by a small circle of filter paper just large enough to touch the wall of the chromatography column. The sand and filter paper prevent the upper layer of support from being disturbed during subsequent operations (this is important). Small irregularities at the top of the column produce large distortions in the shape of each band of sample by the time it reaches the bottom of the column and may cause closely spaced bands to overlap.

The sample is normally added in a solution as concentrated as possible so that narrow bands can be formed. When the solution is ready, the bulb is removed from the base of the column, and the excess solvent is allowed to drain. At the moment the solvent level reaches the top of the packing, the solution of the sample is added and allowed to penetrate into the column. After the sample has penetrated and before the column can become dry, the residue adhering to the walls is washed down with a few drops of solvent. Enough eluting solvent (eluent) is then added to fill the tube.

As the eluent flows through the column, the sample is separated gradually

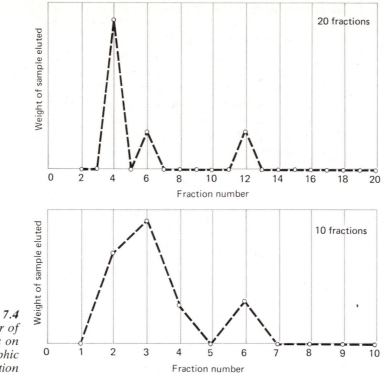

FIGURE 7.4
Effect of Number of
Fractions on
Chromatographic
Resolution

into bands that descend at different rates. It will be necessary from time to time to refill the tube with more eluent to keep the column from running dry. If all of the components of the sample are colored, it will be obvious when to change receivers for each component. When one or more of the components are colorless, it is necessary to collect fixed volumes of eluent in tared flasks, evaporate the solvent, and reweigh the flasks. A graph of the weights of sample eluted in successive fractions plotted against the accumulated volume of eluent reveals information about the number of components and the degree of separation. As shown in Figure 7.4, a larger number of fractions yield more detailed information about the number of components.

Pencil Columns. When you are working with only a few milligrams of sample, the column just described is much too large. TLC could be used but an interesting option is to do column chromatography with a Pasteur pipet for the column. A small wad of glass wool is pushed into the constricted neck of the pipet, followed by enough adsorbent to produce a column about 3–5 cm high. The sample and solvent are added in the way described previously. Frequently, the solvent will not flow through the column on its own and must be forced through (slowly) with a rubber bulb.

Flash Chromatography. Because of the superior resolution of mixtures using TLC, there have been numerous attempts to scale up the procedure to achieve separations of preparative size samples. In general such "thick-layer" methods do not separate as well, and the apparatus becomes unmanageably large for samples of more than about 0.5 g. The recently introduced technique of "flash" chromatography appears to combine the resolving ability of TLC with the large sample capacity of column chromatography and is rapidly becoming one of the most powerful separation techniques available to the organic chemist.[3]

The secret of flash chromatography is to use a very fine and uniformly sized silica gel (typically 32–63 μm particles), similar to that used in preparing TLC plates. The small particle size provides a large surface area and the uniformity enhances the resolution. In order to achieve the desired flow rates through a column packed with such a fine silica gel it is necessary apply a pressure of about 5–15 psi. For safety reasons this, in turn, mandates the use of heavy-walled columns. Threaded glass columns (Ace glass) are well-suited for this application and their use will be described next.

The first step is to assemble the chromatographic column using either an 11- or 15-mm inside diameter tube to which is attached by means of a nylon coupler a fitting bearing a porous plastic plug and a drip tip (see Figure 7.5). A stopcock is not needed. Attach the column firmly to a ring stand with clamps at both the bottom and the top of the column. Position the column on the stand to allow receiving flasks to be slid underneath the drip tip. Now add slowly, with gentle tapping, 15 g (35 mL) of 32–63 μm particle-size silica gel. To the top of the column carefully add about a 1-cm layer of clean sand. Place a 100-mL Erlenmeyer flask underneath the drip tip.

Fill the column with the chosen developing solvent and, after making sure that you are wearing safety goggles,[4] apply 15 lb of air pressure from a rotary vane pump (described in Section 3.2).[5] The solvent should flow down the column at a rate of 2–4 mL/min. Save the solvent for later use. When the liquid level just reaches the top of the sand, release the air pressure and transfer your sample solution to the column. In order to minimize the disturbance of the column surface, the first few milliliters of solution are best added with a long pipet that directs the stream along the walls of the columns.

Adjust the vane pump to deliver 5–10 lb of pressure and force the sample solution onto the column. When the last portion has just penetrated the sand layer, interrupt the pressure and carefully rinse the walls of column with few milliliters of solvent. Reapply 5–10 lb of pressure until the solvent reaches the sand layer. Now filled the column with solvent and apply pressure again. Collect a series of fractions, as described in Section 7.4. Additional solvent may have to be added to the column in order to elute all of the sample components.

---

[3] Still, Khan, and Mitra, *J. Org. Chem.*, **43**, 2923 (1978).

[4] In the laboratory whenever operations are carried out under pressure or vacuum, you should wear safety goggles rather than the less protective safety glasses.

[5] The pressure can be applied conveniently by attaching a large one-holed rubber stopper to the end of the hose from the pump and holding the flat surface of the stopper against the top of the tube.

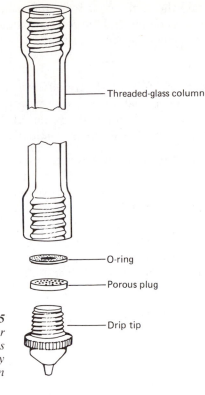

FIGURE 7.5
Fittings for
Threaded-Glass
Chromatography
Column

Threaded-glass column

O-ring

Porous plug

Drip tip

An extraordinary advantage of flash chromatography is that for similar adsorbents the R_f values obtained correspond closely with those obtained on a TLC plate. This correspondence permits one to carry out several trial separations using TLC and then to transfer the best conditions to a flash chromatography separation of a preparative size sample.

The correspondence between the R_f values of the two techniques also can be used advantageously to calculate the volume of solvent required to elute a given component from the column. From the definition of R_f as the ratio of migration distance of the sample to the solvent front, it follows that the volume of solvent required to bring a sample to the bottom of the column is just $1/R_f$ times the volume of solvent required initially to just wet the entire column. For example, consider the case where a particular component has an R_f value of 0.2 and a volume of solvent V is required to just wet the column. After one volume of solvent has passed down the column, the component in question will be 0.2 of the way down; no solvent will yet have emerged. After a second volume V has been added, the component will have moved to a point 0.4 of the way down the column; one volume of solvent will have emerged. After five volumes of solvent have passed down the column, the component will have reached the bottom of the column; four volumes of solvent will have emerged. An additional small volume of solvent will elute the component from the column. Because of the spread in the sample along the column and the consequent lack of a unique value for R_f, one should collect the emerging solvent a little before and a little after the main peak.

Gas–Liquid Chromatography (GLC). Many different styles of GLC instruments are used in undergraduate laboratories; however, the discussion that follows is applicable to most of them with minor alterations. The major components of a gas chromatograph are shown schematically in Figure 7.6.

Recorder. The strip chart recorder makes a permanent record of the signal coming from the sample detector on a paper chart that unrolls at a constant speed while the pen slides back and forth on it in response to the applied signal. Laboratory recorders are classified by size (6″ and 10″ are the two most common), full-scale signal sensitivity (1–10 mV dc is used for GLC work), the minimum response time for the pen to travel the width of the chart (l s for GLC work), and the speed at which the chart paper unrolls ($\frac{1}{2}$″/min is convenient; other speeds are available). The electronic amplifier inside the case has

FIGURE 7.6 Apparatus for Gas–Liquid Chromatography

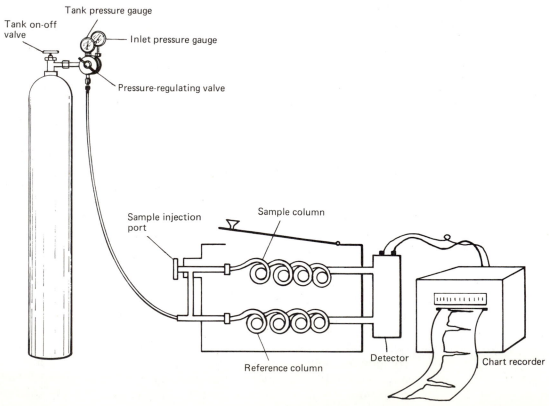

adjustable gain and damping controls, but once set these controls should not be altered by anyone other than service personnel. The recorder has an ON/OFF switch for power, which should be turned on at least 15 min before the recorder is used and left on during the laboratory period. The chart drive switch controls the flow of paper and should be turned on shortly before a sample is injected and turned off when the last peak of interest has been recorded.

Since the signal being recorded is proportional to the amount of sample passing through the detector at any moment, the area under the peak on the chart paper is proportional to the total quantity of sample. Unfortunately, the proportionality constant between chart area and sample size differs slightly with the structure of the compound, so that for quantitative work standard mixtures of known composition must be analyzed first. For qualitative work the relative peak areas are approximately in the ratio of the weight of each component.

The areas under the recorded peaks can be determined by a number of methods: electronic integration, mechanical integration with a planimeter, or cut-and-weigh (this assumes that the density of the chart paper is uniform). In addition to these direct methods, the relative areas of the peaks can be approximated by drawing tangent lines through the sides (see Figure 7.1) and then equating the peak area to the area of the triangle thus constructed: area $= \frac{1}{2}$ (base width) $\times$ (triangle height). Unfortunately, the triangle height is very sensitive to the precision with which the tangent lines are drawn and is difficult to reproduce. Another method of approximating the peak area, which avoids the problem of uncertain height, is to draw the tangent lines to obtain the base width but to take the triangle height to be the height of the peak recorded on the chart paper. Because the quantities of interest are the ratios of areas, the mathematical errors introduced in this hybrid approximation tend to cancel and are more than compensated for by the increased precision of the measurement. If done carefully, this method is reproducible to about 1–2% and is as precise as any other method except electronic integration, which is good to about 0.5%.

Detector. The most widely used detector for undergraduate laboratory work is the thermal conductivity detector, which consists of an electrically heated filament (similar to the one in a light bulb) immersed in the flowing gas stream. The temperature of the filament is determined by the (fixed) electrical current passing through it and the thermal losses to the cooler walls of the detector block from thermal conduction of the flowing gas. The carrier gas conducts heat at a rate characteristic of the gas, its pressure, and the rate at which it flows past the filament. If organic vapor is mixed with the carrier gas, the thermal conductivity of the mixture decreases. The main source of conductivity is from kinetic energy transfer to which the relatively massive organic molecules contribute little. The lowered thermal conductivity results in an increase in the filament temperature, which in turn leads to an increase in the electrical resistance of the filament wire. It is this increase in electrical resistance that is measured and passed on to the recorder as an electrical voltage proportional to the percentage of organic compound present in the detector gas stream.

Because of uncontrollable fluctuations in the gas flow rate and variations in the temperature of the detector walls, the background signal in the absence of organic vapor tends to fluctuate and drift. This can be controlled in practice by splitting the carrier gas into two streams, one of which goes to the injection port and the other to a second reference filament in the detector block. If the electrical signal passed to the recorder is taken as the difference between the sample signal and the reference signal, the distortions introduced by variations in the temperature and gas flow largely cancel.

The electrical current passing through the filament can be changed by the operator, but in practice it is best left untouched once adjusted, since each time it is changed it takes a while for the filaments to stabilize at their new equilibrium temperature. The balance control determines the electrical balance between the reference and sample filaments and is used to bring the recorder to zero when no sample is present. The "zero balance" should be adjusted before each sample is injected. The attenuator switch is a variable resistance network that determines what fraction of the detector signal is passed on to the recorder. The graduations on the dial give the factor by which the signal is reduced, thus a setting of "1" is the maximum sensitivity and "∞" gives no signal at all. The attenuator can be altered during an analysis to keep both the strong and weak signals on the recorder chart scale. Each time the attenuator is altered, some indication of the setting should be noted on the chart paper.

Another kind of detector, which is encountered frequently in research laboratories, is the flame ionization detector. The carrier gas is a mixture of hydrogen and nitrogen, which after passage through the column may also contain sample vapors. This gas mixture is fed to the detector where it is mixed with oxygen and burned in a small burner. The flame is subjected to a strong electric field of several hundred volts that extracts electrons from it (flames contain a small number of electrons from thermal ionization of combustion intermediates). When the sample comes through the column and enters the burner, additional electrons are produced. The electronics of the instrument are sufficiently sensitive to measure this small increase. As with thermal detectors, one must balance out the background signal and control the sensitivity to produce suitable chart readings.

Carrier gas. Helium, the most commonly used carrier gas for thermal conductivity detectors, comes in large tanks pressurized as high as 2600 psi. Because the GLC apparatus operates with a helium pressure of about 20–70 psi, a pressure-reducing and regulating value is required between the tank and the apparatus. The reducing value has two gauges: the one closest to the tank measures the tank pressure, and the other measures the pressure being delivered to the apparatus. The applied pressure, and consequently the gas flow, can be controlled by turning a bar handle that comes out of the body of the pressure regulator. This handle should not be confused with the round ON/OFF tank valve (see Figure 7.6).

The flow rate through $\frac{1}{8}''$ columns should be about 30 mL/min and 60 mL/min for $\frac{1}{4}''$ columns. It can be measured by attaching the bubble flowmeter to the column exit tube on the detector and adjusted by turning the pressure-regulating valve (not the main tank valve) until the time measured with a stopwatch for a soap bubble to rise through a 10-mL volume of the flowmeter is close to 20 sec ($\frac{1}{8}''$ columns) or 10 sec ($\frac{1}{4}''$ columns). The flowmeter is removed after the adjustment.

Columns. Selection of a liquid phase for achieving a desired separation can be quite tedious, since there are literally dozens of combinations of liquid phases and solid supports that can be tried. The information given in Table 7.3 provides a starting point. Columns are connected to the instrument by Swagelok fittings. The diagrams provided by the manufacturer should be studied carefully before attempting to connect a column. The fittings are easily ruined if cross-threaded or overtightened. Loose fittings will leak carrier gas; overtightened ones don't leak at the time but once removed they will leak no matter how they are tightened. Whenever a column in a thermal-detector instrument is changed, it is imperative that the detector current be turned off first and then the helium flow stopped since the filaments will overheat and possibly burn out if they are operated for any length of time without a stream of gas flowing over them.

The HETP[6] of column depends on several factors that change with the velocity of the carrier gas through the column. If the gas flow is too fast, the moving sample does not have time to equilibrate with the stationary liquid phase and the HETP increases. If the gas flow is too slow, the sample spends too much time in the column and diffuses to give broad peaks, which form equation (7.3) corresponds to increased HETP. These two velocity-dependent contributions to HETP were combined by van Deemter with the contribution inherent in the choice of column material to give the equation HETP as a function of velocity.

van Deemter equation: $$\text{HETP} = A + \frac{B}{V} + CV \qquad (7.3)$$

where V is the velocity of the carrier gas and A, B, and C are the parameters characteristic of the column and the factors described previously. The most important consequence of the van Deemter equation and the physical laws it represents is that there is an optimum flow rate for the lower HETP. In critical separations, this optimum must be found; in ordinary work the standard carrier gas flow rates of 30mL/min for $\frac{1}{8}''$ columns and 60 mL/min for $\frac{1}{4}''$ columns can be used.

The temperature of the oven and the column in it affects the quality of separation. Usually, the lower the temperature, the greater the difference will be in relative volatilities of the components. Lower temperatures also lead to peak broadening (increased time for diffusion), so that the increase in separation is not as marked as the difference in retention times might suggest.

Sample injection. The sample is injected by means of a syringe through a rubber septum into a heated injection block where it is vaporized and carried into

---

[6] HETP = column length/number of plates. Broadened peaks correspond to a lowered number of plates.

the column by the flowing carrier gas. Typical syringes used in analytical GLC work hold up to 10 μL of sample.

The syringe is filled by pushing in the plunger to expel the air and then drawing it back slowly while the needle is immersed in the liquid to be analyzed. This operation may have to be repeated if the column of liquid drawn up is not continuous. The syringe is held vertically with the needle up, and the plunger is pressed in until the required sample volume is read on the graduated syringe barrel. The needle is wiped with a tissue to remove excess liquid, and then the plunger is pulled back slightly to draw some air into the syringe needle.

The syringe is held with both hands. One guides the needle into the septum, and the other holds the barrel and prevents the plunger from being forced out by the gas pressure inside the injection block (be careful not to touch the injector port, it can be very hot). The needle is guided through the septum as far as it will go, the plunger is pressed in completely, and the syringe is withdrawn from the injection port. These three operations should be performed quickly but deliberately. If too much time is taken, the resulting peaks will be broadened; if the procedure is done too quickly, there is a high probability that the needle or plunger will be bent. Until you gain experience, it is better to err on the slow side.

The rubber septum is meant to be self-sealing, but after a dozen or so injections it will begin to leak, particularly if a large gauge needle is being used on the syringe. A leaky septum will result in distorted GLC traces. One test for a leaky septum is to place a few drops of water on the suspected septum; if it is leaking helium gas, bubbles will form. If the septum must be replaced, remember to turn off the detector current first so that the filaments will not burn out when the gas stream is interrupted. Also be careful in handling the injector fittings, since they are likely to be very hot.

The syringe should be cleaned after each use by drawing in and ejecting several consecutive samples of acetone and then pushing the plunger back and forth several times in the air to evaporate the acetone.

Calculation of Theoretical GLC Plates. When a sample is injected onto a GLC column, it occupies only a small portion of the column. As the vaporized sample is swept down the column by the carrier gas, it diffuses gradually so that by the time it reaches the detector, the distribution of sample resembles the peak shown in Figure 7.1. It is useful to speak of the "effective" number of times the sample is absorbed in the liquid phase and revaporized during its passage through the column. It can be shown that the total effective number of plates provided by a column is approximately

$$\text{Number of GLC plates} = 16\left(\frac{t_{\text{ret}}}{t_{\text{with}}}\right)^2 \tag{7.4}$$

where t_{ret} is the retention time of the peak maximum and t_{width} of the peak base expressed in the same–units of time. Comparison of equation (7.4) with equation (7.1) reveals the similar theoretical basis of gas–liquid and liquid–chromatography.

7.7 Chromatographic Separations

The separations presented here are designed to illustrate several of the more commonly used chromatographic techniques. The materials to be separated in most of these initial experiments are highly colored because this enables you to see the band development and to determine immediately the consequences of careless or faulty technique.

(A) Separation of Ink Pigments by Thin-Layer Chromatography

Prepare two 2×10-cm thin-layer plates[7] by drawing two horizontal pencil lines across each plate 7 mm from each end. On the bottom line of each plate, about 5 mm from the left-hand edge, make a single sharp dot of ink from a black Flair pen; in the center of the line make a second spot about 2 mm in diameter by momentarily holding the pen tip on the plate; on the right-hand side of the line, about 5 mm from the edge, make a third spot about 5 mm in diameter. Add sufficient acetone to an 8-oz wide-mouth screw-cap bottle containing a filter paper lining (see Section 7.6) until a 3-mm-deep layer is produced. Center one of the spotted plates in the bottle with the upper edge leaning against the side and screw the cap tightly onto the bottle. When the solvent front reaches the upper pencil line, remove the plate and allow the solvent to evaporate. While the first plate is developing, repeat the process with the other plate and a second 8-oz bottle using a $1:1$ mixture of acetone and 95% ethanol.

Determine and record the R_f values for all of the colored spots. Determine which spots, if any, are UV active. Determine which spots are stained by I_2. Make a sketch of the two plates in your laboratory notebook showing the location and shape of the spots with side notes on their response to UV and I_2.

||➡ *CAUTION*   Avoid looking directly at the UV light source since unfiltered light can damage your eyes.

The experiment can be repeated with other colors of Flair pens to determine if the same dyes are used that were found in the analysis of the black Flair-pen ink.

(B) Separation of Plant Pigments by Thin-Layer Chromatography

In a mortar place 1 g of spinach, 1 g of clean sand, 5 mL of acetone, and 5 mL of petroleum ether. Grind the spinach until the green chlorophyll appears to have been extracted completely. Decant the solution into a small beaker.

---

[7] Eastman Chromagram Sheet, Type 6060 or 6061, is suitable. It can be cut conveniently with ordinary scissors. In humid climates it is desirable to activate the plates by heating them in an oven at 100% for 15–30 min.

Prepare two thin-layer plates as described in Part (A) and in the center of each bottom line place a microdrop (see Section 7.5) of the chlorophyll extract. Blow gently on the spot so that the solvent evaporates quickly. Repeat the addition of the extract several times until a distinct green spot is visible. The additions should superpose as closely as possible.

Develop one plate with 1 : 4 (v : v) mixture of acetone and petroleum ether[8] as described in Part (A). Develop the second plate with a 1 : 6 : 1 (v : v : v) mixture of acetone, petroleum ether, and 95% ethanol. Mark the spots with a pencil immediately, since the colors fade (because of air oxidation) fairly quickly.

(C) Analysis of Analgesics by Thin-Layer Chromatography

The colorless components of common analgesics (Contac, APC tablets, etc.) can be identified by comparison of their R_f values with those of the pure components. Common analgesics are usually a mixture of several drugs. Most contain aspirin and caffeine, but several contain other components as well. Because the components are colorless, the spots must be made visible (*visualized*) by either ultraviolet fluorescence quenching or iodine staining.

Select a sample and identify it in your notebook. Pulverize your sample to a fine powder with the back of a laboratory spoon. Place a small wad of glass wool in the tip of a pipet and transfer the powder into the pipet to form a column. With a second pipet, drain 5 mL of 95% ethanol through the column and collect the extract in a small test tube.

Prepare six thin-layer plates that have been impregnated with fluorescent dye by lightly drawing on each of them two pencil lines 7 mm from each end. The five possible components shown below are available as 2% (w : v) solutions in 95% ethanol on the sideshelf. Spot three of the plates in an identical fashion with the extract and three of the knowns (a total of four spots on each plate). A single application of each known solution is sufficient, but two or more applications of the unknown solution will be required to give it adequate spot intensity. Each of the remaining three plates should be spotted with the extract and the remaining two knowns. In placing the spots on the plates it is probably best to spot the extract near the center. When done you will have three pairs of plates; be sure you can identify which plate is which.

Develop one pair of plates in 1 : 12 (v : v) mixture of glacial acetic acid and 1,2-dichloroethane until the solvent has migrated to the top pencil line. Remove the plate, allow the solvent to evaporate, and then examine the plate under an ultraviolet lamp. Circle the dark spots lightly with a pencil and then compute the R_f values of the reference spots on either side of the central spots. After the analysis is complete, place the plate in an iodine chamber and compare the brown spots developed in this way with the circled spots found by UV.

---

[8] Lower percentages of acetone tend to improve the separation of chlorophylls a and b at the expense of poorer separation from the yellow xanthophylls.

▶ *CAUTION* Avoid looking directly at the UV light source since unfiltered light can damage your eyes.

Clean out the development jars to avoid cross contamination of the solvents and repeat the analysis with 5 : 1 : 1 : 2 (v : v : v : v) mixture of 1,2-dichloroethane, acetone, 95% ethanol, and petroleum ether. Finally, after cleaning the jars again, repeat the analysis with 25 : 1 (v : v) mixture of ethyl acetate and acetic acid.[9]

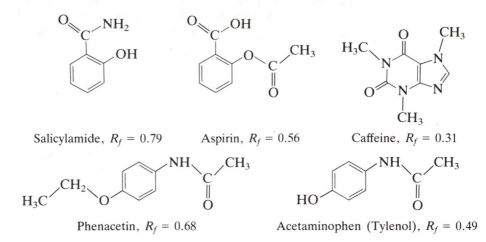

Salicylamide, $R_f = 0.79$ Aspirin, $R_f = 0.56$ Caffeine, $R_f = 0.31$

Phenacetin, $R_f = 0.68$ Acetaminophen (Tylenol), $R_f = 0.49$

(D) Separation of a Dye Mixture

Insert a small wad of glass wool into the constricted end of a 30-cm length of 10-mm diameter tubing and clamp the tube in an upright position (see Figure 7.3). Add a 5-mm layer of coarse sand to the tube. In a 100-mL beaker, prepare a slurry of 6 g of aluminium oxide[10] in 10 mL of warm water, and transfer the slurry in small batches to the tube (swirl between additions). The water that filters through the sand and glass wool should be collected and used to transfer any column material that remains in the beaker. After the packing has settled, add a second 5-mm layer of sand, followed by a small filter paper circle.[11]

When the last drop of water penetrates the column, add 10 drops of the dye solution[12] to the top of the column. When the dye solution has penetrated, add a few drops of water to wash down any dye adhering to the walls. After the wash water has penetrated, fill the tube with water and allow the chromatogram to develop.

[9] The R_f values of aspirin and acetaminophen are particularly sensitive to the activity of the plate and the proportion of acetic acid in the solvent.

[10] Fisher Certified Aluminium Oxide, A-591 is suitable.

[11] These circles can be prepared conveniently from a larger piece of filter paper using a cork borer as a cutter and a large cork as a cutting surface.

[12] A suitable dye solution can be prepared by dissolving 0.1 g each of crystal violet auramine hydrochloride, and malachite green in 100 mL of water. Alternatively, one of the dark-colored commercial food colors may be analyzed.

(E) Flash Chromatography

Assemble a chromatography column, as described in Section 7.6, and pack it with 15 g (35 mL) of 32–63 μm silica gel topped by a 1-cm layer of sand. Prepare 5 mL of leaf extract as in Procedure B, and place it in a 10-mL graduated cylinder. Add 5 mL of water and stir the mixture with a glass rod (to extract most of the highly polar acetone in the sample). Transfer the upper dark green petroleum ether layer to the column by means of a Pasteur pipet. Force the sample onto the column with pressure, as described in Section 7.6, and develop it with 4:1 petroleum ether–acetone.

Compare the pattern of bands with the spots obtained by thin-layer chromatography. Individual bands can be isolated and their purity determined by thin-layer chromatography.

(F) Gas–Liquid Chromatography

After familiarizing yourself with the GLC apparatus (compare with Figure 7.6), check to see that the carrier gas and filament current are on. With the attenuator set at "1", adjust the zero-balance control so that the recorder pen is on zero. Turn the attenuator knob to "8", and start the chart drive.

Draw up 0.5 μL of a standard hydrocarbon mixture[13] in a l- or 10-μL syringe, and inject it into the injection port. Make a mark on the chart paper at the point of injection.[14] Calculate the relative retention times and compare them with the data in Table 7.2. Calculate the number of theoretical plates using equation (7.4) for each of the peaks, and compare their relative areas with the known composition.

Determine the effect of altered environment on relative volatility by injecting a mixture containing three components of similar boiling points but different polarities.[15] Identify the components by their relative areas with the known composition.

Questions 1. Arrange the following compounds in the order of their elution from a silica gel column, with benzene as eluent.

$$CH_3(CH_2)_{10}CH_3 \qquad CH_3CO_2H \qquad CH_3CH_2CH_2OH \qquad CH_3COCH_2CH_3$$

2. Suggest suitable liquid phases for separation of carboxylic acids by liquid–liquid chromatography. (Hint: Consider the solubility relationships discussed in Chapter 6 on extraction.)

---

[13] A suitable mixture is 1 part by volume of *n*-pentane, 2 parts *n*-hexane, and 3 parts *n*-heptane, which on a DC-200 column have relative retention times of about 1.0, 2.0 and 4.0, respectively. Because of the broadening of the slower moving peaks, the specified relative quantities will give peaks about the same height.

[14] A convenient technique for making GLC charts is to momentarily hold your finger over the column exit port. Be certain that the exit port is not too hot.

[15] A suitable mixture is 1 part by volume of 2-propanol (bp 82°), 2 parts ethyl acetate (bp 77°), and 3 parts cyclohexane (bp 81°).

3. Table 7.1 lists several classes of solvents arranged in order of increasing eluting ability. Give a practical example of each class.

4. Figure 7.4 illustrates the effect of the number of fractions on chromatographic resolution obtained on analysis of the same sample. Show that if the 20-fraction chromatrogram samples are combined in consecutive pairs, the 10-fraction chromatogram will be obtained.

5. Silicone oil exhibits approximately ideal behavior as a liquid phase in gas–liquid chromatography, so that relative volatility values, α, obtained from fractional distillation can be employed in estimating the number of GLC theoretical plates required.

   (a) If a mixture of two liquids requires 50 theoretical plates for adequate separation by fractional distillation, how many GLC plates would be required for separation using silicone oil as the liquid phase?

   (b) With silicone oil as the liquid phase, what would be the expected elution order of acetone, n-butyl alcohol, benzene, and pentane?

8 Accessory Laboratory Operations

This chapter describes a number of general laboratory procedures that are used in many of the remaining experiments of this manual.

8.1 Drying Agents

The removal of admixed or dissolved water from starting materials and finished preparations is an important feature of organic laboratory work. In general, water must be regarded as an objectionable impurity since it may bring about an undesired hydrolytic reaction or exert an unfavorable catalytic effect. The presence of water sometimes retards a desired reaction and may inhibit it completely, as in the formation of Grignard reagents. On the other hand, it would be superfluous to remove the last traces of water from a substance that is to be brought into contact with aqueous reagents.

An organic solid may be dried by spreading it in a thin layer exposed to the air at room temperature, but this method usually allows at least a small amount of moisture to remain. More effective drying is obtained by heating the substance in a thin layer in a drying oven at a temperature (*below the melting or decomposition point!*), or by placing it in a desiccator over drying agents such as anhydrous calcium chloride, solid sodium hydroxide, or phosphorus pentoxide. The use of concentrated sulfuric acid in desiccators is dangerous.

An organic liquid, or an organic solid dissolved in an organic solvent, is usually dried by placing the liquid in direct contact with a solid inorganic drying agent. After it is mixed thoroughly and the mixture is allowed to stand, the dried liquid is filtered to remove the spent drying agent and may then be distilled, and so on. If, in any drying operation, sufficient water is present to cause the separation of an aqueous layer, the organic liquid should be separated and treated with a fresh portion of the drying agent.

Selection of a Drying Agent. The selection of an appropriate drying agent involves consideration of the properties of the substance to be dried and the characteristics of the various drying agents. The latter must remove water efficiently and must not dissolve in the liquid or react with it in any way.

The efficiency of a drying agent is a composite of three factors: *capacity* (the amount of water that can be removed per gram of drying agent), *speed* (the rate at which the water is taken up), and *intensity* (the amount of water present even with an excess of drying agent). Ratings[1] for these three factors of a number of common drying agents are presented in Table 8.1. It can be seen from Table 8.1 that the three most efficient drying agents are calcium chloride, magnesium sulfate, and 4 Å molecular sieves. Their use and restrictions will now be described.

Calcium chloride is available in pellet form and is an extremely effective drying agent. Unfortunately, it has two limitations. Calcium chloride binds strongly not only to water but also to alcohols, amines, phenols, and, to a lesser extent, molecules containing other polar functional groups. For this reason it is used primarily for drying hydrocarbons and halogenated hydrocarbons. Also there is the practical problem that any pellets spilled on the desk will eventually absorb enough water to form puddles of calcium chloride solution. Since such puddles are both messy and corrosive, it is imperative that spilled calcium chloride be removed promptly.

Anhydrous magnesium sulfate is a fluffy white solid that settles slowly in an organic solution that has been swirled. Wetted drying agent, by contrast, is dense and tends to clump at the bottom of the flask. The usual practice is to add a small portion and after a few minutes observe its behavior on swirling the flask. Additional portions are added until the newly added drying agent remains fluffy and settles slowly.

Molecular sieves possess a large number of pores; the 4 Å size allows water to penetrate but rejects almost all organic molecules, which are much larger. The

---

[1] The comparisons are based on a study by Pearson and Ollerenshaw, *Chemistry and Industry*, **1966**, p. 370.

TABLE 8.1
Common Drying Agents for Organic Compounds

Drying agent	Capacity	Speed	Intensity
Calcium chloride	High	Medium	High
Calcium sulfate	Low	Very high	Very high
Magnesium sulfate	High	High	Medium high
Molecular sieves, 4 Å	High	High	High
Potassium carbonate	Medium	Medium	Medium
Sodium sulfate	Very high	Low	Low

4 Å molecular sieves are available as a powder, as $\frac{1}{16}''$ and $\frac{1}{8}''$ pellets, and as beads. The $\frac{1}{16}''$ pellets are particularly convenient. Fresh molecular sieves are conditioned by heating at 320° for 3 hr; used sieves can be reconditioned after rinsing them with acetone and allowing all of the solvent to evaporate.

The most common application of molecular sieves is to dry organic solvents either by addition to the solvent or sometimes by passing the solvent down a 2-ft column packed with $\frac{1}{16}''$ pellets. The one inconvenience of the sieves is that unlike magnesium sulfate they give no indication of whether their drying capacity has been saturated.

8.2 Cooling Baths

The common cooling bath is a slush of crushed ice and water. Because of the inversion of density of water at 4°, an ice bath should be stirred well if it is desired to maintain the whole bath at 0°. Temperatures below 0° can be obtained by mixing inorganic salts with ice or cold water. Table 8.2 lists the proportions of ingredients to be mixed to obtain the stated temperature. It is important in using ice–salt baths that the mixture be stirred well. Temperatures down to about −80° can be maintained by addition of pieces of dry ice to isopropanol or other low-melting heat-transfer liquids.

8.3 Refluxing

In preparative organic work it is frequently necessary to maintain a reaction at an approximately constant temperature for a long period of time with a minimum of attention. The simplest procedure for reactions carried out in the solution is to boil the solution and condense the vapors so that they are returned to the reaction flask (refluxing). The temperature in the flask remains nearly constant at approximately the boiling point of the solvent (the actual temperature is elevated above the solvent boiling temperature to an extent governed by the concentration of nonvolatile solutes). This operation is so common and important that the principle

TABLE 8.2
Cooling Baths

Ingredients	Lowest temperature obtained, °C
1 part sodium chloride 3 parts crushed ice	−20
1 part ammonium chloride 1 part sodium nitrate 1 part cold water	−20
3 parts powdered calcium chloride 2 parts crushed ice	−50

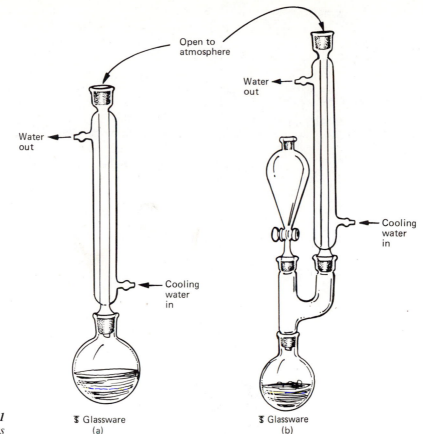

FIGURE 8.1
Reflux Assemblies

and technique should be understood clearly. The apparatus for a simple reflux operation is pictured at the left in Figure 8.1.

Another common requirement in preparative chemistry is to add a liquid to a reaction mixture while the mixture is refluxing. An assembly suitable for this purpose is shown at the right in Figure 8.1. In both assemblies it is *essential* that there be some openings to the atmosphere to prevent pressure buildup. If the reaction is sensitive to atmospheric moisture and must be protected from it, a drying tube is inserted in the top of the reflux condenser; *under no circumstances should a solid stopper be used.*

8.4 Gas Absorption Traps

In some organic preparations, noxious gases are liberated that must not be allowed to escape into the laboratory. Two common methods[2] for trapping

[2] Horodniak and Indicator, *J. Chem. Educ.*, **47**, 568 (1970).

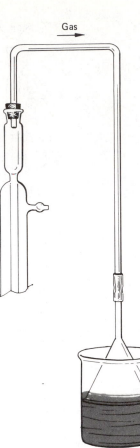

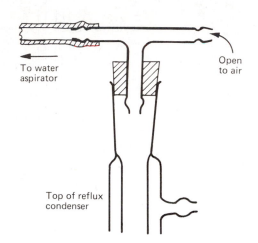

FIGURE 8.2
Gas Absorption Traps

water-soluble gases are pictured in Figure 8.2. The principal precaution to be observed in trapping gases by the inverted funnel method is to construct the apparatus so that a drop of pressure in the reaction flask will not suck water back into the reaction flask. The lower edge of the funnel should not dip into the water *more than 1 or 2 mm*. If the possibility of sucking water back into the reaction flask poses a hazard, a safety flask (see Figure 3.2) should be connected between the reaction vessel and the trap.

In the water-aspirator method, the flow of air sucked in by the aspirator is normally sufficient to overcome all but the most vigorous evolution of gas.

8.5 Stirring

In the organic laboratory, stirring is often needed to hasten solution or reaction by bringing a solid and liquid, or two immiscible liquids, into good contact. Stirring of a *homogeneous* solution is advantageous only when one desires to bring the liquid into good contact with the walls of the vessel to make external heating or cooling more effective and to insure uniform temperature throughout the liquid.

For ordinary laboratory preparations, the movement of a refluxing liquid is usually sufficient to give good mixing. When a solution is not boiling, satisfactory mixing is generally accomplished by intermittent shaking by hand; a circular swirling motion is most effective and reduces the danger of splashing. Occasionally rather violent shaking is required, as when a heavy solid like metallic zinc or iron must be brought into contact with an organic liquid. For material in a beaker, sufficient mixing can usually be obtained by stirring by hand with a glass rod or wooden paddle.

Mechanical stirring is usually needed for large-scale preparations and in any operation where continuous stirring is required for a long time. This is particularly true of heterogeneous reactions in which heat transfer from the central portions of the liquid can be slow enough to cause an excessive temperature rise. Inexpensive motors of the brush type give off sparks and are exceedingly hazardous in an organic laboratory containing vapors of volatile organic chemicals. Nonsparking motors of the induction type normally have high rotation speeds that make them unsuitable; they have low torque at speeds slower than their design speed so that attempts to slow down induction motors by adding a friction device usually are not successful. Geared induction motors are satisfactory but expensive.

Another particularly convenient stirring device is the magnetic stirrer, which consists of an enclosed electric motor attached to a large bar magnet. In practice, a second magnet ("spin bar") is placed in the reaction flask, which rests on the stirrer enclosure; when the motor is turned on, the enclosed magnet produces a rotating magnetic field that forces the spin bar to rotate and stir the contents. The rotating magnetic field is usually sufficiently strong to be effective even if the reaction flask is being heated by a Glass-Col heater or an oil bath. To prevent contamination of the reaction mixture, the spin bar is normally coated with Teflon or glass. Several sizes are available. The magnetic stirrer is more convenient to use than a mechanical stirrer, but it does not have as much turning force and thus is ineffective with viscous reaction mixtures.

8.6 Rotary Evaporation

A common operation in the organic laboratory is the removal of a volatile solvent from a reaction mixture. This can be done by simple distillation, but a faster technique is to use a rotary evaporator, which consists of an electrical motor with an elongated hollow shaft. One end of the shaft is machined to accept joints; the other end is connected through a ball joint to a vacuum pump or aspirator. In some evaporators the vacuum connection is made through a sleeve that fits around the shaft. The apparatus is assembled with the shaft at about 45° angle to the desk top. In operation the motor turns the flask so that a thin film of the solution is continuously being exposed on the upper walls of the flask where it evaporates quickly. In practice the evaporation occurs so quickly that the flask becomes quite cold unless a warm water bath is applied. The advantage of rotary evaporation over simple distillation is the speed of the operation and its simplicity.

II

Identification of Organic Compounds

9

Identification by Chemical Methods

9.1 Introduction

Identification of the molecular structure of organic compounds is a daily activity of practicing organic chemists. Whether the task is to identify the milligram amounts of toxin present in "red-tide" algae or the gallons of distillate from a pilot plant catalytic hydrogenator, the challenge is to translate observations of chemical and physical behavior into a unique structure. Unlike many intellectual problems, structure elucidation usually does not yield to straightforward deductive logic but requires a more intricate interplay of facts and hypotheses. It is this complexity that makes the task always interesting and sometimes just plain fun.

There are two quite different approaches to structure determination. The traditional one, which depends on chemical reactions to identify functional groups and convert them to known derivatives, is the subject of this chapter. The modern method, involving spectrometry, is described in the Chapter 10. Spectrometric methods are used extensively today because they are faster and are capable of dealing with smaller amounts of compounds with more complex structures. Although the traditional methods now are seldom used alone, they are described here for a number of reasons. On occasion, parts of the traditional scheme are still quite useful. Also, the required techniques strongly reinforce fundamental chemical and physical principles and expose the beginning student to the making of chemical judgments, an essential skill for productive research. The time invested in learning how to interpret chemical and physical behavior will be repaid many times over in future work.

Structure elucidation of a completely unknown substance by the traditional methods begins with the isolation of the material in a pure state followed by qualitative tests to disclose the presence of elements such as nitrogen, sulfur, or the halogens. Quantitative analysis furnishes the weight composition of the substance and permits the calculation of an empirical formula, which gives the atomic ratios of the elements present. Determination of the molecular weight permits the assignment of a definite molecular formula that expresses the actual numbers of atoms of each element present in the compound.

The next stage in structure elucidation is to identify the functional groups and other characteristic structural features. If you suspect the compound has been previously prepared and characterized, you would convert it into one or more derivatives that could be compared with the properties reported in the literature. If the compound is new, the partial information obtained from the chemical tests must be pieced together to give a total structure consistent with all available data. Where possible, the tentatively assigned structure is confirmed by an independent synthesis. Alternatively, the compound may be selectively degraded to simpler known substances. The structures of millions of organic compounds were assigned by these same methods.

The full procedure outlined above is much too demanding for beginning students. What will be described in this chapter is a procedure for structure determination that has been simplified by limiting the range of possible functional groups and by restricting the unknowns to previously identified and well characterized compounds. The sequence of steps to be followed is

1. Preliminary examination.
2. Purification of the unknown sample.
3. Physical constant measurements.
4. Element identification.
5. Solubility classification.
6. Functional group identification.
7. Derivative preparation.

9.2 Preliminary Examination

The first step in the identification procedure is to answer some simple but important questions about the unknown. It should be emphasized from the beginning that all observations are to be recorded in the notebook immediately. So many observations are made in the course of a structure identification that some are certain to be forgotten if notebook recording is put off to later.

Physical Appearance. As a first step, note and record whether your sample is a liquid or a solid; the tables that will be consulted in the last stages of the analysis are subdivided into liquid and solid compounds.

Next, note the color of your unknown. Simple aliphatic and aromatic compounds with single functional groups tend to be colorless, whereas aromatic compounds with both electron-donating and -withdrawing groups in conjugation with each other tend to be strongly colored. Many organic compounds, particularly aromatic ones, decompose when heated to form extensively conjugated and hence colored impurities. It is particularly important to note any changes in color as the sample is purified.

Another physical property worth being aware of is the odor of the compound. Although many compounds have characteristic odors, the basis for odor is not well understood, so specific structural conclusions cannot be drawn from the odor. There are, however, some useful generalizations. For example, low-molecular-weight amines have an unmistakable (dead) fish odor; esters are fruity, and so on. There are also many surprises; benzaldehyde and hydrogen cyanide, structurally quite different species, have a similar bitter almond odor. It is a good idea to become acquainted with the odors of the common solvents. It is useful, for example, to be able to recognize quickly the presence of any residual solvent impurities in an unknown.

One should be extremely cautious in observing odors since the vapors could be obnoxious or toxic (see Section 1.1). The proper technique is to open the container well away from your nose and gently waft the vapors toward you. If the odor is not too strong, the container can be brought closer but never right up to the nose. Any odor worth noting will be detected several inches away.

Ignition Test. From the behavior of a small sample when heated in a flame, you can determine if a solid has an accessible melting point, and whether a solid or liquid is volatile, forms volatile decomposition products, or is explosive. Combustion of the sample to give a sooty flame indicates the presence of unsaturation, aromatic groups, or long aliphatic chains. A residue indicates the presence of a metal, usually as the salt of an acid.

Ignition tests can be carried out on a laboratory spatula or on an inverted porcelain crucible cover supported by a wire triangle and ring stand using a small burner. Only a few milligrams of sample should be burned.

9.3 Purification of the Unknown Sample

Your instructor may assign a pure enough unknown that you can go directly to the next step in the identification procedure. If not, you will have to take a small detour. With a solid unknown determine its melting point range. If it is not sharp (less than 2°), recrystallize the solid according to the procedures described in Section 5.2. Determine the melting point again and see if it is now sharp enough to proceed. Additional recrystallizations are required until either the melting point becomes sharp or it has a constant range.

An impure liquid unknown will have to be distilled and a constant boiling fraction collected. Caution is called for here. If the ignition test indicated that the sample decomposes on heating, a vacuum distillation will be necessary. Be sure to record both the pressure and the boiling-point range of the purified sample.

An independent measure of purity for solids and liquids is by thin-layer chromatography. The developing solvent should be chosen so that the R_f of the main component is about 0.5.

9.4 Physical Constants

Melting points and boiling points are characteristic properties of pure materials. The boiling point of a liquid is approximately related to its molecular weight; the melting point of a solid is determined partly by molecular weight but more importantly by the presence or absence of polar groups that interact strongly in the crystal lattice. With liquids the refractive index and density are characteristic properties frequently recorded in handbooks.

If you purified the sample, the melting point or boiling point is already known and you can proceed to the next section on element detection. The determination of boiling points by simple distillation as described in Section 2.4 requires that at least 3 mL of the liquid be available. Boiling points of much smaller samples, even a few drops, can be determined by the inverted capillary technique (Figure 9.1).

Boiling Points of Micro Samples. Place 2–5 drops of the sample in a boiler tube prepared by sealing one end of a Pasteur pipet (or a 5-cm length of 4-mm glass tubing). Seal one end of a melting-point capillary tube and break off the tube about 25 mm from the seal. Drop the capillary tube, open end down, into the boiler tube (Figure 9.1). Attach the boiler tube to a thermometer by means

FIGURE 9.1
Apparatus for Boiling
Points of Micro
Samples

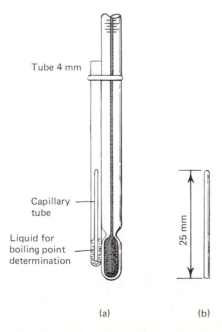

Tube 4 mm

Capillary tube

Liquid for boiling point determination

25 mm

(a)   (b)

of a small rubber band,[1] and support the assembly in an oil bath (a Thiele tube apparatus used for melting points is ideal) so that the sample is at least 10 mm below the bath level. The bath is heated gradually until a rapid stream of bubbles emerges from the capillary. The temperature at which rapid bubbling occurs is a few degrees above the boiling point of the sample. Immediately discontinue heating, which causes the bubbling to cease. When the temperature reaches the boiling point of the sample, and as the temperature continues to fall, the liquid is drawn up into the capillary. The cycle of heating and cooling replaces most of the air in the capillary by the vapor of the sample. At this point resume heating, only more cautiously, so that the temperature rises at a rate of about 2°/min until bubbles once more emerge. Remove the heat and note the exact temperature at which bubbling ceases. This is the boiling point of the liquid, since it is the temperature at which the vapor pressure inside the capillary equals the external atmospheric pressure exerted on the top of the surface of the liquid in the boiler tube.

9.5 Element Detection

Carbon, Hydrogen, and Oxygen. Normally one assumes that a sample obtained in an organic laboratory contains at least carbon and hydrogen. If there is any doubt, you can detect their presence by heating a sample in a tube with dry, powdered copper oxide, whereby carbon dioxide and water are formed. Carbon is detected by passing the evolved gases into an aqueous solution of calcium or barium hydroxide, in which the carbon dioxide produces a precipitate of the carbonate. Hydrogen is detected by the condensation of droplets of water in the cool upper portion of the reaction tube.

There is no satisfactory qualitative test for the presence of oxygen in organic compounds. To determine if oxygen is present quantitative analysis is required. If the sum of the percentages of all known constituent elements does not amount to 100%, the deficit is taken as the percentage of oxygen.

Nitrogen, Halogens, and Sulfur. The qualitative detection of these elements in organic compounds is more difficult than in inorganic compounds because most organic compounds are not appreciably ionized in solution. Since the tests used in qualitative analysis are based upon ionic reactions, they cannot be applied directly to organic compounds. For example, sodium chloride or bromide gives an immediate precipitate of silver halide when treated with an aqueous solution of silver nitrate. Carbon tetrachloride, bromobenzene, and most organic halides do not produce a silver halide precipitate when treated with aqueous silver nitrate solution because they do not furnish an appreciable amount of halide ion in solution.

For qualitative detection it is necessary, therefore, to first convert elements such as nitrogen, sulfur, and halogens into ionized substances. This conversion

---

[1] Tiny rubber bands can be made by cutting off 2-mm lengths of soft rubber tubing.

may be accomplished by several methods, of which the most general is fusion with metallic sodium, which produces sodium cyanide, sodium halides, sodium sulfide, and so forth, as indicated in the following reaction scheme. The resulting anions may then be identified by applying the usual inorganic tests.

Organic compound
containing C, H, $+ Na \xrightarrow[\text{temperature}]{\text{high}} NaCN + NaCl + Na_2S + NaOH +$ etc.
O, N, S, Cl

Sulfur. A fresh portion of the filtered alkaline solution is acidified with acetic acid and treated with an aqueous solution of lead acetate. If sulfide is present, a dark-brown precipitate of lead sulfide results. The acetic acid neutralizes the base to prevent precipitation of lead hydroxide.

$$Na_2S + Pb(OAc)_2 \longrightarrow PbS + 2NaOAc$$
$$\text{(dark brown)}$$

Nitrogen. Two tests for nitrogen are presented here. In the traditional test, a portion of the filtered alkaline solution is treated with aqueous ferrous sulfate and ferric chloride, boiled for a few moments, and acidified with hydrochloric acid. If nitrogen is present, a precipitate of prussian blue results.

$$2 NaCN + FeSO_4 \longrightarrow Fe(CN)_2 + Na_2SO_4$$
$$4 NaCN + Fe(CN)_2 \longrightarrow Na_4Fe(CN)_6$$
$$3 Na_4Fe(CN)_6 + 4 FeCl_3 \longrightarrow Fe_4(Fe(CN)_6)_3 + 12 NaCl$$
$$\text{(prussian blue)}$$

The second test for nitrogen depends on the fact that cyanide ion is a catalyst for the benzoin condensation of 2-pyridinecarboxaldehyde to give a dimeric product that precipitates as a copious bright yellow flocculent solid. The new test is more sensitive and reliable than the traditional one, but suffers from the disadvantage that the carboxaldehyde reagent must be freshly prepared and stored in a refrigerator.

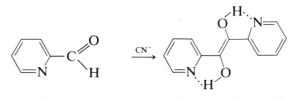

2-Pyridinecarboxaldehyde   (bright yellow solid)

Halogens. A fresh portion of the filtered solution is acidified with nitric acid and boiled for a short time to expel any hydrogen cyanide or hydrogen sulfide that may be present. The resulting solution, containing free nitric acid, is treated with aqueous silver nitrate, and, if halides are present, a precipitate of silver halide results. The individual halides can be distinguished by oxidation to the elemental halogen and observing the color of a methylene chloride extract.

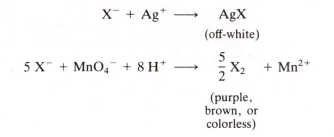

$$X^- + Ag^+ \longrightarrow AgX$$
(off-white)

$$5\,X^- + MnO_4^- + 8\,H^+ \longrightarrow \frac{5}{2}\,X_2 + Mn^{2+}$$
(purple, brown, or colorless)

Sodium Fusion. The sodium fusion should be carried out in the hood. Support a small, Pyrex test tube (about 75×10 mm) by inserting it through a small hole in a piece of transite board so that the tube is held by its rim as in Figure 9.2.

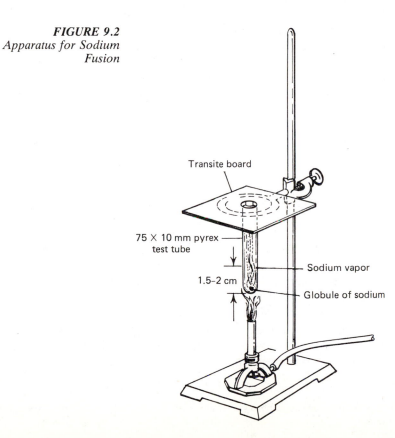

FIGURE 9.2
Apparatus for Sodium
Fusion

Transite board

75 X 10 mm pyrex test tube

1.5–2 cm

Sodium vapor

Globule of sodium

Place a small portion of the sample[2] in the bottom of the test tube. If the sample is a solid, use only about 10 mg; if it is a liquid, use only 2 drops.

Drop a small piece of bright sodium metal no larger than a pea (about 3–4 mm on each edge) into the tube. Some organic compounds will react with the sodium metal but most will not. After any reaction has subsided, heat the bottom of the tube gently with a microburner until the sodium melts and its vapors fill the lower part of the tube. Remove the flame momentarily and then drop another portion of the sample (10 mg of solid; 2 drops of liquid) directly on the molten sodium.

IIII➡ *CAUTION*   Carry out the sodium fusion with great care; be particularly cautious in decomposing the fused mass. You *must wear safety goggles* while carrying out this procedure.

A spontaneous exothermic reaction takes place, frequently accompanied by a flash of fire. Heat the tube again until the vapors start to rise and then remove the flame and add another small portion of the sample. Adjust the burner to produce a hot flame and heat the tube to redness in order to complete the reactions. Extinguish the burner and allow the tube to cool to about room temperature. Slowly add about 1 mL of methanol; after the reaction has subsided, heat the tube gently to evaporate the residual methanol. The methanol frequently ignites during the evaporation but the fire can be extinguished easily by placing a transite board over the top of the test tube.

Allow the tube to cool and then crush it with a clean mortar and pestle. Add about 5–6 mL of distilled water to the mortar and grind the mixture to dissolve the fusion solids. Filter the solution through a small filter paper or wad of glass wool and apply the following tests to portions of the solution.

IIII➡ *CAUTION*   During the crushing and grinding operation cover the mortar and pestle with a laboratory towel.

1. Sulfur. Acidify a 1-mL portion of the fusion solution with acetic acid (use pH paper) and add a few drops of 5% lead acetate solution. A black precipitate indicates sulfide.

2. Nitrogen. *Traditional test*: To a 1-mL portion of the fusion filtrate add 2 drops of ferrous ammonium sulfate solution (saturated) and 2 drops of potassium fluoride solution (10%). Bring the mixture to a gentle boil and then cool it to room temperature. Acidify the mixture carefully with dilute sulfuric acid (20–30%) until the precipitate of iron hydroxide just dissolves. Avoid excess acid. A precipitate of prussian blue indicates the presence of cyanide. A faint precipitate can be detected by allowing the solution to stand for a short time and then filter it through a small filter paper.

---

[2] A stronger test for nitrogen can be obtained by mixing the sample with about half its volume of confectioner's sugar.

Catalytic test: To 1 mL of 1 M aqueous 2-pyridinecarboxaldehyde add a 1-mL portion of the fusion filtrate and allow the mixture to stand for 10 min. If a bright-yellow precipitate forms, the test is positive for nitrogen (as cyanide ion). The test reagent does not store well at room temperature and should be prepared fresh each day, although it will keep for several days if stored in a refrigerator.

3. Halogens. Acidify a 1-mL portion of the fusion filtrate with dilute nitric acid (1 volume of concentrated acid to 1 volume of water) and if nitrogen or sulfur is present, as shown by tests 1 and 2, boil gently for 5-10 min to remove any hydrogen sulfide or hydrogen cyanide (**Caution**—*toxic vapor!*) that may have been formed. Add about 1 mL of a dilute solution of silver nitrate (5–10%) and boil gently for a few minutes. A heavy precipitate indicates the presence of halide; if there is only a faint turbidity, it is probably due to the presence of impurities in the reagents.

The halides chloride, bromide, and iodide can be distinguished by the following procedure. Acidify a 1-mL portion of the fusion solution with dilute nitric acid as above. To this solution add 10 drops of 1% potassium permanganate and shake the test tube for about 1 min. Add *just enough* oxalic acid (about 20 to 30 mg) to discharge the color of the excess permanganate and then add 1 mL of methylene chloride. Shake the test tube and observe the color of the methylene chloride layer. A purple color indicates iodine, a brown color indicates bromine, and the absence of color indicates chlorine. The presence of bromine can be confirmed by the addition of several drops of allyl alcohol, which will react quickly with bromine but not with iodine.

9.6 Solubility Classification

The classification of an unknown according to its solubility behavior, when combined with a knowledge of the elements present, greatly limits the number of functional groups that need be considered. These may be further differentiated either by chemical tests or, as will be discussed in Chapter 10, by spectrometry. Systematic solubility classification was introduced by Kamm[3] and his scheme forms the basis of the flowchart present in Figure 9.3. The classes of molecules included in this chart are restricted to the more commonly encountered functional groups[4].

The principle behind the scheme is that while all hydrocarbons are insoluble in water, the attachment of polar functional groups produces favorable interactions with the polar water molecules such that the substituted hydrocarbon

---

[3] Kamm, *Qualitiative Organic Analysis* (New York: Wiley, 1922).
[4] For more exhaustive treatments of solubility classification and lists of molecular classes see Cheronis, Entrikin, and Hodnett, *Semimicro Qualitative Organic Analysis*, 3rd ed. (New York: Interscience, 1965), or Shriner, Fuson, Curtin, and Morill, *The Systematic Identification of Organic Compounds*, 6th ed. (New York: Wiley, 1980).

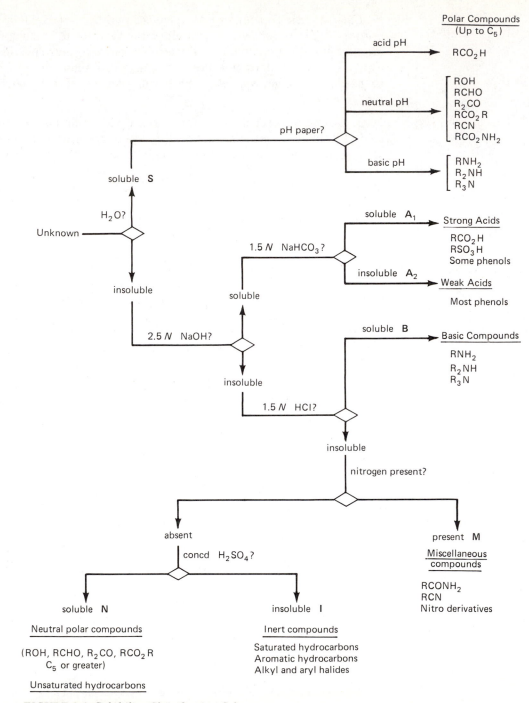

FIGURE 9.3 *Solubility Classification Scheme*

becomes water soluble. The stronger the interactions, the larger the hydro-carbon chain that can be solubilized. By convention a substance is designated "soluble" in a solvent if it dissolves to the extent of 3% or more. This definition is arbitrary, but it provides a practical classification scheme.

In the abbreviated scheme shown here the primary classification solvents are water, 2.5 N NaOH (10%) and 1.5 N HCl (12.5%), which are used to categorize the unknown as water soluble (**S**), acidic (**A**), or basic (**B**). The water-soluble compounds are further divided into those that are soluble in methylene chloride (most simple organic compounds) and those that are not (organic salts and sugars). The acidic compounds are further divided into strong and weak acids according to their solubility in the weak base sodium bicarbonate. Those molecules that are not soluble in any of these aqueous solvents can be sub-divided by their elemental composition and behavior toward concentrated sulfuric acid as shown in the scheme and discussed below.

Most molecules belonging to solubility class **S** have four or fewer carbon atoms combined with a polar functional group. Particularly effective water-solubilizing groups such as carboxylate anions or quaternary ammonium ions confer water solubility to molecules with as many as 20–25 carbon atoms; polyfunctional molecules may have proportionally greater numbers of carbon atoms. The carboxylate anions and ammonium salts will not dissolve in methylene chloride. Any functional group, except the halogens, may be present in a molecule belonging to class **S**. Compounds with five carbons have borderline solubility. Branching increases water solubility; many cyclic compounds with six carbons are soluble.

The following molecules are examples of class **S** compounds; each has a polar functional group and four or fewer carbon atoms.

Water-soluble (class **S***) compounds*

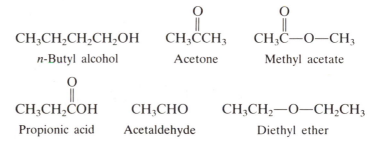

Note particularly the inclusion of diethyl ether, which is widely used to extract organic materials *from* water. There is no contradiction. If one shakes 100 mL of diethyl ether with 100 mL of water, two layers will form. The apparent insolubility is only superficial; about 6 mL of the ether will dissolve in the water and about 1 mL of water will dissolve in the remaining 94 mL of ether. One must not confuse solubility (defined here as 3% solubility) with miscibility (complete solubility), terms that tend to be used interchangeably.

Acidic species with fewer than 20–25 carbon atoms will be soluble in 2.5 N sodium hydroxide because of the strong solubilizing effect of the anion formed. Strong acids, class A_1 (carboxylic acids and some phenols bearing electron-withdrawing groups), will also be soluble in 1.5 N aqueous sodium bicarbonate but weak acids, class A_2 (most phenols, enols, primary and secondary nitro compounds, primary sulfonamides), will not dissolve in this weak base. Examples of A_1 and A_2 compounds are shown below.

Base-soluble (*class* **A**) *molecules*

$$CH_3(CH_2)_{14}CO_2H \qquad (CH_3)_3C-\!\!\langle\bigcirc\rangle\!\!-OH$$

Hexadecanoic acid (A_1)   *p-tert*-Butylphenol (A_2)

Molecules soluble in 1.5 N hydrochloric acid, class **B**, will be primary, secondary, or tertiary amines.

Acid-soluble (*class* **B**) *molecules*

$$CH_3(CH_2)_8NH_2 \qquad (CH_3CH_2CH_2CH_2)_2NH \qquad (CH_3CH_2CH_2)_3N$$

Nonylamine   Di-*n*-butylamine   Tri-*n*-propylamine

In classifying the remaining molecules, which are insoluble in all of the aqueous test solvents, it is useful to distinguish between those that contain nitrogen or sulfur and those that do not. Molecules containing these elements are assigned to class **M**, which includes nitro compounds, amides, mercaptans, sulfides, and sulfonyl derivatives.

Nitrogen- and sulfur-containing (*class* **M**) *molecules*

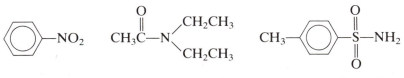

Nitrobenzene   *N,N*-Diethylacetamide   *p*-Toluenesulfonamide

When nitrogen and sulfur are absent, the remaining molecules are classified according to their solubility in cold, concentrated sulfuric acid. Soluble molecules, class **N**, may contain any of the functional groups found in class **S** since each of these groups is capable of being protonated to produce an ionic compound that is soluble in the acid. Another group of molecules falling into class **N** are the alkenes and alkynes, which react with concentrated sulfuric acid to produce polar species that are soluble in that solvent. Compounds that react with sulfuric acid, as evidenced by darkening or heat evolution, should be assigned to class **N** even if an insoluble precipitate is formed. Class **N** molecules will have more than four carbons, and nitrogen and sulfur will be absent as shown in the examples below.

*Class **N** molecules*

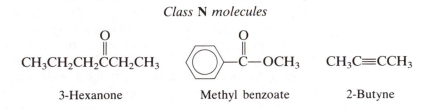

3-Hexanone   Methyl benzoate   2-Butyne

The inert molecules, class **I**, do not dissolve in any of the classification solvents and include the saturated hydrocarbons, aromatic hydrocarbons (weakly reactive toward sulfuric acid because of their delocalization energy), and alkyl and aryl halides. Some examples follow.

*Class **I** molecules*

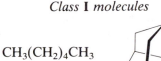

$CH_3(CH_2)_4CH_3$

n-Hexane   Adamantane

Solubility Classification Tests. In a small test tube place 0.1 mL of a liquid or 0.1 g of a solid compound[5] and add, in small portions, a total of 3 mL of water. Between each addition stir the sample vigorously with a rounded stirring rod; with solids it is desirable to crush the crystals to increase their surface area. If the sample has dissolved test the aqueous solution with a wide-range pH paper to determine if the solution is acidic, neutral, or basic. Record your observations directly in your notebook.

Follow the solubility scheme illustrated in Figure 9.3 using fresh samples and 3-mL portions of the appropriate solvents. Keep a careful record of each test as you make it. Note that in following the scheme all of the tests solvents need not be tried, only those that are logically required to classify the sample. Also note that since strongly basic and acidic solutions are quite corrosive, they should be cleaned up immediately if spilled.

The most serious difficulty encountered in these tests is with compounds of borderline solubility. Observe the sample carefully and note whether any of it appears to dissolve. If partial solubility is observed, give the sample a longer time to dissolve since the rate of solution of borderline compounds can be quite slow.

---

[5] It is desirable to use precise amounts of sample. A convenient procedure for solids is to weigh out 0.4 g on a weighing paper and divide it visually into four equal portions. With liquids one can use a dropper that delivers a known number of drops per mL.

9.7 Functional Group Identification

Any compound other than a saturated hydrocarbon has at least one functional group, which can be identified by carrying out a series of "classification tests" that serve to narrow the range of possibilities until only one remains. When the functional group is identified, an appropriate table of characterized compounds containing this group is consulted, and those compounds having chemical and physical properties consistent with the sample are selected. In favorable cases only a few compounds will be found; rarely will there be more than 10.

The lists of compounds containing each functional group give not only the physical properties of the molecules but also the properties of solid substances (derivatives) that can be prepared from it by tested procedures. Since the melting points of these derivatives are usually distinctive, the combination of properties of the original substance and of its derivatives is sufficient to identify it.

Table 9.1 lists the different functional groups that will be considered in this text with the classification tests and derivatives appropriate for each. The list of functional groups is restricted but does include the most commonly encountered types. If your instructor wishes to broaden the range, advanced texts devoted to organic qualitative analysis will have to be consulted.[4] Discussions of the chemistry of each test and its structural significance, as well as the experimental details, are given for each of the functional groups in this section. Once the functional group has been identified, the final identification of the compound can be made by the preparation of derivatives as described in Section 9.8.

In the laboratory it is important to perform the classification tests in a sequence consistent with the accumulated evidence, never at random. A good guide is the solubility classification scheme (Figure 9.3), which lists the possible functional groups for each solubility class. For example, if the elemental analysis reveals nitrogen and the compound falls in solubility class **B** the amine tests should be performed directly.

As a second example, if a neutral compound falls in class **S** or **N** and does not contain nitrogen, sulfur or halogens, the functional group must be one of the following: alcohol, aldehyde, ketone, or ester. In this case, the recommended next step is to test with 2,4-dinitrophenylhydrazine for an aldehyde or ketone. If the test result is positive, further structural distinctions can be made with the tests described in the procedures for aldehydes and ketones. A negative 2,4-dinitrophenylhydrazone test should be followed by the hydroxamate test for esters. If that test is negative, only the alcohol class remains, and this can be confirmed by the classification tests for alcohols. Functional groups of compounds that fall into other solubility classes can have be identified by analogous strategies.

To ensure satisfactory results for the tests, it is recommended that the specified quantities of liquid reagents be measured in a graduated cylinder or a

TABLE 9.1 Classification Tests and Derivatives of Functional Groups

Functional	Classification tests	Derivatives
Alcohol	Oxidation test	p-Nitrobenzoate
	Lucas test	3,5-Dinitrobenzoate
	Iodoform test	Phenylcarbamate
Aldehyde and ketone	Tollens' test	Methone derivatives
	Fuchsin aldehyde test	2,4-Dinitrophenylhydrazone
	2,4-Dinitrophenylhydrazine test	Semicarbazone
	Iodoform test	Oxime
Ester	Hydroxamate test	Saponification equivalent
		Hydrolysis of ester to acid and alcohol, which are suitable derivatives if solid
Carboxylic acid	Solubility classification (pH of solution if **S**)	Acid amides
		Neutralization equivalent
Sulfonic acid	Elemental analysis	S-Benzylthiouronium salts
	Solubility classification (pH of solution if **S**)	Sulfonomides
Acid amide and nitrile	Elemental analysis	Hydrolysis to carboxylic acid and amine or ammonia, which may be identified separately
	Hydroxamate test	
Amine	Elemental analysis	Acetamide
	Solubility classification (pH of solution if **S**)	Benzamide
		Benzenesulfonamide
	Hinsberg test	p-Toluenesulfonamide
		Quarternary ammonium salt
Phenol	Solubility classification	3,5-Dinitrobenzoate
	Ferric chloride test	Aryloxyacetic acid
		Bromination product
Unsaturated hydrocarbons	Solubility classification	No generally applicable derivatives
	Bromine test	Bromine titration
	Permanganate test	
Aromatic and saturated hydrocarbons and halide	Solubility classification	No generally applicable derivatives
	Elemental analysis	
	Alcoholic silver nitrate test	

calibrated dropper. If a test is being done for the first time, it is a good idea to practice on materials of known structure.

Infrared (IR) analysis is a powerful tool for identifying functional groups, since a single IR spectrum reveals much about the nature of all of the fuctional groups present. However, the IR spectrum usually does not provide a total answer and one must resort to either other instrumental techniques or the chemical methods described here.

Alcohols.[6] The chromic acid test is a general test for alcohols or other readily

[6] A suitable set of alcohols with which to practice the alcohol tests is: allyl alcohol, *sec*-butanol, *t*-butanol, and *n*-propanol.

oxidizable functional groups such as aldehydes. The *Lucas test* and the *iodoform test* provide further structural information about the alcohol.

(A) Chromic Acid Oxidation Test

This test, based on the ability of primary and secondary alcohols to be oxidized by chromic acid, distinguishes these alcohols from tertiary alcohols. For low-molecular-weight alcohols, aqueous chromic acid can be used, but for alcohols with large hydrocarbon groups some organic cosolvent is required.

To 5 mL of a 1% solution of sodium dichromate, add 1 drop of concentrated sulfuric acid (**Caution**—*corrosive!*) and mix thoroughly (use a Pasteur pipet and the squirting technique described in Section 6.4; if you insist on shaking the tube, use a cork, not your thumb to stopper the tube!). Add 2 drops of the liquid (or about 40 mg of a solid) to be tested, and warm gently. Observe any change in the color of the solution.

$$H_2Cr_2O_7 + R—CH_2OH \text{ or } R_2CHOH \xrightarrow{H_2SO_4}$$
(orange)

$$Cr_2(SO_4)_3 + R—CO_2H \text{ or } R_2C{=}O$$
(blue-green)

For higher-molecular-weight alcohols that are insoluble in the aqueous reagent, the oxidation test can be carried out by using acetone and chromic acid test reagent. Under these conditions, the blue-green chromic reduction product precipitates and is quite visible in the orange solution.

In 2 mL of acetone in a small test tube, dissolve 2 drops of the liquid (or about 20 mg of a solid) and add 2 drops of *chromic acid test reagent* to the solution.[7] Observe any change that occurs within 5 sec; ignore any change that occurs later. It is advisable to run a control test on a sample of the acetone.

(B) Lucas Test

The reagent used is concentrated hydrochloric acid containing 1 mole of anhydrous zinc chloride to 1 mole of the acid.[8] The Lucas test distinguishes between primary, secondary, and tertiary alcohols and is based upon the rate of formation of the insoluble alkyl chloride. To be reliable the alcohol should be soluble in water (class **S**).

The ease of conversion of alcohol to chloride follows the stability of the

---

[7] Chromic acid test reagent is prepared by dissolving 10 g of chromic anhydride (CrO_3) in 40 mL of 25% sulfuric acid (10 mL of concd sulfuric acid in 30 mL of water). (**Caution:** CrO_3 and H_2SO_4 are *corrosive* and *poisonous!* In contact with organic materials solid CrO_3 may cause combustion!)

[8] Lucas' reagent is prepared by dissolving 34 g of anhydrous (fused) zinc chloride in 27 g of concentrated hydrochloric acid, with stirring and external cooling to avoid loss of hydrogen chloride. The resulting solution has a volume of about 35 mL and is sufficient for about 10 tests. To obtain reliable results, the reagent should be reasonably fresh.

corresponding carbocation, modified by the solubility of the alcohol in the test reagent. Allyl alcohol, $CH_2\!\!=\!\!CH\!-\!CH_2OH$, which yields a stabilized charge-delocalized cation, acts like a tertiary alcohol. Isopropyl alcohol sometimes fails to give a positive test because the chloride product is volatile (36°) and may escape from the solution.

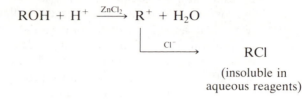

$$ROH + H^+ \xrightarrow{\text{ZnCl}_2} R^+ + H_2O$$

RCl

(insoluble in aqueous reagents)

To 0.5 mL of the alcohol add quickly 3 mL of the hydrochloric acid–zinc chloride reagent at room temperature. Close the tube with a cork and shake it, then allow the mixture to stand. Tertiary alcohols give an immediate separation (emulsion) of the chloride, secondary alcohols require about 5 min, but most primary alcohols do not react significantly in less than an hour. If the result is positive, carry out a second test using concentrated hydrochloric acid alone, instead of the test reagent. This less reactive reagent will give chloride emulsions within 5 min only with tertiary alcohols.

(C) Iodoform Test

This is a test for the specific structural feature $R\!-\!CHOH\!-\!CH_3$ (R may also be H). The test depends on initial oxidation of the alcohol to $R\!-\!CO\!-\!CH_3$, which is iodinated and then cleaved to give a bright yellow precipitate of iodoform.

$$I_2 + NaOH \rightleftharpoons NaOI + NaI + H_2O$$

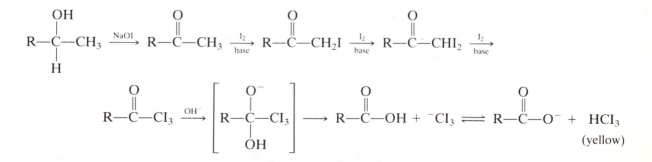

In a clean (acetone-free) 150-mm test tube mix 3 drops of the liquid (or about 50 mg of solid) with 2 mL of water and 2 mL of 10% aqueous sodium hydroxide solution.[9] Add dropwise with shaking a 10% solution of iodine in potassium iodide,[10] until a definite brown color persists (indicating an excess of iodine).

With some compounds a precipitate of iodoform appears almost immediately in the cold. If it does not appear within 5 min, warm the solution to 60° in a beaker of water. If the brown color is discharged, add more of the iodine solution until the iodine color persists for 2 min. Add a few drops of sodium hydroxide solution to remove excess iodine, dilute the mixture with 5 mL of water, and allow it to stand for 5 min at room temperature.

For compounds that are not appreciably soluble in water, the sample may be dissolved in *pure* methanol instead of water. Before starting the test the solvent should be tested to see if iodoform-producing impurities are present.

Iodoform crystallizes as lemon-yellow hexagons having a characteristic odor. Their identity can be confirmed by collecting it with suction, and taking the melting point (119°).

Aldehydes and Ketones.[11] The 2,4-dinitrophenylhydrazone test is positive for both aldehydes and ketones. These may be distinguished by either the *silver mirror test*, which depends on the easy oxidation of aldehydes, or the *Schiff's fuchsin test*, which depends on the ease of formation of SO_2 adducts of aldehydes but not ketones. Another test that will distinguish aldehydes from ketones is the *chromic acid test*, described earlier under alcohols. Aromatic aldehydes take about 60 sec to give a positive test.

The *iodoform test*, also described earlier under alcohols, is specific for molecules containing a methyl group adjacent to a carbonyl group or to any other structure that can form such a methyl carbonyl combination. The only aldehyde that gives a positive iodoform test is acetaldehyde.

(D) 2,4-Dinitrophenylhydrazone Test

Most aldehydes and ketones react with 2,4-dinitrophenylhydrazine reagent to give precipitates of the 2,4-dinitrophenylhydrazones. Esters and amides generally do not respond and can be eliminated on the basis of this test.

The color of the precipitate depends on the degree of conjugation in the aldehyde or ketone. Unconjugated aliphatic carbonyl groups, such as butanal or cyclohexanone give yellow precipitates.

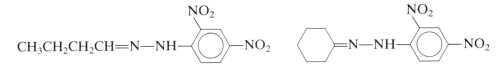

---

[9] This test is suitable for alcohols having significant water solubility. For less-soluble compounds add 1 mL of pure methanol (free of ethanol and acetone).

[10] Iodine–potassium iodide solution is prepared by dissolving 10 g of iodine crystals in a solution of 20 g of potassium iodide in 80 mL of water and stirring until the iodine has dissolved.

[11] A suitable set of aldehydes and ketones with which to practice is: acetone, benzaldehyde, and cyclohexanone.

Conjugated carbonyls, such as benzaldehyde or methyl vinyl ketone, give red precipitates.

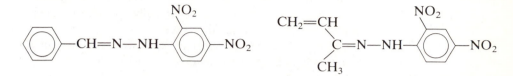

Unfortunately the reagent is orange-red; one should establish that a reddish precipitate is really a new product and not just the starting reagent that has been made insoluble by the addition of the unknown.

In a clean, small test tube, place 1 mL of 2,4-dinitrophenylhydrazine reagent[12] and add a few drops of liquid (or about 50 mg of solid dissolved in the minimum amount of 95% ethanol). A positive test is the formation of a yellow to red precipitate. Most aldehydes and ketones will give a precipitate immediately, although some sterically hindered ones may take longer. If no precipitate appears within 15 min, heat the solution gently for 5 min; examine the test tube after it has cooled to room temperature.

(E) Tollens' Reagent (Silver Mirror Test)

This test involves reduction of an alkaline solution of silver ammonium hydroxide to metallic silver and oxidation of the aldehyde, but *not* a ketone, to the carboxylic acid.

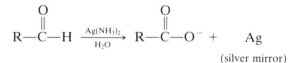

(silver mirror)

This is an extremely mild oxidation and alcohols do not respond. Fehling's or Benedict's solution (alkaline cupric tartrate or citrate) also may be used as a test for aldehydes but the Tollens' test is more sensitive.

In a thoroughly clean 75-mm test tube, place 1 mL of a 5% solution of silver nitrate and add a drop of 10% aqueous sodium hydroxide. Add a very dilute solution of ammonia (about 2%) drop by drop, with constant shaking until the precipitate of silver oxide just dissolves. To obtain a sensitive reagent it is necessary to avoid a large excess of ammonia.

IIII▶ *CAUTION* The silver ammonium hydroxide reagent should be freshly prepared just before use and *should not be stored.* On standing the solution may decompose and deposit an explosive precipitate of silver nitride, Ag_3N.[13]

---

[12] The reagent is a 4% solution of 2,4-dinitrophenylhydrazine in acidified ethanol–water.

[13] As soon as the test has been completed, pour the contents into the "silver waste" container and wash the tube with water. A freshly formed silver mirror can be removed with soap and a test tube brush; residual silver stains can be removed with dilute nitric acid.

Add 2 drops of the unknown to be tested, shake the tube, and allow it to stand for 10 min. If no reaction has occurred in this time, place the tube in a beaker of water that has been heated to about 40° and allow it to stand for 5 min. A positive test is the formation of a silver mirror (if the tube is clean) or a black precipitate of finely divided silver.

Water-insoluble compounds give weak or negative tests. With such unknowns it is helpful to dissolve them in 0.5 mL of AR (Analytical Reagent) grade acetone.

(F) Schiff's Fuchsin Test

The intensely colored triphenylmethane dye fuchsin reacts with bisulfite (a source of SO_2) to produce the colorless "leuco" form of the dye. Aldehydes, but not ketones, react with this "leuco" dye to produce a new triphenylmethane dye possessing a similar fuchsia color.

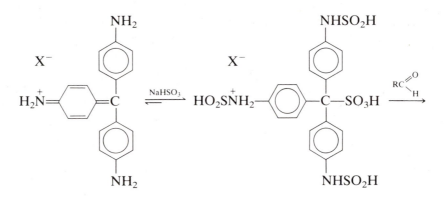

(fuchsia)   (colorless)

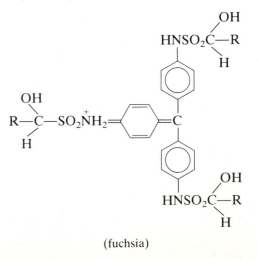

(fuchsia)

To a few drops of the unknown to be tested, in 4–5 mL of water, add about 1 mL of the fuchsin test reagent[14] and observe any development of purple color.

Ketones do not respond to this test *when perfectly pure*, but the color reaction is very sensitive and responds to mere traces of an aldehyde.

Esters. An extremely sensitive test for esters in the formation of the intensely colored ferric hydroxamate derivative. The first step of the test is the base-catalyzed conversion of the ester to a hydroxamic acid. In the next step the hydroxamic acid is treated with ferric chloride, which produces the red-violet octahedral ferric hydroxamate.

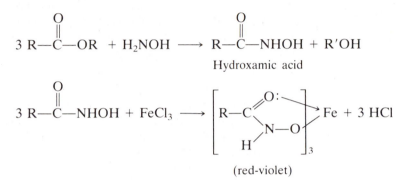

All carboxylic acid esters produce the ferric hydroxamate. Acid chlorides, anhydrides, and imides react with hydroxylamine to give hydroxamic acids and thus they also give positive tests. Amides and nitriles generally do not react sufficiently under the specified reaction conditions to give more than pale colorations. Free carboxylic acids (other than formic acid) give negative tests.

Some phenols react with ferric chloride to give a somewhat similar color. For this reason, if a positive test is obtained, it is necessary to run a "blank" test in which all of the steps are carried out except the addition of the hydroxylamine.

(G) Hydroxamate Test[15]

In a small test tube, place a few drops of a liquid (or about 50 mg of a solid) unknown and 1 mL of 7% methanolic hydroxylamine hydrochloride ($NH_2OH \cdot HCl$) containing a pH indicator.[16] Add 10% methanolic potassium hydroxide

[14] The fuchsin test solution is prepared by dissolving 100 mg of pure fuchsin (*p*-rosaniline hydrochloride) in 100 mL of distilled water and adding 4 mL of saturated sodium bisulfite solution. After about an hour, 5 mL of water to which 2 mL of concentrated hydrochloric acid has been added is added slowly. This produces a practically colorless and sensitive reagent. If the solution is not colorless, shake it with a little decolorizing carbon and filter it.

[15] A suitable set of compounds on which to practice the ester test is: ethyl acetate, benzoic acid, and *p*-cresol.

[16] The indicating reagent is prepared by dissolving 70 g of hydroxylamine hydrochloride, 100 mg of thymolphthalein, and 15 mg of methyl yellow in 1 L of methanol. The solution is neutralized by careful addition of 10% methanolic potassium hydroxide until the rose color just turns orange.

until the mixture just turns blue, then add 0.5 mL more. Heat the solution to boiling and after allowing it to cool slightly, acidify it with 7% methanolic hydrochloric acid[17] until the solution turns rose color. Then add 2 drops of 3% ferric chloride solution. If the color is weak add a few more drops of ferric chloride solition. A positive test is the development of an intense red-violet color.

Carboxylic Acids. The presence of a carboxylic acid functional group is revealed by the solubility behavior of the compound. If the solubility class is **S**, the aqueous solution will be acidic. The only possible confusion here would be with a low-molecular-weight sulfonic acid, which would require a positive elemental analysis for sulfur. If the solubility class is **A**, and sulfur is absent, the molecule is either a carboxylic acid or one of the few phenols that are substituted with strongly electron-withdrawing groups, such that they move from their normal classification as weak acids (A_2). Tests for phenols are described later in this section.

Sulfonic Acids. There is no specific chemical test for detecting a sulfonic acid functional group, although they may be detected readily by IR. The presence of such a group is indicated by a positive elemental analysis for sulfur, combined with solubility evidence: either an acidic solution (if the molecule is water soluble, class **S**), or solubility in sodium carbonate (class A_1). The only confusion that might arise would be in the rare circumstance that the molecule was a carboxylic acid substituted by some neutral, sulfur-containing substituent. Fortunately, good derivatives can be made from sulfonic acids, and its suspected presence can be confirmed.

Carboxamides and Nitriles. These two functional groups can be detected by their reaction with hydroxylamine and ferric chloride to form a red-violet complex. The general structure of the complex from an acid amide or nitrile is the same as that described under the ester test reactions; it differs only by substitution of an imino group for the ester carbonyl group.

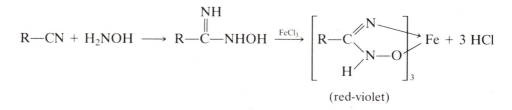

(red-violet)

Amides and nitriles are less reactive than esters so that more vigorous conditions are required for the reaction with hydroxylamine. This is achieved by substituting propylene glycol (bp 187°) for methyl alcohol (bp 65°) as solvent.

[17] This is approximately 2 M acid prepared by dissolving 17 mL of concentrated aqueous hydrochloric acid in 83 mL of methanol.

(H) Amide and Nitrile Test[18]

Add 30–50 mg of the unknown to 2 mL of 1 N hydroxylamine hydrochloride dissolved in propylene glycol. Add 1 mL of 1 N potassium hydroxide in propylene glycol, and boil the mixture gently for 2 min. Cool the solution to room temperature and add 0.5–1 mL of 5% aqueous ferric chloride. A red-violet color is a positive test.

Amines. All classes of alkyl and aryl amines—primary, secondary, and tertiary (1°, 2°, 3°)—have an unshared electron pair on nitrogen and are basic and nucleophilic. The unshared electron pair is responsible for the ability of these amines to form salts with acids and coordination complexes with metal cations and to undergo alkylation with alkyl halides (nucleophilic displacements).

The availability of the unshared pair for combination with an electrophile (measured by base strength or nucleophilicity) is strongly influenced by the nature of the groups attached to the nitrogen atom. In general, alkyl groups, through electron release, enhance the base strength, while electron-withdrawing groups such as aryl and acyl diminish it. Base strength and nucleophilicity are also influenced by the steric bulk of the substituents, both being diminished by large groups.

All of the amines considered here will fall into either the **S** or **B** solubility class and thus are easily recognized by their elemental composition and basicity. The primary and secondary amines differ from tertiary amines in their behavior toward acyl and sulfonyl halides. A particularly useful reagent is benzenesulfonyl chloride and excess base (Hinsberg test).[19] All three classes of amines react with benzenesulfonyl chloride but they differ in how the intermediate products respond to base. With primary amines the sulfonamide formed is acidic and dissolves in the excess base used to yield a solution of the corresponding anion. Addition of excess hydrochloric acid converts the anion into the water-insoluble sulfonamide.

$$RNH_2 + \langle\bigcirc\rangle{-}SO_2Cl \xrightarrow{KOH} \langle\bigcirc\rangle{-}SO_2NHR + KCl + H_2O \underset{\text{excess acid}}{\overset{\text{excess base}}{\rightleftharpoons}}$$

1° amine   water insoluble

$$\langle\bigcirc\rangle{-}SO_2NR^- \ K^+ + H_2O$$

water soluble

Secondary amines react with benzenesulfonyl chloride but the sulfonamide lacks an amide hydrogen and is insoluble in the basic reagent.

[18] A suitable set of compounds with which to practice this test is benzamide and benzonitrile.
[19] For a detailed discussion of the Hinsberg test see Gambill, Roberts, and Shechter, *J. Chem. Educ.*, **49**, 287 (1972).

$$R_2NH + \langle\bigcirc\rangle\!-\!SO_2Cl \xrightarrow{KOH} \langle\bigcirc\rangle\!-\!SO_2NR_2 + KCl + H_2O \xrightarrow{excess\ base} no\ reaction$$

2° amine   water insoluble

Tertiary amines react differently with benzenesulfonyl chloride; the intermediate ammonium ion does not have a proton to lose and reacts rapidly with hydroxide ion to displace the benzenesulfonate anion and regenerate the tertiary amine.

$$R_3N + \langle\bigcirc\rangle\!-\!SO_2Cl \longrightarrow \langle\bigcirc\rangle\!-\!SO_2\overset{+}{N}R_3\ Cl^- \xrightarrow[H_2O]{OH^-}$$

3° amine   water soluble

$$\langle\bigcirc\rangle\!-\!SO_3^- + NR_3 + Cl^-$$

water soluble

The overall reaction amounts to an amine-catalyzed hydrolysis of the benzenesulfonyl chloride. With tertiary amines there can be a side reaction between the amine and the intermediate ammonium ion to produce a complex mixture of water-insoluble products,[19] which could lead to confusion with the results for a secondary amine. This complication can be minimized by keeping the concentration of the amine low (as specified in the test procedure).

(I) Hinsberg Test

To 8–10 drops of the amine in a large test tube, add 10 mL of 10% aqueous potassium hydroxide and 10 drops of benzenesulfonyl chloride (**Caution**—*disagreeable odor!*). Shake the tube thoroughly and note any reaction. Warm the mixture very gently with shaking (*do not boil*) for 10 min. The reaction mixture should still be strongly alkaline at this point. Cool the test tube to room temperature, shake well, and note whether any solid or liquid separates. Do not confuse any separated material with unreacted benzenesulfonyl chloride.

If the mixture has formed two liquid layers (not counting any unreacted benzenesulfonyl chloride layer), separate them and determine if the upper, organic layer is soluble in 5% hydrochloric acid. If the organic material is soluble, it indicates a tertiary amine (acid soluble, unreactive toward benzenesulfonyl chloride). However, if the organic material does not dissolve in the hydrochloric acid, it indicates a secondary amine (acid- and base-insoluble secondary sulfonamide). Add hydrochloric acid to the aqueous phase until the pH is 4 or less. If a precipitate forms, it indicates that the unknown was a primary amine.

Failure of the original basic mixture to separate indicates the presence of a primary sulfonamide. This can be confirmed by adjusting the pH to 4 and noting the formation of a precipitate.

Phenols.[20] Many phenols and related compounds form colored coordination complexes with ferric iron, in which six molecules of a monohydric phenol are combined with one atom of iron to form a complex anion. Most phenols produce red, blue, purple, or green colors. Sterically hindered phenols give negative tests. Aliphatic enols (ethyl acetoacetate, acetylacetone) give a positive test.

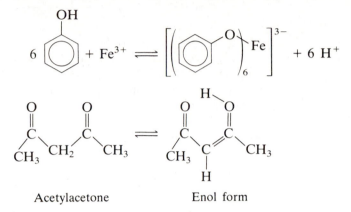

Acetylacetone   Enol form

(J) Ferric Complex

To 2 mL of ethanol in a test tube, add 2 drops of a liquid (or 20 mg of a solid) unknown and a few drops of a 3% aqueous solution of ferric chloride. Shake well and observe the color.

▶ CAUTION Phenol, the cresols, and other phenolic compounds in the pure state or in concentrated solution are toxic and cause painful burns, If any of these come in contact with the skin, wash the area quickly and thoroughly with soap and water.

Hydrocarbons.[21] There are four classes of hydrocarbons: (1) the saturated hydrocarbons, (2) the alkenes (olefins), (3) the alkynes (acetylenes), and (4) the aromatic hydrocarbons. Of these only the alkenes and alkynes will be soluble in cold sulfuric acid (class **N**); the saturated hydrocarbons will fall in class **I**. A test for an aromatic hydrocarbon was described earlier in this section. There are no simple chemical tests for saturated hydrocarbons; these substances must be detected by their failure to give positive tests for either an aromatic ring or unsaturation. Saturated hydrocarbons are best detected by nuclear magnetic resonance, as described in Section 10.3.

The suspected presence of unsaturation can be confirmed by the *cis* hydroxylation with aqueous permanganate (Baeyer test) and by the *trans* addition of bromine in carbon tetrachloride. Almost all alkenes and alkynes react with these reagents.

---

[20] A suitable set of phenols on which to practice this test is: 2-naphthol, *p*-nitrophenol, and salicylic acid.
[21] A suitable set of hydrocarbons on which to practice these tests is: cyclohexane, cyclohexene, and toluene.

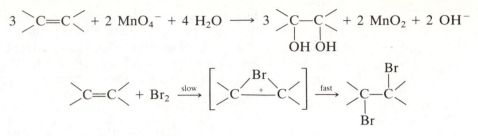

The only exceptions are molecules with a strongly electron-withdrawing group on the multiple bond, which fail to react with bromine because the intermediate bromononium ion is formed too slowly.

Another complication of the bromine test is the tendency of C—H bonds adjacent to a double bond to discharge the bromine color by a free-radical substitution reaction that is accompanied by the evolution of hydrogen bromide.

$$C{=}C{-}C{\overset{H}{\diagdown}} + Br_2 \longrightarrow C{=}C{-}C{\overset{Br}{\diagdown}} + HBr$$

Many phenols and some ketones also substitute bromine and evolve hydrogen bromide, but by an ionic mechanism. These substitution reactions can be detected by the evolved hydrogen bromide vapor, which is not soluble in the carbon tetrachloride solvent and tends to form an "acid fog" when one blows across the top of the reaction vessel.

All aliphatic amines and some pyridine derivatives discharge the bromine color by the reversible formation of colorless bromine adducts.

The Baeyer permanganate test is superior to the bromine test, but it also has complications. All easily oxidized molecules, such as aldehydes and phenols, give positive Baeyer tests. Fortunately, the two tests are largely complementary. It is recommended that the permanganate test be tried first; then, if it is positive, the bromine test should be tried.

(K) Permanganate Test (Baeyer Test)

In a small test tube dissolve 3 drops of the liquid (or 30 mg of a solid) unknown in 1 mL of pure *alcohol-free* acetone. The solvent must be tested beforehand for purity. Add dropwise, with vigorous shaking, a 1% aqueous solution of potassium permanganate. A positive test is the loss within 1 min of the purple permanganate ion color and formation of the insoluble, brown hydrated oxides of manganese. Record the number of drops necessary to develop a persistent purple color; do not be deceived by a slight reaction caused by impurities in the unknown.

(L) Bromine Test

This test should be carried out in the hood. In a small test tube dissolve 3 drops of the liquid (or 30 mg of solid) unknown in 1 mL of carbon tetrachloride and add dropwise, with shaking, a 2% solution of bromine in carbon tetrachloride.

Record the number of drops necessary to develop a persistent (for 1 min) bromine color.

Bromine can cause painful burns. If any of the solution is spilled on the skin, wash the area quickly and thoroughly with water and then apply a dressing soaked in 10% sodium thiosulfate solution; see a physician.

Prolonged exposure to carbon tetrachloride vapor should be avoided because of its toxicity.

Aromatic Hydrocarbons. Molecules falling into solubility class **I** include saturated hydrocarbons, aromatic hydrocarbons, and their derivatives. The flame test carried out in the preliminary examination may have suggested the presence of an aromatic ring by the appearance of a yellow, sooty flame. Confirmation can be obtained from the Friedel–Crafts alkylation test described here.

Aromatic hydrocarbons (and many of their derivatives) react serially with chloroform in the presence of anhydrous aluminum chloride to produce triarylmethanes.

$$ArH + CHCl_3 \xrightarrow{AlCl_3} ArCHCl_2 \xrightarrow[ArH]{AlCl_3} Ar_2CHCl \xrightarrow[ArH]{AlCl_3} Ar_3CH$$

The intermediate chlorohydrocarbons react with aluminum chloride to produce carbocations that abstract a hydride ion from the triarylmethane to yield highly colored triarylmethyl cations. For example,

$$Ar_2CHCl + AlCl_3 \longrightarrow Ar_2CH^+ + AlCl_4^-$$

$$Ar_2CH^+ + Ar_3CH \longrightarrow Ar_2CH_2 + \quad Ar_3C^+$$
(highly colored)

The color depends on the number of rings in the hydrocarbon. Benzene and its derivatives give an orange-red color; naphthalene and phenanthrene as well as their derivatives give blue-purple colors; and anthracene ring produces a green color. In general, the observed color depends on the nature of the substituents, but in the classification scheme described here the substituents will be either alkyl groups or halogens, which do not change the colors significantly.

In carrying out the test it is essential that the aluminum chloride be completely anhydrous. This is accomplished in the test procedure by freshly subliming a sample of aluminum chloride, which drives off any water that may be present.

(M) Friedel–Crafts Test[22]

Place about 100 mg of anhydrous aluminum chloride in a small, dry Pyrex test tube and heat it strongly with the tube held almost horizontally so as to sublime the chloride onto the cooler wall of the tube. While the tube is cooling, prepare

---

[22] A suitable set of hydrocarbons on which to practice this test is: toluene, naphthalene, and cyclohexane.

in the hood in another small test tube a solution of about 20 mg of unknown in 10 drops of chloroform (**Caution**—*chloroform is toxic*; see Section 1.1.) Add this solution to the test tube containing the freshly sublimed aluminum chloride by dropping it directly onto the salt and note the color, if any, where they meet.

Alkyl and Aryl Halides.[23] Alkyl halides can be distinguished from aryl halides by a combination of two tests. The first is with alcoholic silver nitrate, which forms a precipitate of silver halide with alkyl halides that undergo S_N1 reactions. The order of reactivity for R groups is allyl and benzyl > tertiary > secondary $\gg$ primary. The order for the halide leaving group is I > Br > Cl. Secondary and primary halides give no reaction within 5 min; secondary halides react only when the solution is boiled. Primary, aromatic, and vinyl halides usually do not react even after 5 min of heating under reflux.

Primary chlorides and bromides can be distinguished from the aromatic and vinyl halides by the reaction with sodium iodide in acetone. Primary bromides undergo S_N2 displacement reactions within 5 min at room temperature to produce sodium bromide, which is insoluble in acetone. The same reaction occurs with primary chlorides at 50° to produce sodium chloride, which also precipitates.

$$R - X \ (X = Cl, \ Br) + KI \longrightarrow R - I + \underset{\text{white precipitate}}{KX}$$

Secondary and tertiary bromides and some secondary chlorides also react at 50°.

(N) Alcoholic Silver Nitrate

In a small test tube place 2 mL of a 2% solution of silver nitrate in ethanol and add 1 drop of a liquid (or 10 mg of a solid) unknown. A positive test is a precipitate of whitish silver halide within 5 min. If no reaction occurs in that time, boil the solution gently for 5 more min.

If a precipitate forms, either at room temperature or on heating, it is advisable to verify that it is not the silver salt of an organic acid by adding two drops of dilute nitric acid (20:1 water:acid). The acid salts will dissolve; the halides will not.

(O) Sodium Iodide in Acetone

In a small test tube dissolve 2 drops of a liquid (or 20 mg of a solid) unknown in the minimum volume of acetone and add 1 mL of the sodium iodide solution (15 g of sodium iodide in 100 mL of analytical reagent grade acetone). A positive test is a white precipitate within 5 min at room temperature. If no reaction occurs, place the test tube in a beaker of water at 50° and after 5 min cool the test tube to room temperature and note if a precipitate has formed.

---

[23] A suitable set of halides on which to practice these tests is: chlorobenzene, *n*-butyl bromide, *t*-butyl chloride, and *sec*-butyl chloride.

9.8 Derivatization of Functional Groups

At this stage in the analysis the functional group has been identified. It is time to consult a table of characterized molecules containing the functional group and to select those with consistent physical properties. Extensive tables of organic molecules organized by functional group are available;[4,24] abbreviated tables containing many of the more common compounds are included in this section. Your instructor will tell you which set of tables to consult in identifying your unknown.

At least one derivative can be prepared for most functional groups, and, when available, information on the melting points of derivatives is included in the table. For those functional groups without generally applicable derivatives, you must rely on the physical properties of the unknown to make an identification. It is for such compounds that the spectrometric tools discussed in the next chapter are of particular value.

This section is organized by procedures for derivative preparation listed according to functional group. Where more than one derivative is given for a functional group only those derivatives required to distinguish among the possible compounds need be prepared.

Most of the procedures are described for sample sizes of 0.25 g. You may wish to increase the scale of the reactions, but if you do, remember that while you increase the quantities of all reagents the reaction time does not change. Students with more laboratory experience may wish to reduce the scale by a factor of four. The smaller the scale, the greater the attention that must be paid to careful technique; separations are a particular source of trouble on a small scale. If the scale is small enough, it is better to replace the separatory funnel by a pair of test tubes and a Pasteur pipet and to use the squirting technique (see Section 6.4).

Alcohols. Three alcohol derivatives are described here: the esters of *p*-nitrobenzoic acid, of 3,5-dinitrobenzoic acid, and of phenylcarbamic acid (also known as phenylurethans). The first two are formed from the carboxylic acid chloride in the presence of 4-methylpyridine as both a catalyst and a base to neutralize the acid produced. The mechanism, in skeletal form, is believed to be

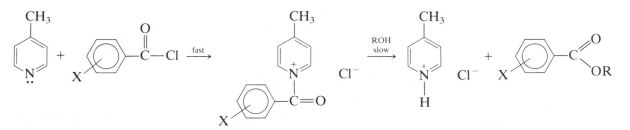

---

[24] Rappoport, Ed., *Handbook of Tables for Organic Compound Identification* (Boca Raton, FL: CRC Press, Inc.).

This reaction works well with primary and secondary alcohols, but with tertiary alcohols the product tends to undergo E2 elimination to give the carboxylic acid and an alkene. This competing reaction is minimized by use of milder conditions (for longer times) and a more powerful catalyst, 4-dimethylaminopyridine. In most published procedures the parent compound, pyridine, is used as the base and catalyst. However, one study of the rates of reaction using different bases gave the following order.

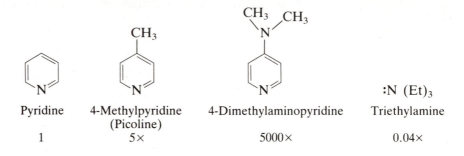

Pyridine	4-Methylpyridine (Picoline)	4-Dimethylaminopyridine	Triethylamine
1	5×	5000×	0.04×

It is interesting to note that triethylamine, although it is the strongest base of the compounds shown, was the poorest catalyst. Perhaps the low catalytic effect of triethylamine results from the greater steric bulk of the three ethyl groups, which would be consistent with the proposed mechanism.

The third derivative, the phenylcarbamate, is prepared from phenyl isocyanate. The mechanism for its formation is

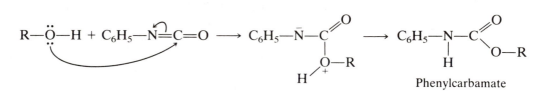

This reaction is base catalyzed, a reflection of the greater nucleophilicity of the alcoholate anion.

$$R—O—H + base \rightleftharpoons R—O^- + H—base^+$$

Alcohol   Alcoholate
                              anion

It has been suggested that 4-dimethylaminopyridine is a superior base for the isocyanate reaction, but the evidence is less well documented than the case for the acid chloride reaction.

Also, with the isocyanate reaction, tertiary alcohols tend to give an elimination reaction. The by-product, water, hydrolyzes an equivalent of phenyl isocyanate to produce phenylcarbamic acid, which is extremely unstable and decomposes by splitting off CO_2 to produce aniline, $C_6H_5—NH_2$. The aniline reacts rapidly with a second equivalent of phenyl isocyanate to produce the crystalline, but very insoluble, diphenylurea.

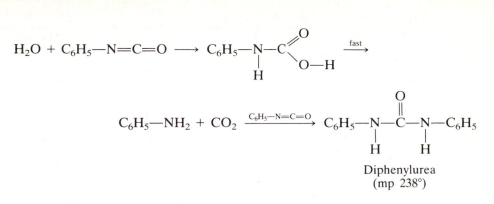

$$H_2O + C_6H_5-N=C=O \longrightarrow C_6H_5-\underset{\underset{H}{|}}{N}-C\underset{O-H}{\overset{O}{\diagup}} \overset{fast}{\longrightarrow}$$

$$C_6H_5-NH_2 + CO_2 \xrightarrow{C_6H_5-N=C=O} C_6H_5-\underset{\underset{H}{|}}{N}-\overset{\overset{O}{\|}}{C}-\underset{\underset{H}{|}}{N}-C_6H_5$$

Diphenylurea
(mp 238°)

(A) 3,5-Dinitrobenzoates and *p*-Nitrobenzoates

(1) Primary alcohols. In a large 150-mm test tube containing a boiling chip place 1 mL of cyclohexane, 0.5 mL of 4-methylpyridine (**Caution**—*lachrymator!*) and 0.25 g of the alcohol. Cautiously add 0.5 g of fresh 3,5-dinitrobenzoyl chloride[25] (or *p*-nitrobenzoyl chloride) and, after the initial reaction has subsided, heat the mixture at a gentle reflux for about 10 min. Allow the mixture to cool for a few minutes and then, while stirring thoroughly, *slowly* pour it into 10 mL of ice-cold 20% hydrochloric acid. If the product does not solidify immediately, cool the mixture in an ice bath for several minutes.

Remove the cyclohexane layer with a Pasteur pipet and place it in a small beaker. If any solid has separated, remove it by filtration and add it to the cyclohexane layer in the beaker. Evaporate the cyclohexane, crush the solid, and stir it well with 5 mL of 10% sodium carbonate solution. Collect the product on a Hirsch filter, and recrystallize it from ethanol–water. It may be necessary to repeat the crystallization to achieve a sharp, precise melting point (Table 9.2).

(2) Tertiary alcohols. Follow the same procedure described for primary and secondary alcohols except substitute 0.5 g of 4-dimethylaminopyridine for the 4-methylpyridine and instead of heating the solution allow it to stand, well stoppered, for at least 24 hr at room temperature.

(B) Phenylcarbamates (Urethans)

In a large test tube mix 0.25 g of the alcohol with 0.3 mL of phenyl isocyanate (**Caution**—*lachrymator!*) and add a few crystals of 4-dimethylaminopyridine as catalyst. With primary and secondary alcohols warm the test tube in a beaker of boiling water for 5–10 min; with tertiary alcohols, stopper the tube and set

---

[25] Carboxylic acid chlorides are rapidly hydrolyzed by moisture. If a bottle of the chloride has been left unsealed for some time, the acid chloride should be recrystallized from cyclohexane and the melting point checked (3,5-dinitrobenzoyl chloride, 70°; *p*-nitrobenzoyl chloride, 75°). The carboxylic acids are essentially insoluble in cyclohexane.

Old bottles of acid chlorides can build up hazardous pressures of hydrogen chloride and should be opened cautiously.

TABLE 9.2
Derivatives of Alcohols

Alcohol	Bp, °C	Melting point of derivative, °C[a]		
		Phenyl-carbamate	3,5-Dinitro-benzoate	p-Nitro-benzoate
Methyl	65	47	108	96
Ethyl	78	52	93	57
Isopropyl	82	88	122–123	110
t-Butyl	83	136	142	116
Allyl	97	70	48–49	(28)
n-Propyl	97	51	74	35
s-Butyl	99	65	76	(26)
t-Pentyl	102	42	116	85
Isobutyl	108	86	87	69
3-Pentanol	116	48	101	(17)
n-Butyl	118	63	64	36
2,3-Dimethyl-2-butanol	118		111	82
2-Pentanol	119		61	
2-Methyl-2-pentanol	121	239	72	70
3-Methyl-3-pentanol	123	50	62; 97	
2-Methoxyethanol	125			51
2-Methyl-1-butanol	129		70	
2-Chloroethanol	131	51	95	
4-Methyl-2-pentanol	132	143	65	(26)
3-Methyl-1-butanol	132	55	61	(21)
2-Ethoxyethanol	135		75	
3-Hexanol	136		77	
2,2-Dimethyl-1-butanol	137		51	
1-Pentanol	138	46	46	(11)
2-Hexanol	139		39	
2,4-Dimethyl-3-pentanol	140			155
Cyclopentanol	141	132	115	62
2-Ethyl-1-butanol	148		52	
2-Methyl-1-pentanol	148		51	
4-Heptanol	156		64	35
1-Hexanol	158	42	58	(5)
2-Heptabol	159		49	
Cyclohexanol	161	82	113	50
2-Furfuryl	172	45	81	76
1-Heptanol	177	68	47	(10)
Tetrahydrofurfuryl	178	61	83–84	46–48
2-Octanol	179	114	32	(28)
1-Octanol	195	74	61	(12)
Benzyl	205	78	113	85
2-Phenylethanol	221	79	108	62
Benzohydrol (mp 66–67°)	297	140	141	132

[a] Two values are given for certain derivatives that may be encountered in polymorphic forms. Blanks mean data are not available. Melting points too low to be useful are enclosed in parentheses.

it aside at room temperature for at least 24 hr. When the reaction is complete, cool the tube in an ice bath and scratch the inside wall with a glass rod to induce crystallization. Dissolve the product in 2 mL of hot ligroin, bp 100–120° (**Caution**—*flammable solvent!*), which leaves any diarylurea as undissolved residue. Filter the hot solution, allow the filtrate to cool, and collect the crystals. Recrystallize the product from hot ligroin and take the melting point (see Table 9.2).

Aldehydes and Ketones. Four carbonyl derivatives are described here. The first, the methone derivative, is suitable only for aldehydes; the others are suitable for both aldehydes and ketones.

Aldehydes undergo a base-catalyzed condensation with the reactive methylene group of 5,5-dimethyl-1,3-cyclohexanedione (also called methone or dimedone) to give an intermediate product that reacts with a second equivalent of the ketone to furnish the crystalline methone derivative (**I**).

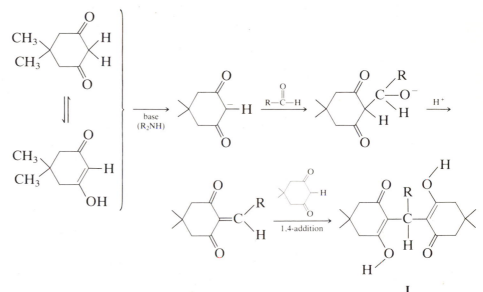

I
Methone derivative

The methone derivatives (but not those of formaldehyde and *o*-hydroxybenzaldehydes) are cyclized readily to xanthenedione derivatives (**II**) by heating with dilute acid in ethanol.

The methone derivatives are particularly good derivatives for small amounts of aldehyde because of the large increase in molecular weight.

Perhaps the single best derivative of both aldehydes and ketones are the 2,4-dinitrophenylhydrazones. They offer a large increase in weight, form quickly, and crystallize easily. The chemistry of 2,4-dinitrophenylhydrazone formation is

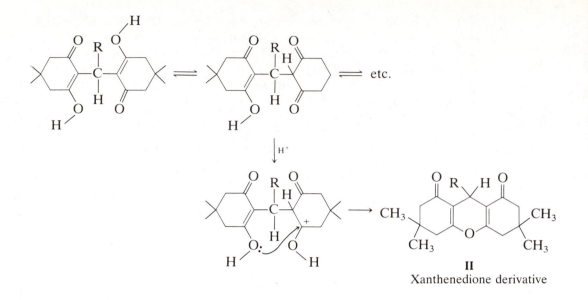

II
Xanthenedione derivative

complex but understandable if it is taken a step at a time. In the following mechanism HA represents added acid catalyst and H$_2$NNHDNPh represents 2,4-dinitrophenylhydrazine.

$$R_2C{=}O + HA \xrightleftharpoons{\text{fast}} \left[R_2C{=}\overset{+}{O}H \longleftrightarrow R_2\overset{+}{C}{-}OH \right]$$

$$\left[R_2C{=}\overset{+}{O}H \longleftrightarrow R_2\overset{+}{C}{-}OH \right] + H_2NNHDNPh \xrightleftharpoons{\text{slow}}$$

$$R_2C\!\!\begin{array}{c} \nearrow OH \\ \searrow \overset{+}{N}H_2{-}NHDNPh \end{array} \xrightleftharpoons{\text{fast}} R_2C\!\!\begin{array}{c} \nearrow OH \\ \searrow NH{-}NHDNPh \end{array} + H^+ \xrightleftharpoons{\text{HA}}$$

$$R_2C\!\!\begin{array}{c} \nearrow \overset{+}{O}H_2 \\ \searrow NH{-}NHDNPh \end{array} \rightleftharpoons R_2C{-}NH{-}NHDNPh \rightleftharpoons R_2C{=}N{-}NHDNPh$$

The rate of reaction is proportional to the concentration of protonated carbonyl compound and *un*protonated 2,4-dinitrophenylhydrazine. The reaction is slow under basic conditions because there is not enough protonated carbonyl compound to give a significant reaction. On the other hand, if too much acid is present, the reaction also slows down because the 2,4-dinitrophenylhydrazine is protonated to form the unreactive phenylhydrazine salt.

The reaction is reversible and in strong aqueous acid, 2,4-dinitrophenyl-hydrazones will hydrolyze because protonation of the resulting 2,4-dinitro-phenylhydrazine pulls the equilibrium toward the carbonyl compound.

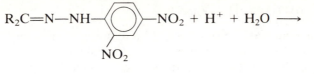

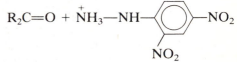

The semicarbazone and oxime derivatives are chemically analogous to the 2,4-dinitrophenylhydrazones.

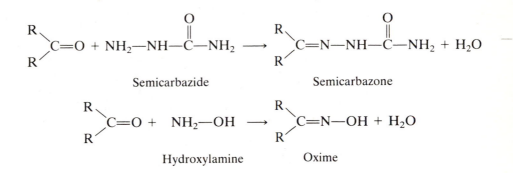

Many aldehydes and ketones give oximes that are liquids at ordinary temperatures (e.g., butanal, 2-butanone), or are very difficult to obtain in crystalline form. For such compounds the semicarbazones or 2,4-dinitrophenylhydrazones are likely to be suitable crystalline derivatives for characterization.

The oximes of aldehydes and of unsymmetrical ketones are capable ot exhibiting geometrical isomerism and may exist in *syn* and *anti* configurations. Benzaldoxime, for example, is known in two forms of different melting points and different chemical properties. The acetate of the α-oxime on warming with sodium carbonate solution gives benzonitrile (C_6H_5—CN) but the acetate of the β-form merely regenerates the oxime of this treatment.[26]

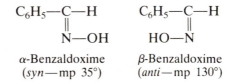

α-Benzaldoxime	β-Benzaldoxime
(*syn*—mp 35°)	(*anti*—mp 130°)

[26] Stereoisomeric ketoximes when subjected to the Beckmann rearrangement give different carboxylic amides; see Blatt, *Chem. Rev.*, **12**, 215 (1933); Popp and McEwen, *ibid.*, **58**, 370 (1958).

(C) Methone Derivative

In a small flask, place 0.50 g (3.6 mmole) of methone, 5 mL of 50% aqueous ethanol, and *not more than* 0.15 g of the aldehyde (~1.5 mmole). Add *one drop* of a secondary amine as catalyst (diethylamine or piperidine), and introduce two small boiling chips. Attach a small reflux condenser, and heat the flask in a beaker about one fourth filled with water. Apply heat gently until the reaction mixture reaches its boiling point. After the mixture has boiled gently for 5–10 min, remove the flask and add water dropwise until a turbidity develops. Allow the solution to cool, with occasional shaking. Methone derivatives often crystallize slowly; if necessary, stopper the flask and allow it to stand overnight or longer. Collect the crystals with suction and wash them with cold 50% aqueous ethanol. The yield is about 0.30 g. After drying the crystals, take the melting point (see Table 9.3).

The methone derivative (except that of formaldehyde)[27] undergoes cyclization very readily. For this purpose dissolve about 0.2 g in 6–10 mL of 80% ethanol, add one drop of concentrated hydrochloric acid, and boil the solution gently for 5 min. Add water dropwise to the hot solution until a faint turbidity develops and allow the liquid to cool. Collect the crystals of the cyclized product (II) and take its melting point (see Table 9.3).

(D) 2,4-Dinitrophenylhydrazones

To prepare the reagent place 0.3 g of moist 2,4-dinitrophenylhydrazine[28] in a 50-mL Erlenmeyer flask and add 1 mL of water, followed by dropwise addition of 1 mL of concentrated sulfuric acid (**Caution**—*corrosive!*), with swirling. Allow the solution to cool and then add 15 mL of 95% ethanol.

Prepare a solution of the carbonyl compound in a 50-mL flask by adding 0.25 g of the compound to 10 mL of 95% alcohol. After the carbonyl compound

---

[27] Cyclization of the formaldehyde derivative requires 6–8 hr heating with 10 times its weight of concentrated sulfuric acid. After pouring into water and neutralizing with sodium carbonate, the product is collected with suction and recrystallized from ethanol.

[28] Transportation regulations require that 2,4-dinitrophenylhydrazine must be shipped in a moist condition, containing 20% of water. A procedure for the preparation of this compound from 2,4-dinitrochlorobenzene is given by Fieser and Williamson, *Organic Experiments*, 4th ed. (Lexington, MA: Heath, 1979).

TABLE 9.3
Methone (I) and
Xanthenedione (II)
Derivatives of
Aldehydes

Aldehyde	Melting point, °C		Aldehyde	Melting point, °C	
	I	II		I	II
Formaldehyde	190–191	(171)[27]	Isovaleraldehyde	154–155	170–172
Acetaldehyde	141–142	176–177	*n*-Hexaldehyde	107–108	
Propionaldehyde	157–158	141–143	Benzaldehyde	194–195	204–205
n-Butyraldehyde	134–135	135–136	Anisaldehyde	142–143	241–243
Isobutyraldehyde	153–154	154–155	Piperonal	177–178	218–220
n-Valeraldehyde	107–109	112–113	2-Furaldehyde	159–160	

has dissolved completely, slowly add the 2,4-dinitrophenylhydrazine test reagent and allow the combined solutions to stand at room temperature. Crystallization of the hydrazone usually occurs within 5–10 min. If the product does not separate in this time, attach a reflux condenser and heat the flask on a steam bath for 15 min. If the hydrazone seems to be too soluble in ethanol, add water, dropwise, at the boiling point of the solution until the product just begins to separate. Allow the solution to cool slowly, since rapid chilling may cause the hydrazone to separate as an oil.

Collect the crystals by suction filtration, and wash them with a little cold ethanol. It is usually not necessary to recrystallize the product. However, if the meeting range is broad, try recrystallization from ethanol. If the hyrazone does not dissolve completely in ethanol, add ethyl acetate dropwise to the hot mixture until solution occurs. The melting points of the 2,4-dinitrophenylhydrazones of many aldehydes and ketones are listed in Table 9.4.

(E) Semicarbazones

In a large test tube prepare a solution of 0.25 g of semicarbazide hydrochloride (NH_2—CO—$NHNH_2$·HCl) and 0.4 g of sodium acetate crystals in 3 mL of water. Add 0.25 g of a carbonyl compound, close the tube with a cork, and shake it *vigorously*. Allow the reaction mixture to stand (shake it occasionally) until the product has crystallized completely. If necessary, cool the tube in an ice bath. Collect the crystals with suction and wash them with a little cold water. After they have dried in the air take their melting point.

Reactive carbonyl compounds (butanal, cyclohexanone, etc.) may be converted to semicarbazones by the method given above. With less-reactive compounds it is advantageous to heat the tube in a beaker of boiling water for a few minutes and allow it to cool slowly.

Dissolve a water-insoluble compound in 3 mL of ethanol; add water, dropwise, until the solution becomes turbid, and then add 0.25 g of semicarbazide hydrochloride and 0.4 g of sodium acetate crystals. Shake the contents of the tube vigorously and warm it to 80–100°.

The melting points of the semicarbazones of many aldehydes and ketones are listed in Table 9.4.

(F) Oximes

In 3 mL of water dissolve 0.5 g of hydroxylamine hydrochloride (NH_2OH·HCl), add 2 mL of 10% aqueous sodium hydroxide, and introduce 0.25 g of the aldehyde or ketone. If the compound does not dissolve completely, add *just enough* ethanol, dropwise with shaking, to obtain a clear or only faintly turbid solution. Heat the solution in a boiling water bath for 10–20 min, cool it in an ice bath, and induce crystallization by scratching the walls of the container with a glass rod. If necessary, allow the solution to stand overnight or longer. Collect the crystals on a small suction filter, and wash them with 1 mL of ice cold water. Recrystallize the oxime from a little water or aqueous ethanol, reserving a few

TABLE 9.4
*Derivatives of
Aldehydes and
Ketones*

	Bp (mp), °C	Melting point of derivative, °C[a]		
		Oxime[b]	Semi-carbazone	2,4-Dinitrophenyl-hydrazone[c]
Aldehydes[d]				
Acetaldehyde	21	47	162	168; 157
Propionaldehyde	50	40	89; 154	154
Isobutryaldehyde	64	oil	125	187
n-Butryaldehyde	74	oil	104	123
Isovaleraldehyde	92	48	107	123
n-Valeraldehyde	103	52	108	98; 107
n-Hexaldehyde	131	51	106	104; 107
n-Heptaldehyde	153	57	109	108
2-Furaldehyde	161	89; 74	202	230; 212
Benzaldehyde	179	35; 130	222	237
Salicyladehyde	197	57	231	252 dec
p-Tolualdehyde	204	179; 110	221	239
Citral	228	oil	164	116
Chloral hydrate	(53)	56	90 dec	131
4-Hydroxy-3-methoxy-benzaldehyde	(81)	117; 122	230; 240 dec	271 dec
Ketones				
Acetone	56	59	187	126
2-Butanone	80	oil	146	117
3-Methyl-2-butanone	94	oil	113	120
2-Pentanone	102		106; 112	143
3-Pentanone	102	69	139	156
Methyl t-butyl	106	75; 79	157	125
4-Methyl-2-pentanone	119	58	134	95
2,4-Dimethyl-3-pentanone	124	34	160	88; 94
2-Hexanone	129		122	106
Cyclopentanone	131	56	203; 210; 216	142; 146
4-Heptanone	145	oil	133	75
2-Heptanone	151	oil	123; 127	74; 89
Cyclohexanone	155	90	166	162
Acetaphenone	200	59	198	240
Benzalacetone	(41)	115	187	223
Benzophenone	(48)	141	164	239
Benzalacetophenone	(58)	116; 75	168; 180	245
Benzil (mono)	(95)	137; 108	175; 182	189
Benzil (di)		237	244	
Benzoin	(133)	151; 99	206 dec	245
dl-Camphor	(176)	118	235	

[a] Two values are given for certain derivatives that may be encountered in polymorphic forms or as *syn* and *anti* geometrical isomers.

[b] Oximes of many aliphatic carbonyl compounds separate as oily liquids that resist efforts to induce crystallization. The simpler oximes are appreciably soluble in water and in organic solvents.

[c] A few dinitrophenylhydrazones exist in red and yellow forms, which have different melting points. Mixtures of the two have lower or intermediate melting points.

[d] Methone derivatives are useful for some aliphatic aldehydes.

tiny crystals for seeding. The melting points of the oximes of many aldehydes and ketones are listed in Table 9.4.

Esters. There are no universal ester derivatives. One identifying characteristic that will serve instead is the saponification equivalent, which is the weight of the ester (in grams) that reacts with 1 mole of alkali. For mono esters the saponification equivalent is the molecular weight of the ester. This is determined by heating a weighed sample of ester with an excess of standardized potassium hydroxide solution (usually in aqueous methanol or ethanol) and titrating the excess alkali with standardized hydrochloric acid, using phenolphthalein as indicator.

$$\underset{\substack{\| \\ \text{R--C--OR}}}{\overset{\text{O}}{}} + {}^-\text{OH} \longrightarrow \underset{\substack{\| \\ \text{R--C--O}^-}}{\overset{\text{O}}{}} + \text{HOR}$$

The saponification equivalent[29] expresses the molecular weight of the ester divided by the number of ester groups in the molecule. For esters of dibasic acids or of bifunctional alcohols it is one half the molecular weight, for trifunctional esters one third the molecular weight, etc.

If either the acid or alcohol fragment of the ester is a solid, the ester can be hydrolyzed (saponified) with base and the solids isolated. These, after purification, are suitable derivatives without further transformation. The boiling and melting points of many common esters are listed in Table 9.5.

(G) Saponification Equivalent of an Ester

Prepare an ethanolic solution of potassium hydroxide by dissolving 1 g of potassium hydroxide pellets in 5 mL of water and adding 20 mL of ethanol; if sediment is present allow the solution to stand until it has settled. Withdraw *carefully*, by means of a pipet, two 10-mL portions of the solution and place them in separate 125-mL Erlenmeyer flasks. Add about a 0.250-g sample of the ester, weighed accurately to ±0.001 g, to one of the flasks. Attach a reflux condenser,[30] add a boiling chip, and boil the solution gently for 30 min. Meanwhile, titrate the 10-mL portion of potassium hydroxide solution in the other flask, using standardized (about 0.2 N) hydrochloric acid, with phenolphthalein as an indicator.

---

[29] The saponification equivalent of an ester is different from the saponification number, commonly used in industry for fats and fatty oils, which is defined as the number of milligrams of potassium hydroxide required to saponify *one gram* of the fat or fatty oil (mixtures of glyceryl esters of higher aliphatic acids).

$$\text{Saponification number} = \frac{\text{mL of 1 } N \text{ alkali} \times 56.1}{\text{grams of ester}}$$

The factor 56.1 is the molecular weight of potassium hydroxide.

[30] The condenser is best attached by means of a cleanly drilled cork. A conically shaped flask with a ground glass joint may be substituted, but the joint must be greased carefully since otherwise it might freeze from contact with traces of base.

TABLE 9.5
Boiling Points of
Liquid Esters and
Melting Points of
Solid Esters

Liquid ester	Bp, °C	Liquid ester	Bp, °C
Methyl formate	32	Pentyl formate	132
Ethyl formate	54	Ethyl 3-methylbutanoate	135
Methyl acetate	57	Isobutyl propanoate	137
Isopropyl formate	68; 71	Isopentyl acetate	142
Ethyl acetate	77	Propyl butanoate	143
Methyl propanonate	80	Ethyl pentanoate	146
Methyl propenoate	80	Butyl propanoate	147
Propyl formate	81	Pentyl acetate	149
Isopropyl acetate	91	Isobutyl 2-methylpropanoate	149
Methyl 2-methylpropanoate	93	Methyl hexanoate	151
sec-Butyl formate	97	Isopentyl propanoate	160
t-Butyl acetate	98	Butyl butanoate	165
Ethyl propanoate	99	Propyl pentanoate	167
Propyl acetate	101	Ethyl hexanoate	168
Methyl butanoate	102	Cyclohexyl acetate	175
Allyl acetate	104	Isopentyl butanoate	178
Ethyl 2-methylpropanoate	110	Pentyl butanoate	185
sec-Butyl acetate	112	Propyl hexanoate	186
Methyl 3-methylbutanoate	117	Butyl pentanoate	186
Isobutyl acetate	117	Ethyl heptanoate	189
Ethyl butanoate	122	Isopentyl 3-methylbutanoate	190
Propyl propanoate	122	Ethylene glycol diacetate	190
Butyl acetate	126	Tetrahydrofurfuryl acetate	194
Diethyl carbonate	127	Methyl octanoate	195
Methyl pentanoate	128	Methyl benzoate	200
Isopropyl butanoate	128	Ethyl benzoate	213

Solid ester	Mp, °C	Solid ester	Mp, °C
d-Bornyl acetate (bp 221)	29	Ethyl 3,5-dinitrobenzoate	93
Ethyl 2-nitrobenzoate	30	Methyl 4-nitrobenzoate	96
Ethyl octadecanoate	33	2-Naphthyl benzoate	107
Methyl cinnamate (bp 261)	36	Isopropyl 4-nitrobenzoate	111
Methyl 4-chlorobenzoate	44	Cyclohexyl 3,5-dinitrobenzoate	112
1-Naphthyl acetate	49	Cholesteryl acetate	114
Ethyl 4-nitrobenzoate	56	Ethyl 4-nitrobenzoate	116
2-Naphthyl acetate	71	*t*-Butyl 4-nitrobenzoate	116
Ethylene glycol dibenzoate	73	Hydroquinone diacetate	124
Propyl 3,5-dinitrobenzoate	74	*t*-Butyl 3,5-dinitrobenzoate	142
Methyl 4-bromobenzoate	81	Hydroquinone dibenzoate	199;

When the saponification has been completed, cool the solution and rinse the condenser tube with 5–10 mL of water (collect the rinsing water directly in the flask). Titrate the alkali remaining in the solution against standardized hydrochloric acid, as in the previous titration. The difference in the volumes of acid required in the two titrations represents the amount of alkali consumed in the saponification. Calculate the saponification equivalent by the formula given below. The result should correspond to the molecular weight (within about 5%) if the sample of ester was fairly pure.

$$\text{Saponification equivalent} = \frac{\text{weight of ester (in grams)}}{\text{mL of NaOH consumed} \times N \text{ of the NaOH}} \times 1000$$

(H) Hydrolysis of an Ester

In a 100-mL round-bottomed flask provided with a reflux condenser (greased joints), place 1.0 g of an ester. To this add 5 mL of 10% aqueous sodium hydroxide and about 10 mL of water (if the ester boils above 150°, substitute 10 mL of ethylene glycol). Add two boiling chips and boil for about 2 hr. Cool the solution in the flask and carefully transfer it to a small separatory funnel. Extract the strongly basic solution twice with 10-mL portions of methylene chloride and combine the extracts (containing the alcohol) in a clean flask. If the alcohol is a solid, it can be isolated by concentration of the methylene chloride solution by distillation followed by evaporation of any residual solvent. Recrystallization of the alcohol may be desirable. Isolation of a liquid alcohol in the quantity specified here requires a Hickman distillation apparatus.

The acid can be isolated from the residual basic solution by acidification with dilute sulfuric acid (about 10%). The acid is a solid; it can be isolated as described above for the alcohol fragment of the ester.

Carboxylic Acids. Amides and N-substituted amides of carboxylic acid make excellent derivatives because of their high melting points and easy preparation and purification. The preparation of any of these derivatives starts with conversion of the carboxylic acid to the acid chloride using thionyl chloride catalyzed by dimethylformamide (DMF).

$$\underset{\substack{\| \\ \text{O}}}{\text{R}-\text{C}-\text{OH}} + \underset{\substack{\| \\ \text{O} \\ \text{Cl} \quad \text{Cl}}}{\text{S}} \xrightarrow{\text{DMF}} \underset{\substack{\| \\ \text{O}}}{\text{R}-\text{C}-\text{Cl}} + \text{SO}_2 + \text{HCl}$$

The DMF catalyst is particularly effective. The proposed mechanism involves formation of a chloroimmonium ion.

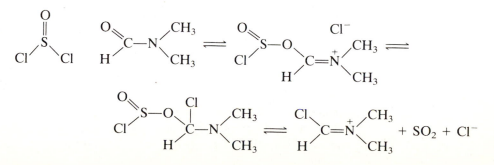

It is this chloroimmonium ion that is the active chlorine-transfer agent.

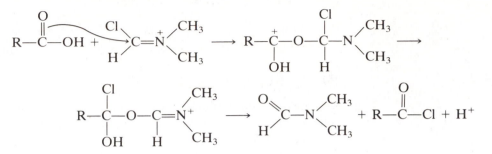

Through all of these transformations the dimethylamino group serves as a reversible electron source, releasing electrons to stabilize the adjacent cationic center and taking them back up again when that center encounters better electron sources (e.g., nucleophiles and the C=O group). In the overall sequence the DMF is regenerated and thus is a true catalyst.

Acid chlorides react rapidly with ammonia, amines, alcohols, and *water*. In the procedure described here the reaction is moderated by adding the inert solvent methylene chloride, chosen for its easy removal. One equivalent of HCl is liberated, which produces copious white fumes on contact with ammonia and water. The reaction should be carried out in a hood.

$$R-\overset{\overset{\text{O}}{\|}}{C}-Cl + \overset{..}{N}H_2R \longrightarrow R-\overset{\overset{O^-}{|}}{\underset{\underset{Cl}{|}}{C}}-\overset{+}{N}H_2R \longrightarrow$$

$$R-\overset{\overset{\text{O}}{\|}}{C}-\overset{+}{N}H_2R \rightleftharpoons R-\overset{\overset{\text{O}}{\|}}{C}-NHR$$
Amide

A useful alternative to a solid derivative is to determine the neutralization equivalent (equivalent weight) of the acid, which is defined as the weight of acid required to neutralize 1 mole of base. For mono acids the neutralization equivalent is the molecular weight of the acid; for di or tri acids it is one half or one third of the molecular weight.

(I) Acid Amides, Anilides, and *p*-Toluidides

1. Conversion of acid to acid chloride. The preparation of these derivatives can release hazardous fumes and should be carried out **in a hood**! In a small, dry flask place 0.250 g of the acid, 1 mL of thionyl chloride, and 1 drop of pyridine or dimethylformamide. Add a boiling chip, attach a reflux condenser, and boil the mixture gently in a water bath for 30 min. During this time the acid should react and dissolve completely. Small amounts of sulfur dioxide and hydrogen chloride are evolved during the heating. The excess thionyl chloride could be removed by warming under reduced pressure, but that is not necessary.

⟹ CAUTION Handle thionyl chloride carefully. The liquid burns the skin and the vapor is an irritant and harmful to breathe.

2. Conversion to acid amide. Cool the acid chloride mixture and add 5–10 mL of methylene chloride. Pour the solution of the acid chloride (and excess thionyl chloride) *very cautiously* and *in small portions* into 5 mL of cold, concentrated, aqueous ammonia contained in a small Erlenmeyer flask. The reaction mixture should mixed thoroughly by swirling during the addition. The acid amide may separate as a crystalline precipitate during the reaction.

After allowing the mixture to stand for 20 min, occasionally shaking it, separate the methylene chloride layer and remove the solvent by distillation. Wash the residue with water, collect the crystals by suction filtration, and press them as dry as possible on the filter. Recrystallize the amide from ethanol–water and determine its melting point. The melting points of the amides of the common carboxylic acids are listed in Table 9.6.

3. Conversion to Anilide. Prepare the acid chloride as described in Part 1 above and add 5–10 mL of methylene chloride to the cooled mixture. Dissolve 0.5 g of aniline in 10 mL of methylene chloride in a small Erlenmeyer flask and add the acid chloride solution to it *cautiously* and *in small portions*. The reaction mixture should be mixed thoroughly by swirling between additions.

After allowing the mixture to stand for 20 min, occasionally shaking it, transfer the methylene chloride solution to a separatory funnel and wash it in sequence with 5 mL of water, 5 mL of 5% hydrochloric acid, 5 mL of 5% sodium hydroxide, and finally 5 mL of water. Remove the solvent by distillation, recrystallize the residue from ethanol–water, and determine its melting point. The melting points of the anilides of the common carboxylic acids are listed in Table 9.6.

4. Conversion to p-Toluidide. Use the same procedure described for anilides, except substitute *p*-toluidine for aniline. The melting points of the *p*-toluidides of the common carboxylic acids are listed in Table 9.6.

(J) Neutralization Equivalent of an Acid

Place about 0.250 g of the acid, weighed accurately to ±0.001 g, in a 125-mL Erlenmeyer flask and add 5 mL of water and 20 mL of ethanol. Swirl the flask to dissolve the acid. Titrate the solution using standardized aqueous sodium hydroxide (about 0.2 *N*) with phenolphthalein as an indicator. The neutralization equivalent is calculated as

$$\text{Neutralization equivalent} = \frac{\text{weight of acid (in grams)}}{\text{mL of NaOH consumed} \times N \text{ of the NaOH}} \times 1000$$

TABLE 9.6
Derivatives of
Carboxylic Acids

		Melting point of derivative, °C		
	Bp, °C	Amide	Anilide	*p*-Toluidide
Liquid carboxylic acids				
Methanoic (formic)	101		50	53
Ethanoic (acetic)	118	82	114	153
Propenoic (acrylic)	141	84	104	141
Propanoic (propanoic)	141	81	106	126
2-Methylpropanoic (isobutyric)	155	128	105	109
Butanoic (butyric)	163	115	96	75
3-Methylbutanoic	177	135	110	106
Pentanoic (valeric)	186	106	63	74
2,2-Dichloroethanoic	194	98	118	153
Hexanoic (caproic)	205	100	94	74
Heptanoic (enanthic)	223	96	65; 70	81
Octanoic (caprylic)	239	106; 110	57	70
Nonanoic (pelargonic)	254	99	57	84
	Mp, °C			
Solid carboxylic acids				
Octadecanoic (stearic)	70	109		96
Phenylethanoic	77	156	65	117
2-Benzoylbenzoic	90; 128	165	195	
Pentandioic (glutaric)	97	175	223	218
Ethanedioic (oxalic)	101	219	148	169
2-Methylbenzoic	105	143	125	144
3-Methylbenzoic	112	94	126	118
Benzoic	122.4	130	160	158
trans-Cinnamic	133	147	109; 153	168
Propanedioic (malonic)	135	106	132	156
2-Acetoxybenzoic (aspirin)	135	138	136	
cis-Butenedioic (maleic)	137	172	187; 198	142
2-Chlorobenzoic	140	142; 202	114; 118	131
3-Nitrobenzoic	140	143	154	162
2-Nitrobenzoic	146	176	155	
Diphenylacetic	148	168	180	172
2-Bromobenzoic	150	155	141	
Benzilic	150	153	175	190
Hexanedioic (adipic)	153	125; 230	151; 249	241
2-Hydroxybenzoic (salicylic)	158	142	136	156
2-Iodobenzoic	162	110	141	
4-Methylbenzoic (*p*-toluic)	179	160	144	160; 165
4-Methoxybenzoic (*p*-anisic)	185	162; 167	170	186
2-Naphthoic	186	192	171	192
Phthalic (mono)[a]	200; 230	149	170	150
Phthalic (di)[a]		220	254	201
3,5-Dinitrobenzoic	205	183	234	
4-Nitrobenzoic	241	198	204; 211	192; 204

[a] Phthalic acid is a dicarboxylic acid that forms both mono- and dicarboxylic acid derivatives.

Sulfonic Acids. Alkylsulfonic and arylsulfonic acids and their metallic salts can be characterized by conversion to salts of organic bases, such as the *S*-benzylthiouronium salts, which are crystalline and have suitable melting points.

$$\left[C_6H_5CH_2{-}S{-}C \underset{NH_2}{\overset{NH_2}{\diagup}} \right]^+ \; {}^-[O_3S{-}C_6H_4CH_3]$$

Another, more complicated method involves reaction of the sulfonic acid with phosphorous pentachloride to form the arylsulfonyl chloride, which is treated with ammonia (or an amine) to furnish a crystalline arylsulfonamide. These reactions are analogous to those discussed for formation of amide derivatives of carboxylic acids in Section I of this chapter.

$$RSO_3H + PCl_5 \longrightarrow RSO_2Cl + POCl_3 + HCl$$

$$RSO_2Cl + NH_3 \longrightarrow RSO_2NH_2 + HCl$$
$$\text{Sulfonamide}$$

(K) *S*-Benzylthiouronium Salts

In a small test tube, dissolve 0.25 g of the sodium or potassium salt of the sulfonic acid in a minimum amount of water (warm if necessary). If only the sulfonic acid is available, dissolve it in 0.5 mL of 10% sodium hydroxide, add 1 mL (or more) of water, add a drop of phenolphthalein solution, and neutralize the excess base by adding hydrochloric acid drop by drop.

In a separate test tube prepare a concentrated solution of 0.25 g of *S*-benzylthiouronium chloride[31] in water. Mix the solutions together, shake well, and cool the mixture in an ice–water bath. If crystals do not form within a few minutes, scratch the inside of the tube with a glass rod to promote crystallization. Collect the product on a small suction filter, wash sparingly with cold water, and recrystallize it from 50% aqueous ethanol (reserving a seed crystal for inoculation). The melting points of the *S*-benzylthiouronium salts of the common sulfonic acids are listed in Table 9.7.

---

[31] *S*-Benzylthiouronium chloride is a relatively expensive reagent. Suitable material may be prepared readily from thiourea and benzyl chloride. In a flask fitted with a reflux condenser, a mixture of 25 g (16 mL) of benzyl chloride (**Caution**—*lachrymator!*), 15 g of thiourea, and 40 mL of ethanol (or methanol) is warmed on a steam bath. Soon a vigorous exothermic reaction occurs and the thiourea dissolves completely. The pale-yellow solution is refluxed for 30 min, transferred while hot to a beaker, and cooled in an ice–water bath. The mass of white crystals is collected on a suction filter, washed with several 15-mL portions of cold ethyl acetate or ethanol, and pressed well. The dried product is stored in an amber bottle. The yield is 30–35 g of the salt. The crude product is satisfactory for the preparation of derivatives. It may be purified by recrystallization from ethanol or from 15–20% aqueous hydrochloric acid. The compound is dimorphic: stable form, mp 174–176°; metastable form, mp 142–145° (with reversion to the stable form).

TABLE 9.7
Derivatives of
Sulfonic Acids

| Sulfonic acid | Mp, °C | Melting point of derivative, °C | |
		S-Benzylthiouronium salt	Sulfonamide
Methane-	20		90
3-Nitrobenzene-	48	146	167
Hexadecane-1-	54		97
2-Methylbenzene-	57	170	156
2,4-Dimethylbenzene-	62	146	139
3,4-Dimethylbenzene-	64	208	144
Benzene-	66	148	153
2-Carboxybenzene	68	206	194
Naphthalene-1-	90	137	150
Naphthalene-2-	91	190	217
4-Chlorobenzene-	93	175	144
4-Methylbenzene-	104	181	105; 139
Naphthalene-1,6-di-	125	81	297

(L) Sulfonamides

This reaction can release hazardous vapors and should be carried out in a hood. In a small, round-bottomed flask place 0.25 g of the sulfonic acid (or its sodium or potassium salt) and about 1.0 g of phosphorous pentachloride (**Caution**—*corrosive material; do not inhale vapors*). Attach a reflux condenser and heat the mixture in a sand bath at 150° for 45 min. Cool the mixture to room temperature, add 5–10 mL of methylene chloride, and break up any solid masses present. Filter the solution through a dry filter paper into a small Erlenmeyer flask containing 5 mL of cold, concentrated, aqueous ammonia. Swirl the flask during the addition. Occasionally, the sulfonamide will separate as a crystalline precipitate, but usually it will not.

After allowing the mixture to stand for 20 min, occasionally shaking it, separate the methylene chloride layer and remove the solvent by distillation. Recrystallize the sulfonamide from ethanol or ethanol–water. The melting points of the sulfonamides of the common sulfonic acids are listed in Table 9.7.

Nitriles and Amides. There are no good derivatives of nitriles and amides; the best strategy is to hydrolyze them to the corresponding carboxylic acid and ammonia (or amine). The hydrolysis can be achieved with either acid or base, but only the base procedure will be described here.

$$R\text{—}C\equiv N \xrightarrow[\text{H}_2\text{O}]{\text{NaOH}} R\text{—}\overset{\overset{\displaystyle O}{\|}}{C}\text{—}NH_2 \longrightarrow R\text{—}\overset{\overset{\displaystyle O}{\|}}{C}\text{—}O^- + NH_3$$

$$R\text{—}\overset{\overset{\displaystyle O}{\|}}{C}\text{—}NR_2 \xrightarrow[\text{H}_2\text{O}]{\text{NaOH}} R\text{—}\overset{\overset{\displaystyle O}{\|}}{C}\text{—}O^- + HNR_2$$

Nitrile	Bp, °C	Mp, °C	Amide[b]	Bp, °C	Mp, °C
Acrylonitrile[a]	77		*N,N*-Dimethylformamide	153	
Acetonitrile	81		*N,N*-Diethylformamide	176	
Propionitrile	97	−93	*N*-Methylformamide	185	
Isobutyronitrile	108		*N*-Formylpiperidine	222	
n-Butyronitrile	117		*N,N*-Dimethylbenzamide		41
Benzonitrile	191	−13	*N*-Benzoylpiperidine		48
2-Methylbenzonitrile	205	13	*N*-Propylacetanilide		50
2-Cyanopyridine	212	26	*N*-Benzylacetamide		54
Phenylacetonitrile	234		*N*-Ethylacetanilide		54
3-Cyanopyridine	240	52	*N,N*-Diphenylformamide		73
1,4-Dicyanobutane	295	2	*N*-Methyl-4-acetotoluidide		83
2-Chlorobenzonitrile		41	*N,N*-Diphenylacetamide		101
Diphenylacetonitrile		75	*N*-Methylacetanilide		102
4-Cyanopyridine		80	*N*-Ethyl-4-nitroacetanilide		118
			N-Phenylsuccinimide		156
			N-Phenylphthalimide		205

[a] Cancer suspect agent.
[b] Table 9.6 lists boiling points and melting points of many unsubstituted amides, anilides, and 2-methylanilides. Table 9.9 lists these data for a number of *N*-substituted acetamides, benzamides, benzenesulfonamides, and 4-methylbenzenesulfonamides.

For purely physical reasons one must use a different procedure for substituted (on nitrogen) and unsubstituted amides. With nitriles and unsubstituted amides, only ammonia is produced and this need not be characterized. With substituted amides for which the amine fragment is *volatile*, a trap containing acid must be attached to the apparatus in order to collect the amine for identification. With substituted amides for which the amine fragment is *nonvolatile*, no trap is required. The amine can be recovered from the basic hydrolysis mixture by extraction; acidification of the mixture then yields the acid. The boiling and melting points of many nitriles and amides are listed in Table 9.8.

(M) Base Hydrolysis of Nitriles and Unsubstituted Amides

In a small flask place 2 g of potassium hydroxide, 5 mL of ethylene glycol, and 0.5 g of the nitrile or unsubstituted amide. Add a boiling chip, attach a reflux condenser, and heat the mixture under reflux for 1 hr. Cool the mixture to room temperature, dilute with 5 mL of water, and extract twice with 5-mL portions of methylene chloride to remove any unreacted nitrile or amide. The methylene chloride layer should be set aside. Acidify the aqueous solution with 6 *N* (50%) hydrochloric acid until the pH is about 1 to 2, and extract again with methylene chloride. This extract, should be evaporated or distilled to recover the acid.

(N) Base Hydrolysis of Substituted Amides

1. Volatile amines. The procedure is the same as that described above for unsubstituted amides, except that a glass tube, leading to an Erlenmeyer flask containing 5 mL of 6 N hydrochloric acid, should be attached to the top of the reflux condenser to trap the volatile amines. At the end of the reaction, the acid should be neutralized with 10% sodium hydroxide solution and extracted several times with 5-mL portions of methylene chloride. The amine can be recovered from the methylene chloride by distillation.

2. Nonvolatile amines. The procedure is identical to that described for unsubstituted amides. The nonvolatile amine will be in the first methylene chloride extract (of the basic solution) and the acid will be in the second methylene chloride extract (of the acidic solution).

Amines. The most effective derivative of primary and secondary amines are amides. Four are described here. The chemistry of their formation is essentially the same as that described in connection with the tests for amines. They have in common an acyl or sulfonyl group that is activated by using a chloride or anhydride leaving group in place of the hydroxyl group of the acid. When a chloride is used some base must be present to neutralize the hydrogen chloride produced, which would otherwise react with amine to produce an unreactive hydrochloride salt.

$$
\underset{\substack{\parallel \\ O}}{Ar-C-X} + NHR_2 \longrightarrow \underset{\substack{\parallel \\ O}}{Ar-C-NR_2} + HX
$$

$$
ArSO_2X + NHR_2 \longrightarrow Ar-SO_2NR_2 + HX
$$

The melting points of these derivatives are listed in Table 9.9.

The arylsulfonyl derivatives of aliphatic and aromatic amines are usually the better crystalline derivatives for identification purposes. If the benzenesulfonamide is a liquid, the *p*-toluenesulfonamide may serve. For the nitroanilines, halogenated anilines, and similar weak bases, which frequently give poor results in an aqueous medium, the use of pyridine in an anhydrous system is advantageous.

With tertiary amines, amide derivatives are impossible and one must turn to formation of a quarternary ammonium salt using either methyl or ethyl iodide. The iodide salts are nicely crystalline but tend to decompose near their melting points or on prolonged exposure to light. Many rapidly absorb moisture from the air. Melting points of the salts of a few tertiary amines are listed in Table 9.10.

(O) Acetylation with Acetic Anhydride in Water. The Lumiere–Barbier Method

In a large test tube dissolve 0.25 g of the 1° or 2° amine in 5 mL of 5% hydrochloric acid. Prepare a solution of 0.5 g sodium acetate crystals (trihydrate) and set this aside.

Warm the solution of the amine hydrochloride to 50°, add 0.3 mL of acetic anhydride, and swirl the mixture to dissolve the anhydride. *At once* add

| Amine | Bp °C | Melting point of amide, °C | | | |
		Acetyl	Benzoyl	Benzene-sulfonyl	*p*-Toluene-sulfonyl
Methylamine	−6	oil	80	30	75
Ethylamine	17	oil	71	58	63
Isopropylamine	33		100	26	
n-Propylamine	49	47	84	36	52
t-Butylamine	45	98	134		
sec-Butylamine	63		76	70	55
Isobutylamine	69	107	57	53	78
n-Butylamine	77	oil	42		44
Ethylenediamine	116	172 (di)	244 (di)	168 (di)	160 (di)
Cyclohexylamine	134	104	149	89	
Benzylamine	185	60	105	88	116
Aniline	185	114	163	112	103
2-Methylaniline	199	112	144	124	108
4-Methylaniline (mp 45°)	200	148	158	120	117
3-Methylaniline	203	66	125	95	114
2-Chloroaniline	208	87	99	129	105; 193
2-Ethylaniline (mp 47°)	210	111	147		
2,4-Dimethylaniline	212	130	192	129	
2,5-Dimethylaniline	215	139	140	139	119
2,6-Dimethylaniline	216		168		
2,4-Dimethylaniline	217		192	130	
N-Ethyl-3-methylaniline	221		72		
2-Methoxyaniline	225	85	60	89	127
4-Chloroaniline (mp 70°)	232	172; 179	192	122	95; 119
4-Methoxyaniline (mp 57°)	243	130	157	95	114
2-Ethoxyaniline	229	79	104	102	164
4-Ethoxyaniline	254	135	173	143	107
3-Nitroaniline (mp 114°)		152	155	136	
4-Nitroaniline (mp 147°)			199	139	
Dimethylamine	7	oil	41	47	79
Diethylamine	55	oil	42	42	60
Diisopropylamine	84			94	
Piperidine	105	oil	48	93	96
Di-*n*-propylamine	109			51	
Morpholine	130	oil	75	118	147
N-Methylaniline	192	102	63	79	94
N-Ethylaniline	205	54	60	oil	87
N-Methyl-2-methylaniline	208	83	53	64	60

TABLE 9.10
*Derivatives of Tertiary
Amines*

| Tertiary amine | Mp, °C | Melting point of derivative, °C | |
		Methiodide	Ethiodide
Triethylamine	89	280	
Pyridine	116	117	90
2-Methylpyridine (α-picoline)	129	230	123
2,6-Dimethylpyridine (2,6-lutidine)	142	233	
3-Methylpyridine (β-picoline)	143	92	
4-Methylpyridine (γ-picoline)	143	167	
Tripropylamine	157	207	238 (dec)
Dimethylaniline	193	220 (dec)	136
Tributylamine	216	180; 186	
Diethylaniline	216	102	
Quinoline	237	133	

the sodium acetate solution and mix the reactants thoroughly. After a few minutes, cool the reaction mixture and stir it vigorously to induce crystallization. Collect the crystals by suction filtration.

(P) Acylation with Acyl Halides and Aqueous Sodium Hydroxide (Schotten–Baumann Method)

To 10 mL of 5% aqueous sodium hydroxide in a large test tube, add 0.25 g of the amine and 0.5 g (0.6 mL of a liquid) of the acyl or sulfonyl chloride (**Caution**—*irritating vapor*). Stopper the tube firmly and shake it vigorously; release any internal pressure by *cautiously* removing the stopper. Continue to shake for about 10 min. Collect the precipitate by suction filtration and wash it with water, followed by a little dilute hydrochloric acid. Recrystallize the derivative from methanol or aqueous ethanol; a few crystals should be reserved for seeding.

(Q) Acylation in Pyridine

In a small, round-bottomed flask, place 0.25 g (0.6 mL of a liquid) of the acyl or sulfonyl chloride (**Caution**—*irritating vapor*), and 2 mL of pyridine (**Caution**—*disagreeable odor*). Reflux the mixture gently for 30 min and then pour it *cautiously* into 10–15 mL of cold water. Stir the product until it crystallizes, collect it on a suction filter, and recrystallize it from aqueous ethanol.

(R) Quaternary Ammonium Salts

In a small flask mix 0.3 mL (0.25 g) of the 3° amine and add, in the hood, 0.3 mL of methyl iodide (bp 43°).

||||▶ *CAUTION*   Methyl iodide is a cancer suspect agent and toxic.

Attach a reflux condenser, warm the flask so that the methyl iodide boils gently (to avoid the loss of methyl iodide) for about 5 min. Allow the flask to cool for 10 min, and add 2 mL of methylene chloride. Collect the crystals of the methiodide with suction and wash them with a little fresh methylene chloride. Without delay, determine the melting point of the product and place the excess in a stoppered vial. If the product is impure, it may be recrystallized from methanol or ethanol.

The same procedure can be used with ethyl iodide to prepare the ethiodide.

Phenols. Phenols possess two functional groups joined together: a benzene ring and a hydroxyl group. The chemistry of phenols is largely the chemistry of these two groups. However, because of the interaction of the π-electrons, their properties are somewhat modified. The proton of the hydroxyl group is more acidic than the proton of an aliphatic alcohol; the benzene ring is more susceptible to attack by electrophilic reagents.

Of the three derivatives described here, the first depends on the alcohol-like properties of a phenol and the chemistry is the same as that described in the alcohol-derivative section.

The aryloxyacetic acid derivative formation depends on the enhanced acidity of the phenol proton, which easily yields the highly nucleophilic phenolate anion.

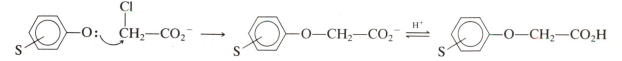

Phenolate anion

This strong nucleophile rapidly displaces the Cl of chloroacetate anion in an S_N2 reaction to yield the aryloxyacetate anion, which upon acidification gives the aryloxyacetic acid.

$$ \text{S} - \text{O:} \quad \underset{|}{\overset{\text{Cl}}{\text{CH}_2}} - \text{CO}_2^- \longrightarrow \text{S} - \text{O} - \text{CH}_2 - \text{CO}_2^- \underset{}{\overset{\text{H}^+}{\rightleftharpoons}} \text{S} - \text{O} - \text{CH}_2 - \text{CO}_2\text{H} $$

The bromination of phenol is very fast in any available *ortho* or *para* position. Phenol itself gives a tribromide; phenols with one or more of these positions occupied will give correspondingly less bromine uptake. The mechanism involves electrophilic aromatic substitution with a bromonium ion or its equivalent (such as Br_3^+).

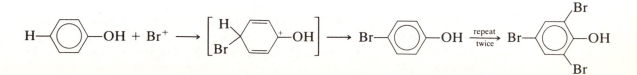

TABLE 9.11 *Derivatives of Phenols*

| Phenol | Mp, °C | Bp, °C | Melting point of derivative, °C | | |
			3,5-Dinitrobenzoate	Aryloxyacetic acid	Bromo derivative
5-Isopropyl-2-methyl- (carvacrol)	1	238	77; 83	151	46
2-Chloro-	7	175	143	145	48 (mono)
					76 (di)
3-Methyl- (*m*-cresol)	10	203	165	103	84 (tri)
2,4-Dimethyl-	26	212	164	141	
2-Methoxy- (guaiacol)	28	205	141	116	116 (tri)
2-Methyl-	32	192	138	152	56 (di)
4-Methyl-	35	203	188	135	49 (di)
					108 (tetra)
2-Nitro-	45	216	155	158	95 (tri)
2,6-Dimethyl-	48	212	158	139	79
2-Isopropyl-5-methyl- (thymol)	50	232	103	149	55
4-Methoxy-	57	243		110	
3,4-Dimethyl-	63	225	181	162	171 (tri)
3,5-Dimethyl-	64	219	195	81; 111	166 (tri)
4-Bromo-	66	222	191	157	95 (tri)
2,5-Dimethyl-	73	212	137	118	178 (tri)
1-Naphthol	95	278	217	193	105 (di)
3-Nitro-	96		159	156	91 (di)
4-*t*-Butyl-	98	237	156	86	50 (mono)
					67 (di)
1,2-Dihydroxy- (catechol)	105	246	152 (dil)	137	192 (tetra)
1,3-Dihydroxy- (resorcinol)	110	281	201 (di)	175; 195	112 (tri)
4-Nitro-	112		186	187	142 (di)
2-Naphthol	123	285	210	154	84
1,4-Dihydroxy- (hydroquinone)	171	286	317 (di)	250	186 (di)

(S) 3,5-Dinitrobenzoates

Follow the procedure given for the preparation of 3,5-dinitrobenzoates of alcohols. The melting points of the 3,5-dinitrobenzoates of many common phenols are listed in Table 9.11.

(T) Aryloxyacetic Acids

In a large test tube dissolve 0.25 g of the unknown phenol in 5 mL of 10% aqueous sodium hydroxide and add 0.25 g of chloroacetic acid with vigorous shaking. If any of the sodium salt of the phenol separates, add another few milliliters of water to dissolve it. Heat the solution in a gently boiling water bath for 1 hr and, after cooling, add 1 mL of water. Acidify the solution to pH 4 (universal pH paper), and extract it with two 10-mL portions of diethyl ether (**Caution**—*flammable solvent*). Wash the combined ether layers with cold water.

To isolate the aryloxyacetic acid, extract the ether solution with 5 mL of 5% sodium carbonate solution. Draw off the aqueous layer and add it to a flask containing 2 mL of concentrated hydrochloric acid diluted with 10 mL of water. Filter the product with suction and recrystallize it from water or 95% ethanol. The melting points of the aryloxyacetic acid derivatives of many common phenols are listed in Table 9.11.

(U) Bromo Derivatives

In a large test tube dissolve 0.25 g of the phenol in 3 mL of methanol and add 3 mL of water. Add dropwise, with swirling, aqueous brominating solution[32] until a yellow color persists even after thorough mixing. Add 15 mL of water to precipitate the product and collect it on a small suction filter and wash it thoroughly with about 10 mL of 5% aqueous sodium bisulfite solution. Finally, wash the bromophenol with about 10 mL of water and press it as dry as possible on the filter.

Dissolve the crude product in about 25 mL of hot ethanol, by warming on a steam bath, and filter the hot solution through a fluted filter into a clean 125-mL Erlenmeyer flask. To the hot filtrate add about 50 mL of water in small portions until the bromophenol begins to separate. Mix the solution thoroughly, heat to redissolve the solid, and set the solution aside to cool undisturbed. After the product has crystallized, collect it by suction filtration, wash it with a few milliliters of cold 50% aqueous ethanol, and spread it on a filter paper to dry.

Hydrocarbons and Halohydrocarbons. There are no universal derivatives for hydrocarbons; the recommended procedure is usually to functionalize the compound in some way and then identify the product. For example, alkenes can be acetylated under Friedel–Crafts conditions and the resulting ketone identified in the usual manner. Aromatic hydrocarbons can be nitrated and the nitro product reduced to an amine, which can be derivatized. The situation is no better with halohydrocarbons, which usually must be converted by a Grignard reaction to a carboxylic acid. This approach is lengthy, it frequently gives mixtures, and the yields can be low. With the development of NMR spectroscopy, it has become easy to identify ordinary hydrocarbons and chlorocarbons spectrometrically. For this reason, the cumbersome chemical procedures will not be described. The boiling points of representative hydrocarbons and chlorocarbons are listed in Tables 9.12 through 9.15.

---

[32] Aqueous brominating solution is prepared by adding 2 mL (6 g) of bromine (**Caution**—*severe hazard!* See Section 1.1) to a solution of 8 g of potassium bromide in 50 mL of water.

TABLE 9.12
Boiling Points of Saturated Hydrocarbons

	Bp, °C		Bp, °C
Pentane	36	2,2,4-Trimethylpentane	99
Cyclopentane	49	*trans*-1,4-Dimethylcyclohexane	119
2,2-Dimethylbutane	50	Octane	126
2,3-Dimethylbutane	58	Nonane	151
2-Methylpentane	60	Decane	174
3-Methylpentane	63	Eicosane (mp 37°)	343
Hexane	69	Norbornane (mp 87°, subl)	—
Cyclohexane	81	Adamantane (mp 268°, sealed)	—
Heptane	98		

TABLE 9.13
Boiling Points of Unsaturated Hydrocarbons

	Bp, °C		Bp, °C
1-Pentene	30	3-Hexane	82
2-Methyl-1,3-butadiene		Cyclohexene	84
(isoprene)	34	2-Hexyne	84
trans-2-Pentene	36	1-Heptene	94
cis-2-Pentene	37	1-Heptyne	100
2-Methyl-2-butene	39	2,4,4-Trimethyl-1-pentene	102
Cyclopentadiene	41	2,4,4-Trimethyl-2-pentene	104
1,3-Pentadiene	41	1-Octene	123
(piperylene)		Cyclooctene	146
3,3-Dimethyl-1-butene	41	1,5-Cyclooctadiene	150
1-Hexene	63	*d,l*-α-Pinene	155
cis-3-Hexene	66	(−)-β-Pinene	167
trans-3-Hexene	67	Limonene	176
1-Hexyne	71	1-Decene	181
1,3-Cyclohexadiene	80		

TABLE 9.14
Boiling and Melting Points of Aromatic Hydrocarbons

Liquids	Bp, °C	Solids	Mp, °C
Benzene	80	Tetrahydronaphthalene	206
Toluene	111	Diphenylmethane	25
Ethylbenzene	136	Bibenzyl	53
p-Xylene	138	Biphenyl	69
m-Xylene	139	Naphthalene	80
o-Xylene	144	Triphenylmethane	92
Isopropylbenzene	152	Acenaphthene	96
n-Propylbenzene	159	Phenanthrene	96
1,3,5-Trimethylbenzene	165	Fluorene	116
t-Butylbenzene	169	Pyrene	148
Isobutylbenzene	173	1,1′-Binaphthyl	160
sec-Butylbenzene	173	Hexamethylbenzene	165
Indene	182	Anthracene	216
n-Butylbenzene	183		

TABLE 9.15
Boiling Points of Halogenated Hydrocarbons

	Bp, °C		Bp, °C
Chlorides		**Bromides**	
n-Propyl	47	Ethyl	38
t-Butyl	51	Isopropyl	60
sec-Butyl	68	Propyl	71
Isobutyl	69	*t*-Butyl	72
n-Butyl	78	Isobutyl	91
Neopentyl	85	*sec*-Butyl	91
t-Pentyl	86	Putyl	101
Cyclohexyl	143	*t*-Pentyl	108
Chlorobenzene	132	Neopentyl	109
2-Chlorotoluene	159	Bromobenzene	156
1,4-Dichlorobenzene (mp 53°)	173	1-Bromoheptane	174; 180
Hexachloroethane (mp 187° subl)	185	1,4-Dibromobenzene (mp 89°)	219
2,4-Dichlorotoluene	200		
1-Chloronaphthalene	259	**Iodides**	
Triphenylmethyl (mp 113°)		Methyl	43
		Ethyl	72
		Isopropyl	90
		Propyl	102

Questions

1. Predict the solubility class of each compound.
   (a) aniline
   (b) benzamide
   (c) butanoic acid amide
   (d) citric acid
   (e) cyclohexanone
   (f) diethyl ether
   (g) methylene chloride
   (h) *p*-nitrobenzoic acid
   (i) *n*-pentane
   (j) 2-propanol

2. Using the solubility classification solvents, indicate how you would separate each of the following binary mixtures and recover the components.
   (a) benzoic acid and benzamide
   (b) *n*-butyl bromide and di-*n*-butyl ether
   (c) nitrobenzene and aniline
   (d) 1-pentene and 2-pentanone

3. A white solid, mp 110–112°, was insoluble in water but soluble in both aqueous sodium hydroxide and sodium carbonate. Elemental analysis was negative for N, S, and halogens. What is a reasonable structure for the solid?

4. A colorless liquid, bp 177–179°, fell into solubility class **N**. Treatment with phenyl isocyanate gave a solid derivative, mp 112–114°. What is a reasonable structure for the liquid?

5. What test could be used to distinguish between each pair of compounds?
   (a) 1-butanol and 2-butanol
   (b) 2-methyl-2-propanol and 1-butanol
   (c) butanal and 2-butanone

(d) acetic acid and acetamide

(e) triethylamine and dibutylamine

(f) cyclohexene and cyclohexane

6. What derivative would you prepare to distinguish between each pair of compounds?

(a) isobutyl alcohol and 3-pentanol

(b) 2-pentanone and 3-pentanone

(c) propanedioic acid and 2-acetoxybenzoic acid

(d) benzenesulfonic acid and 2-carboxybenzenesulfonic acid

(e) 4-methylaniline and 3-methylaniline

10 Identification by Spectrometric Methods

The classical methods for organic structure determination involve a series of test reaction and chemical transformations. The newer spectrometric methods probe the sample with electron beams (mass spectrometry), with electromagnetic radiation (infrared, ultraviolet, and X-ray spectroscopy), or with a combination of electromagnetic radiation and a strong magnetic field (nuclear magnetic resonance).[1] The classical methods are destructive; that is, the sample is gone once the reaction is complete. So is mass spectrometry, but on such a small scale (micrograms or less) as to be usually ignored. The other spectroscopic methods are essentially nondestructive. Historically, it was this feature as much as anything that led organic chemists to welcome spectroscopy when the instruments first became commercially available. It soon became apparent, however, that spectroscopy offered much more; with this technique the enviroment of the atoms, the bonds, and combinations of bonds could also be probed with an ever-increasing variety and specificity denied to chemical reactions.

Spectrometry has entered a new era with the exploitation of microelectronic devices and computers. Complex sequences and combinations of nondestructive electromagnetic probes can be applied automatically with the massive outpouring of data reduced and analyzed by a computer. Elaborate computer programs are available that compare an observed spectrum with a large file of spectra. In a matter of minutes, a file of 20,000–100,000 spectra is searched and any match reported; even when no matching spectrum is found, the program reports the structures of the compounds with the nearest fitting spectra—information that frequently is sufficient to identify the unknown.

[1] This list does not include the less common techniques such as photoelectron spectroscopy, circular dichroism, electron spin resonance, ion cyclotron resonance, and electron diffraction.

The discussion of spectrometry presented here is much less ambitious. Only the fundamentals of the four most widely used methods will be considered. Even though abbreviated, the material presented is suitable for the many routine applications of spectrometry in the organic laboratory.

10.1 Mass Spectrometry

The mass spectrometer bombards a stream of vaporized sample with a high-energy electron beam, which, on collision with a molecule of the sample, ejects an electron from it to produce a vibrationally excited radical-cation species.

$$\text{Molecule} \xrightarrow[\text{radical cation}]{\text{electron beam}} [\text{molecule}]^{\cdot +} + \text{ejected electron}$$
$$\text{radical cation}$$

$$\text{neutral and ionized fragments}$$

The collisions are so energetic (typically 1400 kcal/mole) that the radical cations formed behave as though they have been heated to thousands of degrees and begin to decompose almost immediately to form a variety of neutral and ionized fragments. The collection of ionized fragments along with the remaining parent ions are drawn out of the ionizing chamber by a set of charged electrodes and toward an ion-collecting target that generates a signal in proportion to the number of ions reaching it.

A simplified diagram of a mass spectrometer is shown in Figure 10.1. The

FIGURE 10.1
Diagram of Mass
Spectrometer

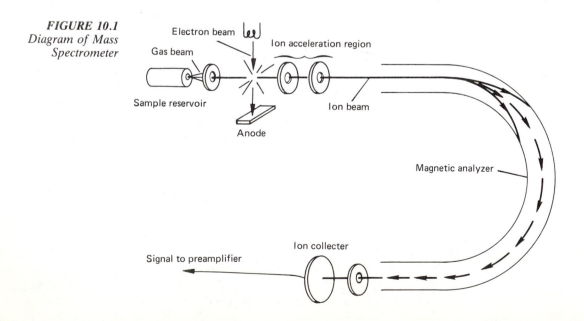

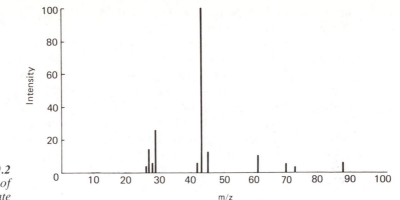

FIGURE 10.2
*Mass Spectrum of
Ethyl Acetate*

ionization chamber is at the left of the diagram; the target (detector) is at the bottom. Along the way to the target the rushing flow of ions passes through a strong magnetic field that bends the ion beam by an amount that depends on both the mass of the ion and the strength of the magnetic field. For a particular magnetic field strength only those ions possessing a characteristic mass-to-charge ratio (conventionally abbreviated m/z) will reach the narrow opening of the target. As the magnetic field is altered, species with different m/z will be focused on the target. Once the calibration between m/z and magnetic field strength has been determined, it is possible to record a mass spectrum that shows the relative abundance of each ionic species produced in the fragmentation process. A representative mass spectrum, that for ethyl acetate, is shown in Figure 10.2.

Mass spectra are both fingerprints of compounds, and, if interpreted in terms of stability and ease of forming different ions, sets of clues to the structures. One of the most valuable pieces of information provided by mass spectrometry is the molecular weight of the parent ion, which will normally be observed as one of the group of peaks at highest m/z. For ethyl acetate (Figure 10.2) this is 88. Not visible in Figure 10.2 are low-intensity peaks at m/z 89 and 90, which arise from trace amounts of ^{13}C in the original sample. There is also a weak peak at m/z 87, which arises from the loss of a hydrogen atom from the parent ion.

From close examination of the peaks around the parent peak, it is possible to obtain some idea of the elemental composition. Ordinary organic chemicals are not isotopically pure because of the 1.08% natural abundance of ^{13}C. For the saturated hydrocarbon $C_{10}H_{22}$, this means that there is a $10 \times 1.08 = 11\%$ probability that a particular molecule will contain one ^{13}C atom. The mass spectrum of the C_{10} hydrocarbon will show parent peaks at m/z 142 and 143 in the ratio of 100:11. The appearance of isotopic peaks is quite helpful in identifying the presence of chlorine, since the ^{35}Cl and ^{37}Cl isotopes occur naturally in an easily reconized ratio of 3:1. The appearance of two peaks in this ratio, separated by two mass units, is strong evidence for one chlorine atom in the molecule. If two chlorine atoms were present, three peaks would occur in the ratio of 9:6:1 (Why?) with each peak separated from the next by two mass units. Naturally occurring bromine contains ^{79}Br and ^{81}Br in the ratio of $1:1$; fluorine and iodine are monoisotopic.

With special mass spectrometers of extremely high mass resolution, it is possible to make use of the nonintegral values of atomic weights of even the pure isotopes (^{12}C = 12.00000 basis, ^{1}H = 1.007825, ^{14}N = 14.00307, ^{16}O = 15.99491, etc.) to determine empirical formulas directly. For example, ethyl acetate with an empirical formula $C_4H_8O_2$ has a molecular weight of 88.0524. Ethyl n-propyl ether, $C_5H_{12}O$, has a molecular weight of 88.0888. These two compounds, although they have the same low-resolution molecular weight of 88, could easily be distinguished by a high-resolution mass spectrometer, which will give molecular weights to ±0.002 mass unit.

Interpretation of the clues provided by the fragment ions requires considerable practice and study but can provide complete structures, particularly if auxiliary data from other spectroscopic sources or chemical studies are available. The details of the process will not be described here; the interested reader is referred to the excellent book by McLafferty.[2] The mass spectrum of ethyl acetate reproduced in Figure 10.2 gives an idea of the structural information available by this technique. The largest peak in the spectrum (*base peak*) at m/z 43 is the fragment ion [CH$_3$CO]$^{\cdot+}$ that arises from the parent peak (m/z 88) by loss of a neutral CH$_3$CH$_2$O radical. The peak at m/z 45 arises from the same bond cleavage, but this time with the loss of a neutral CH$_3$CO· radical. The peak at m/z 29 is from the [CH$_3$CH$_2$]$^+$ ion. The other significant peaks arise from more complex rearrangements and eliminations. The relative abundance of the peaks reflect the relative stabilities of the cations and radicals formed.

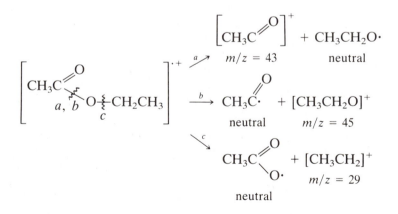

A particularly powerful organic analytical technique is the computer-controlled gas-chromatograph–mass-spectrometer combination with which microgram or even nanogram quantities of complex mixtures can be analyzed as rapidly as the peaks come off the chromatograph. Such instruments form the backbone of environmental-pollution analysis.

---

[2] McLafferty, *Interpretation of Mass Spectra*, 3rd ed. (New York : Wiley, 1981).

10.2 Infrared Spectroscopy

The electromagnetic spectrum runs continuously from X-rays, through light, to radio waves. Electromagnetic radiation consists of bundles of energy (photons) surrounded by oscillating electric and magnetic fields. At the X-ray end of the spectrum, the frequency and energy are high, the wavelength is short; at the radiowave end of the spectrum the converse is true. Molecules respond in various ways to electromagnetic radiation, depending on the frequency. Figure 10.3 displays the electromagnetic spectrum and several of the regions of particular interest to organic chemists.

A useful mechanical model for discussing the infrared (IR) region of absorption (wavelengths of about 1–50 μm) is to treat the atoms as masses connected by springs (bonds). A property of mechanical systems is their characteristic vibrational frequencies (resonances) that absorb energy from applied oscillatory forces. Small masses and stiff springs give rise to high vibrational frequencies; it is a consequence of the small weights of atoms and the tightness of bonds that molecular vibrations occur in the infrared region (6×10^{12} to 3×10^{14} Hz). The probability of absorption depends on the change of dipole moments during the vibration; a large change in dipole moment accompanies intense absorption.

There are two basis types, or *modes*, of motion of the nuclei relative to each other: stretching and bending.

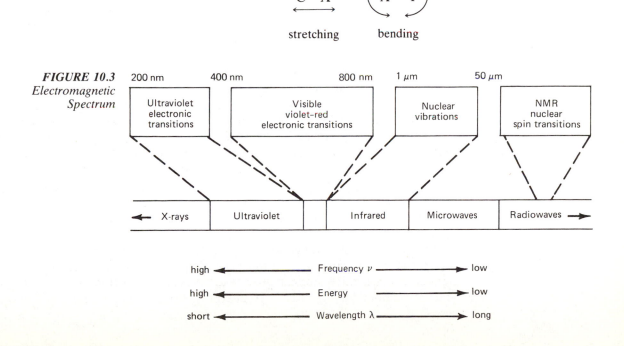

FIGURE 10.3
Electromagnetic Spectrum

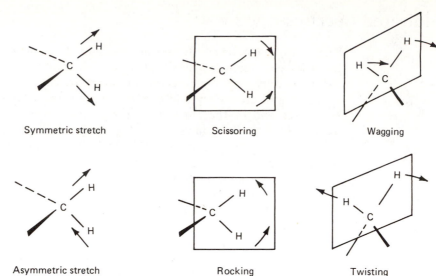

FIGURE 10.4
*Vibrational Modes of
the Methylene Group*

Symmetric stretch   Scissoring   Wagging

Asymmetric stretch   Rocking   Twisting

These basic modes combine to form more complex vibrational modes. For example, a methylene group has six modes, each with a characteristic frequency, as shown in Figure 10.4.

Even for simple molecules the number of vibrational frequencies is large (for nonlinear molecules, $3N - 6$ vibrations where N is the number of atoms). Not all of these will appear in the spectrum because many are highly symmetrical vibrations that produce small changes in dipole moment during the vibration. A further complication is that overtones (analogous to those of a squeaky violin) and combination bands occur with moderate intensity. In view of this complex situation, it is fortunate that certain functional groups and structural units have characteristic absorption frequencies that change little from molecule to molecule. For many structural units, the frequency shifts that do occur can be related to variations in the neighboring structure. Table 10.1 lists a few structural units and their characteristic absorption ranges in the infrared region. A more complete listing is given in the Appendix. The frequencies listed in Table 10.1 arise from stretching modes. The stronger the bond joining the atoms, the higher the frequency. The apparent exceptions are the various X—H bonds, which have normal single bond strengths but absorb at high frequencies because of the lightness of the hydrogen atom (the analogy here is to the vocal cords of a soprano compared to those of a basso).

A more detailed analysis of the characteristic frequencies can be found in specialized texts.

Frequenices in the range of 1430–625 cm^{-1} (7–16 microns), the so-called fingerprint region, contain most of the bending vibrations. The vibrational motions occurring in this region can be quite complex mixtures of the basic modes, and you are cautioned not to attempt to interpret every band or wiggle found in this region.

TABLE 10.1
Infrared Fundamental Absorption Frequencies

	Frequency, cm⁻¹	Wavelength	Intensity[a]
C=O stretching vibrations			
Aldehydes	1740–1720	5.75–5.81	s
Ketones			
Saturated	1725–1705	5.80–5.87	s
Aromatic	1700–1680	5.88–5.95	s
Carboxylic acids	1725–1720	5.80–5.95	s
Esters	1700–1630	5.71–5.81	s
Amides		5.81–6.13	s
C—H, N—H, and O—H stretching vibrations			
C—H			
Alkanes	2960–2850	3.38–3.51	m–s
Alkanes	3095–3010	3.23–3.32	m
Alkynes	near 3300	near 3.03	s
Aromatic	near 3030	near 3.30	v
Aldehyde	2960–2820 and	3.45–3.55 and	w
	2775–2700	3.60–3.70	w
N—H			
Amine	3500–3300	2.86–3.03	m
Amide	3500–3140	2.86–3.18	m
O—H			
Alcohols and phenols			
Not hydrogen bonded	3650–3590	2.74–2.79	v, sharp
Hydrogen bonded	3550–3200	2.82–3.13	v, broad
Carboxylic acids	2700–2500	3.70–4.00	w
Triple-bond stretching vibrations			
Nitriles	2260–2215	4.42–4.51	m
Alkynes	2260–2100	4.42–4.76	v
C=C stretching vibrations			
Alkenes	1680–1620	5.95–6.17	v
Conjugated diene	near 1650 and	near 6.06 and	w
	near 1600	near 6.25	w
Aromatic	near 1600 and	near 6.25	v
	near 1500	near 6.67	v

[a] w = weak absorption; m = medium absorption; s = strong absorption; v = variable intensity absorption.

Assignment of Principal Functional Groups.

This section presents a scheme for the systematic identification of the principal functional groups. A flowchart of the scheme is given in Figure 10.5; the circled numbers on the chart correspond to the **boldface** numbered steps that follow. Further details for selected functional groups along with representative spectra are presented in the next section.

The assignment scheme starts the carbonyl group since, when present, the carbonyl stretch of aldehydes, ketones, esters, acid chlorides, and amides is

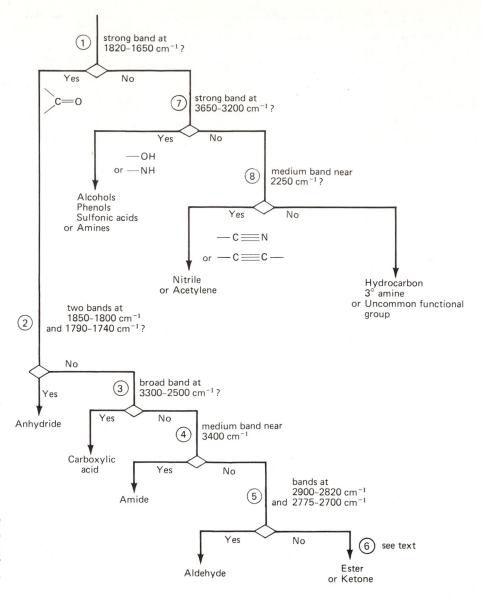

FIGURE 10.5
*Infrared Assignment
Scheme for the
Principal Functional
Groups*

usually the strongest band in the spectrum. The next most characteristic absorptions are those for O—H groups and N—H bonds. Triple bonds, though less common and frequently weak, also have characteristic frequencies. The scheme is not infallible, but it is much superior to random guessing; although the conclusions of the scheme are stated firmly, it should be understood that confirmation is always desirable.

1. Is there strong absorption in the region of 1650–1820 cm^{-1}? If not, jump to step **7.** If there is strong absorption (see Figures 10.10 and 10.11 for

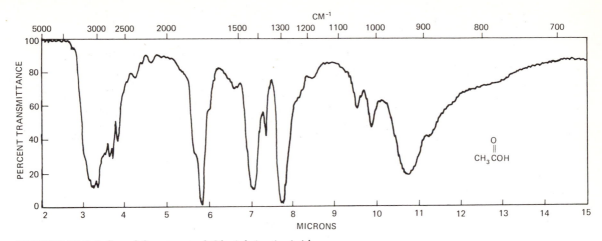

FIGURE 10.6 *Infrared Spectrum of Glacial Acetic Acid*

examples), the molecule contains a carbonyl group, which can be further specified by looking for other characteristic absorptions starting with step **2**. The strong carbonyl band arises from the large change in dipole moment as the carbon–oxygen bond stretches and contracts.

2. Is there a pair of strong bands at 1800–1850 and 1740–1790 cm^{-1}? If not, proceed to step **3**. If a pair of bands is present, the molecule probably contains an anhydride functional group.

3. Is there broad absorption in the region of 2500–3300 cm^{-1} (Figure 10.6 shows an example)? If not, proceed to **4**. If this is present, the molecule is a carboxylic acid. The absorption arises from O—H stretching vibration and is broad because the different molecules of a sample of a carboxylic acid occur with varied degrees of hydrogen, each with its characteristic absorption frequency.

4. Is there a medium-intensity absorption near 3400 cm^{-1} (an example appears in Figure 10.7)? If not, proceed to step **5**. If the 3400 cm^{-1} band is present, the molecule is a primary or secondary amide. Primary amides show two bands (~3500 and ~3400 cm^{-1}), whereas secondary amides show only one band near 3400 cm^{-1}. These absorptions are associated with N—H stretching vibrations. Tertiary amides do not absorb in this region.

5. Is there a pair of weak bands in the 2820–2900 and 2700–2775 cm^{-1} regions ? If not, proceed to step **6**. If such a pair of bands is present, the molecule is probably an aldehyde. The bands have been associated with the mixing of the C—H and C—O vibrations. The identification of aldehydes and ketones is discussed more completely in the following section.

6. For the restricted classes of functional groups being considered here only two possibilities remain: esters and ketones.

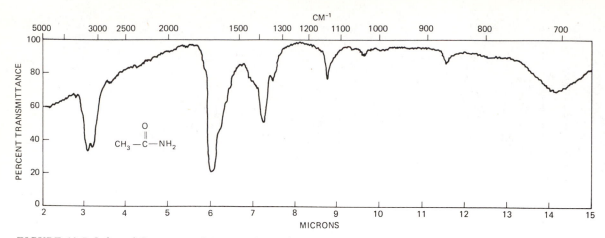

FIGURE 10.7 *Infrared Spectrum of Acetamide (Melt)*

Saturated esters absorb in the range of 1735–1750 cm^{-1} (see Figure 10.8), which increases to 1760–1820 cm^{-1} with strained cyclic esters. Aryl and α,β-unsaturated esters absorb at 1717–1730 cm^{-1}.

Saturated ketones absorb in the range of 1705–1725 cm^{-1}. Ketone groups adjacent to α,β-unsaturation, phenyl groups, or cyclopropane rings absorb at 1665–1695 cm^{-1}.

7. If a carbonyl group is absent, look for hydroxyl or amine absorption in the range of 3200–3650 cm^{-1}. If none is present, go to step **8**. Alcohols and phenols absorb strongly in this range; primary and secondary amine show medium-strength absorptions. The amines can sometimes be distinguished by the presence of weak absorptions near 1600 cm^{-1} arising from

FIGURE 10.8 *Infrared Spectrum of Neat n-Butyl Acetate*

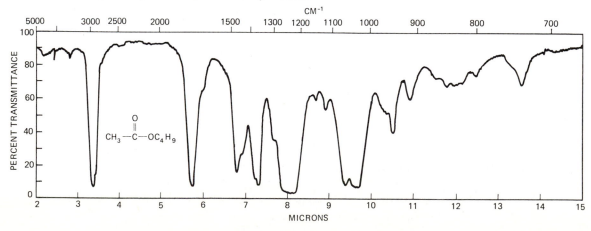

N—H bending vibrations. Hydroxyl groups show O—H bending vibrations in the fingerprint region. The spectroscopic identification of alcohols and amines is discussed more fully in the section that follows this scheme.

8. If O—H and N—H absorptions are absent, look near 2250 cm^{-1}. If no absorption is present, go to step **9**. Nitriles show a medium intensity, sharp absorption near 2200 cm^{-1} characteristic of the C—N stretch. Alkynes absorb near 2150 cm^{-1}, but the band is usually weak.

9. If none of the above infrared bands is present, the molecule either is a hydrocarbon or contains one of the less common functional groups. The spectroscopic identification of hydrocarbons is discussed in the next section, as are sulfonic acids and their derivatives. If another functional group is suspected, you will have to consult a specialized text on infrared spectroscopy.

Spectroscopic Details for Selected Functional Groups

Alcohols. The infrared spectra of alcohols typically show absorption in the 3200–3600 cm^{-1} range from the O—H stretching of the species formed by intermolecular hydrogen bonding. Under conditions of high dilution, the hydrogen-bonded species dissociate and a weaker band at higher energy (3600–3650 cm^{-1}) appears for the free hydroxyl. Strong absorptions near 1050–1150 cm^{-1} and 1250–1410 cm^{-1} are characteristic of C—O stretching vibrations. These features can be seen in the IR spectrum of neat (pure liquid) *n*-butanol reproduced in Figure 10.9.

Aldehydes and ketones. The carbonyl group is one of the most easily identified peaks in an infrared spectrum. Saturated acyclic ketones absorb strongly in the 1705–1725 cm^{-1} range (see the spectrum of cyclohexanone,

FIGURE 10.9 *Infrared Spectrum of Neat n-Butanol*

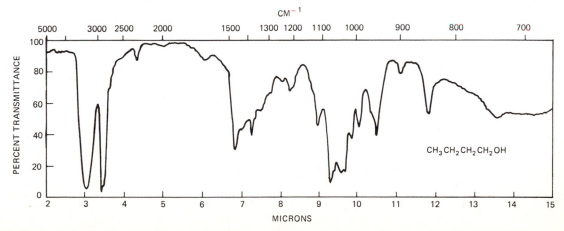

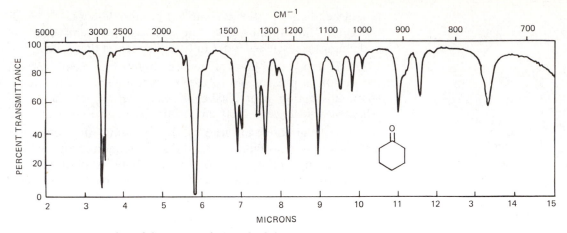

FIGURE 10.10 *Infrared Spectrum of Neat Cyclohexanone*

Figure 10.10). If α,β-unsaturation is present, the peak shifts to lower energy, 1680–1685 cm^{-1}. Cyclic ketones with five-membered or smaller rings are shifted to higher energies: five-membered, 1740–1750 cm^{-1}; four-membered, 1775 cm^{-1}.

Aldehydes absorb in the 1720–1740 cm^{-1} range. As with ketones, conjugation shifts the peaks to lower energies: α,β-unsaturation, 1680–1705 cm^{-1}, and aromatic, 1695–1715 cm^{-1}. Aldehydes show two additional weak bands at 2700–2775 and 2820–2900 cm^{-1}. These features can be seen in the spectrum of benzaldehyde (Figure 10.11).

Amines. Primary amines, which have *two* N—H bonds, show *two* stretching bands in the 3300–3500 cm^{-1} range; secondary amines show *one* band. Tertiary

FIGURE 10.11 *Infrared Spectrum of Neat Benzaldehyde*

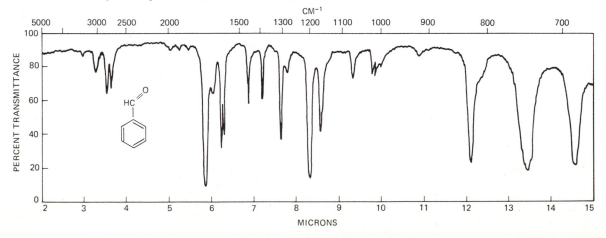

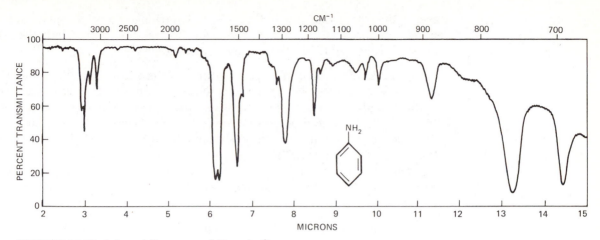

FIGURE 10.12 *Infrared Spectrum of Neat Aniline*

amines have no N—H stretching absorption. Primary amines show a NH_2 scissoring mode in the 1560–1640 cm^{-1} range. Secondary amines absorb near 1500 cm^{-1}. Aromatic amines, for example aniline (Figure 10.12), show strong C—N stretching absorption in the 1250–1370 cm^{-1} range.

Hydrocarbons. The infrared spectra of all hydrocarbons show strong C—H stretching vibrations in the 2850–3300 cm^{-1} range. If a spectrometer with high resolution is used, the subclasses of hydrocarbons can be distinguished, but with ordinary instruments other regions of the spectrum are usually more revealing. Alkenes show a medium intensity C=C stretching absorption near 1670 cm^{-1}. Alkynes have a distinctive C≡C stretching absorption near 2200 cm^{-1}. Aromatic hydrocarbons do not show truly distinctive absorptions.

Figure 10.13 shows the spectrum on neat cyclohexene in which the C—H and C=C bands are visible. It shows also, as is normal, many other longer

FIGURE 10.13 *Infrared Spectrum of Neat Cyclohexene*

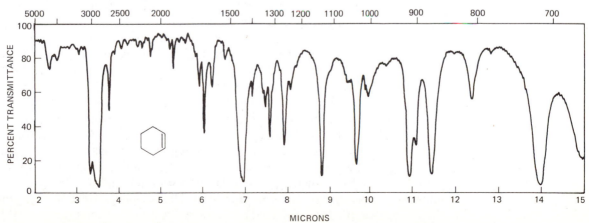

wavelength absorptions related to C—C stretching vibrations and various bending vibrations. Even though these are not readily interpreted, they do provide a "fingerprint" characteristic of cyclohexene.

Sulfonic acids and derivatives. Sulfonic acids show a medium-intensity band in the 3100–3400 cm^{-1} range associated with O—H stretching. The anhydrous acids give strong S—O absorptions at 1340–1350 cm^{-1} and 1150–1165 cm^{-1}; the hydrated acids show an additional band at 1120–1230 cm^{-1}, which has been attributed to S—O double-bond stretch of the hydronium sulfonate salt (RSO_3^- H_3O^+).

Sulfones (R_2SO_2), sulfonamides (RSO_2NR_2), and covalent sulfonates (RSO_3R) show a similar pair of strong S—O absorptions at 1300–1375 cm^{-1} and 1120–1195 cm^{-1}.

Infrared Sampling Techniques.

With pure liquid samples, the simplest technique is to prepare a thin film between a pair of polished NaCl salt plates. A drop of sample is placed in the center of one plate, and the second plate is placed gently on top. The sandwich is pressed together to produce an *even* film of liquid. Salt plates are soft and easily scratched. They also cleave readily if uneven pressure is applied. Another problem is that they dissolve or etch when exposed to water or hydroxylic solvents. Even the moisture on one's fingers is sufficient to damage them; the polished faces should never be touched. The best solvents for cleaning salt plates are carbon tetrachloride and chloroform. Because of the toxicity of these solvents with prolonged exposure, they should be handled only in the hood or other well-ventilated area. The plates are best handled with rubber-tipped tweezers, and they should be stored in a desiccator.

Another common device used for solutions and liquids that are not too viscous is the solution cell, made from salt plates separated by a thin spacer and held together in a metal frame. Solution cells, because of their precision construction, are relatively expensive. A typical cell has a path length of 0.2 mm and holds about 0.5 mL of solution. With such a cell 5% solutions yield good spectra. If infrared spectra are determined on solutions, it is normal practice to place a solution cell filled with pure solvent in the reference beam to cancel (approximately) any absorptions of the solvent.

With solid samples, two other techniques are used. The *mull* method requires that a 5- to 10-mg sample be ground vigorously in an agate or mullite mortar until the sample is glossy and cakes to the side of the mortar (an indication that the particle size is in the micrometer range). A drop of Nujol (commercial mineral oil) is added and the sample reground to disperse the solid in the Nujol. The mixture should resemble Vaseline in viscosity. If it is too thick, another drop of Nujol is added and the sample ground again. If the solid particles are too coarse, the sample will scatter the infrared light at short wavelengths and yield a distorted spectrum. The mull is transferred to a salt plate with a microspatula; a second plate is put on top and rotated until all air bubbles are squeezed from the window area. The two plates are held together by the viscous sample and supported in the IR beam by a V-shaped sample holder.

The spectra of solids can also be determined by the potassium bromide *pellet* method. Approximately 1–2 mg of solid is ground *briefly* with about 200 mg of spectral grade potassium bromide (too much grinding produces a powder that absorbs water rapidly from the atmosphere). The powder is transferred to a pellet press[3] and pressure applied to produce a fused translucent wafer that is then mounted in a special holder.

10.3 Nuclear Magnetic Resonance

In the previous section it was shown how functional groups can be identified readily by infrared absorption spectroscopy. This section will describe the complementary technique of nuclear magnetic resonance (NMR), which is useful for identifying the hydrocarbon skeleton to which the functional group is attached. Proton NMR allows the qualitative and quantitative characterization of the different types of hydrogen in the molecule (aromatic, aliphatic, etc.). The various ^{13}C NMR techniques identify the carbons of the skeleton. Although IR and NMR by themselves are valuable structural tools, the combination of NMR and IR is even more powerful and sufficient to identify all but the most complex molecules.

Theory of ^{1}H and ^{13}C NMR.[4] The nucleus of a hydrogen atom is surrounded by a weak magnetic field associated with the nuclear spin. When hydrogen atoms or materials containing hydrogen atoms are placed in a strong external magnetic field, the weak nuclear magnetic moments line up either parallel or antiparallel to the direction of the magnetic lines of force. The two possible orientations differ in energy by an amount proportional to the strength of the applied magnetic field; if the nuclei are simultaneously exposed to electromagnetic radiation of the correct frequency (energy = $h\nu$), the parallel moments absorb energy, flip over, and become antiparallel moments. It is this flipping process, called *nuclear magnetic resonance*, that gives rise to the absorption spectrum.

The condition for proton resonance absorption is that $\nu = 4.2577 \times 10^3 \mathbf{B}$, where ν is the radiating frequency and $\mathbf{B}$ is the magnetic field strength at the nucleus measured in gauss.[5] If all of the protons were exposed to the same magnetic field, they would absorb at the same frequency. Actually, when a molecule is subjected to a magnetic field, the bonding electrons circulate in such a way as to generate additional weak magnetic fields of their own. In general, these new magnetic fields oppose the large, fixed magnetic field, which has the effect of partially shielding the underlying proton nuclei. The greater the

---

[3] The KBr Mini Press, available from Wilks Scientific Corporation, Norwalk, CT, is a simple, effective, and inexpensive press. The Mini Press also serves as a holder for the wafer. The press should be *cleaned thoroughly and dried carefully* after each use.

[4] This section describes the origin of proton NMR signals. It applies also to ^{13}C NMR except that the condition for resonance is different.

[5] For ^{13}C nuclei the condition is $\nu = 1.0705 \times 10^3$ $\mathbf{B}$.

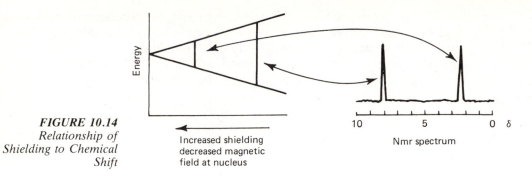

FIGURE 10.14
Relationship of
Shielding to Chemical
Shift

shielding, the lower the frequency required to produce resonance. To a first approximation the amount of shielding is proportional to the electron density around the proton (there are secondary effects, which will be discussed later) and thus protons adjacent to electron-donating groups will absorb at lower frequencies. The relationship of varied shielding to NMR absorption peak shift is illustrated in Figure 10.14.

Chemical Shift. Because of the correlation of electron density with structure, proton NMR is a powerful tool for determining the structural environment of hydrogen atoms in organic molecules. By common usage the reference standard for practical proton NMR measurements is the highly symmetrical tetramethyl-silane (TMS, $(CH_3)_4Si$), an inert, volatile liquid that is easily separated from the sample after the measurement. The protons of TMS are more highly shielded than the protons of most other organic compounds. In recording NMR spectra, the most widely used measure is the chemical shift, δ, defined as

$$\delta = \frac{\Delta \nu}{\nu_0} \times 10^6 \text{ ppm}$$

where $\Delta \nu$ is the difference in absorption frequencies of TMS and the proton being examined, and ν_0 is the radiation frequency of the NMR machine.[6] The chart paper used in a calibrated NMR instrument allows both δ and $\Delta \nu$ to be read directly.[7]

Some typical proton chemical shifts are listed in Table 10.2. These shifts portray fairly well the range of shifts observed in practice. A more detailed listing of ^{1}H NMR shifts is given in the Appendix.

---

[6] An older unit, now all but gone from the literature, is the tau unit defined as $\tau = 10.0 - \delta$.

[7] In practice, NMR machines are designed to operate at a constant frequency, and it is the magnetic field that is varied. However, because of the proportional relationship between absorption frequency and magnetic field, the change in magnetic field can be expressed as an equivalent change in frequency. The protons of molecules less shielded than those found in TMS are referred to as deshielded, as occurring downfield from TMS, or as having positive chemical shifts.

	Structural environment	Chemical shift from TMS, ppm
TABLE 10.2		

TABLE 10.2
Nuclear Magnetic Resonance Chemical Shifts

Structural environment	Chemical shift from TMS, ppm
$-\overset{\mid}{\underset{\mid}{C}}-CH_3$ (saturated)	0.8–1.1
$-CH_2-$ (saturated)	1.2–1.3
$-\overset{\mid}{CH}-$ (saturated)	1.4–1.6
$X-\overset{\mid}{\underset{\mid}{C}}-CH_3$ (X = halogen, —O, $>$N)	1.0–2.0
$>C=C<_{CH_3}$	1.6–1.9
$ArCH_3$	2.1–2.5
$O=\overset{\mid}{C}-CH_3$	2.1–2.6
$-C\equiv C-H$	2.4–3.1
$-O-CH_3$	3.5–3.8
$>C=CH_2$ (nonconjugated)	4.6–5.0
$>C=C<^H$	5.2–5.7
$Ar-H$	6.6–8.0
$-C\overset{O}{\underset{H}{<}}$	9.8–10.8

The ^1H NMR spectrum of methyl *p*-bromobenzoate is reproduced in Figure 10.15, which shows the downfield aromatic absorptions and (relative to the aromatic protons) the upfield methyl absorptions. The reference peak of TMS, which is added to the sample, appears at $\delta = 0.0$

The major determinant of proton chemical shifts is the inductive electron withdrawal by electronegative groups and is approximately additive so that, for example, the chemical shift (in parts per million) for the protons in methane is 0.8; in methyl chloride, 3.0; in methylene chloride, 5.3; and in chloroform, 7.3. The other important factor, magnetic anisotropy of neighboring bonds, is deeply interwoven and difficult to separate from the inductively caused chemical shifts. However, two situations clearly reveal its presence. In aromatic compounds the ring protons typically absorb at $\delta = 6.6–8.0$, a region well below typical alkene absorption ($\delta = 5.1–6.3$) even though both employ sp^2-hybridized carbon atoms to bond to hydrogen and must therefore have similar inductive effects and electron densities at the protons. In acetylenes, the carbon atoms use sp-hybrids for the C—H bonds and are more electronegative than the sp^2-hybridized

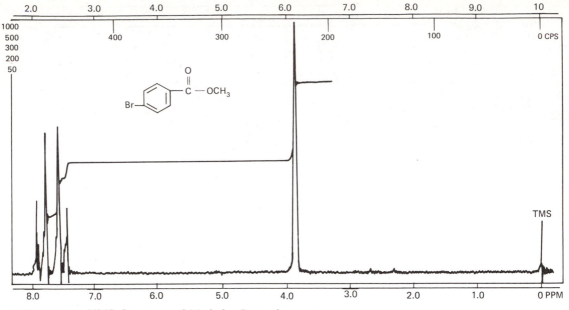

FIGURE 10.15 *NMR Spectrum of Methyl p-Bromobenzoate*

carbon atoms of aromatics or alkenes. Nevertheless, the protons of acetylenes occur upfield (more shielded) at $\delta = 2.4$–3.1. These anomalous shifts are attributed to magnetic anisotropy that arises whenever the electrons of a bond, under the influence of the applied magnetic field, can circulate with particular ease in one direction. In aromatic compounds this easy circulation occurs in the plane of the benzene ring (hence the name "ring current"); in acetylenes it occurs around the axis of the cylindrically symmetric triple bond. These extra currents act as little electromagnets, which produce extra magnetic fields that alter the magnetic field strengths at nearby protons. Figure 10.16 depicts the magnetic fields induced in aromatics and acetylenes and shows how the local fields at the hydrogen nuclei are altered.

If the NMR machine is operated at low radiation levels, the areas under the peaks are proportional to the numbers of protons giving rise to the absorptions. The machine can be operated in an integral mode such that the pen displacement on the chart is proportional to the accumulated area under each peak. Figure 10.17 reproduces the NMR absorption spectrum of 1,2,2-trichloropropane on which is superposed the integral for each peak. The integral heights correspond to the expected area ratios of $3:2$ with the methylene hydrogens appearing further downfield than the methyl hydrogens.

Exchange. Table 10.2 does not contain an entry for hydroxyl protons of carboxylic acids or alcohols because these chemical shifts are highly dependent on the solvent. One of the novel features of NMR spectroscopy is that when

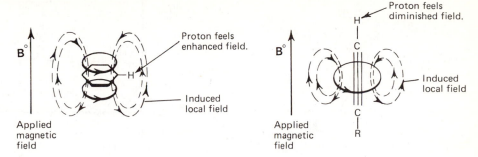

FIGURE 10.16
Alteration of Magnetic
Field at the Proton Due
to Induced Local Field

protons with different chemical shifts exchange positions, the NMR response depends on the rate of exchange relative to the time for one cycle of the applied radiation (at 100 mHz one cycle is 10^{-8} s). If exchange is rapid on this "NMR time scale," only a single sharp peak will be observed at an averaged position proportional to the concentrations of each kind of exchanging proton. If the exchange rate is similar to the NMR cycle time, the peaks of the exchanging protons are broadened and span the entire gap between the characteristic chemical shift positions. If that gap happens to be large, the absorption peaks can be so broad that they become lost in the background noise. The exchange phenomenon, one of the few directly observable consequences of the uncertainty principle in quantum mechanics, can be analyzed rigorously and used to

FIGURE 10.17 *NMR Spectrum of 1,2,2-Trichloropropane with Superposed Intensity*

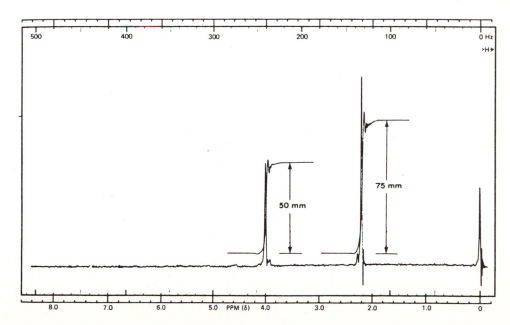

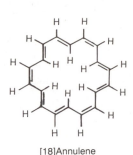

[18]Annulene

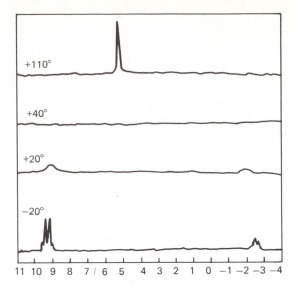

FIGURE 10.18
NMR Spectra of
[18]Annulene at
ferent Temperatures

calculate rates of rapid proton exchanges. In principle the same phenomenon applies to protons that exchange positions by rotation around bonds; if the rotation rate happens to approximate the NMR cycle time, broadening is observed. With some NMR machines, the temperature of the sample and consequently the rate of exchange can be controlled to make the broadening phenomenon appear or become more distinct.

A striking example of a temperature-dependent exchange process is shown by the NMR spectra of [18]annulene (see Figure 10.18). At −20° two sets of peaks are present centered at 8.94 (12 outer protons) and −2.50 δ (6 inner protons).[8] At a high temperature (+110°) the ring rapidly flips inside out and the entire set of 18 protons appears as a single sharp peak at 5.45 δ. At intermediate temperatures the peaks are broad; at 40° the peaks are so broad as to be indistinguishable from the baseline noise.

Spin-Spin Interactions. In the preceding paragraphs we have described how the application of a magnetic field to isolated protons causes their magnetic moments (spins) to align themselves either with the field (α spins) or against it (β spins). The difference in energy between the two states is proportional to the strength of the magnetic field.[5] If two protons are within a few angstroms of each other, their weak magnetic moments (spins) can interact to change the energies of the spin states, and this results in a change in the chemical shifts of the two protons. Since a sample of a substance contains many molecules (6×10^{23}/mole) and these can have all of the possible different combinations of proton spins, spin–spin interaction can lead to a complex pattern of overlapping

---

[8] The unusual chemical shift of the inner protons results from the large ring current acting in [18]annulene (see Figure 10.16 and related discussion).

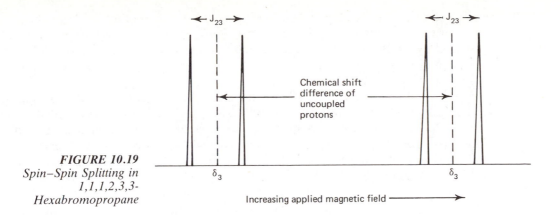

FIGURE 10.19
Spin–Spin Splitting in
1,1,1,2,3,3-
Hexabromopropane

absorption spectra. Fortunately, many of the widely occurring patterns are simple and can be readily used to help identify an unknown.

For example, 1,1,1,2,3,3-hexabromopropane ($Br_2CHCHBrCBr_3$) has two different kinds of protons and might have been expected to give rise to two peaks in the area ratio of 1 : 1. In fact, four peaks are observed (see Figure 10.19) as a consequence of the interaction between the spins and different ways the neighboring spins can be aligned. The pair of peaks associated with the proton on carbon-2 have an *average* value close to the chemical shift that would be expected in the absence of neighboring spins. The same is true for the pair of peaks associated with the proton of carbon-3. The difference in chemical shifts (splitting) between either pair of peaks is called the coupling constant, conventionally designated as J. The splitting is the same for both pairs of peaks since it arises from the same interaction. It is common practice to add subscripts to identify the protons that are coupled (e.g., J_{23} for the hexabromopropane).

The simple picture of spin–spin splitting applies as long as the coupling constant is small compared to the difference in chemical shifts between the centers of the doublets. When this condition is not met, additional quantum-mechanical effects enter; the result is that for two-proton systems the inner peaks in the spectrum become stronger at the expense of the outer peaks and the observed spacing between the peaks is no longer simply J. In the limit of no chemical shift difference (e.g., $Br_2CH—CHBr_2$), the outer peaks have zero intensity. This is why the protons of a CH_2 group generally show only a single sharp peak even though the protons are strongly coupled to each other. The important determinant is the ratio of the coupling constant to the chemical-shift difference. The spectra of molecules having small ratios of coupling constants to chemical-shift differences are referred to as first-order spectra; those displaying the additional quantum-mechanical effects are designated second-order spectra.

Molecules containing two magnetically nonequivalent protons are known as AB systems. Molecules with three nonequivalent protons are called ABC systems, if each proton has a different chemical shift. In ABC systems, three coupling constants J_{AB}, and J_{AC}, and J_{BC} must be considered. If all three are small compared to the chemical-shift differences between protons A, B, and C, a first-order spectrum results, consisting of 12 peaks. The A protons are split into a doublet by the B protons, and each member of this doublet is split into another doublet by coupling with the C protons. The same splitting pattern applies to the B and C proton absorptions.

If two of the three protons have the same chemical shift and are coupled identically to the third proton (e.g., $Br_2CH—CH_2Br$), the system is called an A_2B. Because J_{AA} is large compared to the zero difference in chemical shift of the two A protons, the two magnetically equivalent nuclei give no observable splitting of each other. If J_{AB} is large compared to the difference in chemical shifts of the A and B protons, the spectrum appears as a triplet for the B protons (total area of 1 unit with components in the ratio of $1:2:1$ separated by an amount J_{AB}) and a doublet for the A protons (total area of 2 units with two components in the ratio of $1:1$ separated by an amount J_{AB}).

For the general case of A_nB the B protons will occur as $n + 1$ peaks with a total area of 1 unit separated by J_{AB} Hz. The relative areas of the components within the multiplet are the coefficients of the different powers of x and y obtained by expanding the expression $(x + y)^n$. The B protons occur as a doublet of equal height with a total area of n units separated by J_{AB} Hz.

Although second-order spectra of ABC and larger spins are quite complex, the quantum-mechanical equations describing them can be solved exactly. If the chemical shifts and coupling constants are known or can be estimated, computer programs are available that will produce the spectrum for systems containing up to seven interacting spins. The reverse process of extracting chemical shifts and coupling constants from an experimental spectrum is much more tedious. Table 10.3 gives some characteristic coupling constants.

As an example of the use of spin–spin coupling in the interpretation of NMR spectra, consider the spectrum in Figure 10.20, obtained from a compound with an empirical formula $C_9H_{10}O$. Three groups of peaks are apparent. The relative peak areas, obtained by determining the spectrum in the "integral mode," are in the ratio of $5:2:3$. From the chemical shift data in Table 10.2 the complex group

TABLE 10.3
Spin–Spin Coupling Constants

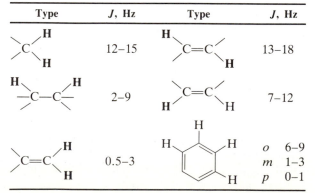

Type	J, Hz	Type	J, Hz
$\diagdown C \diagup \begin{smallmatrix}H\\ \\H\end{smallmatrix}$	12–15	$\diagup C {=} C \diagdown \begin{smallmatrix}H\\ \\H\end{smallmatrix}$	13–18
H—C—C—H	2–9	C=C (H,H)	7–12
C=C (H,H)	0.5–3	(benzene ring)	o 6–9 / m 1–3 / p 0–1

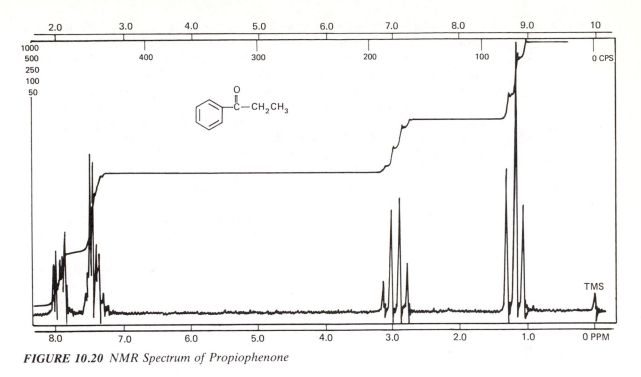

FIGURE 10.20 *NMR Spectrum of Propiophenone*

of peaks of area 5 centered near $\delta = 7.5$ ppm can be assigned to a phenyl group. The complexity of this set of peaks arises because it is a second-order A_2B_2C spin system. The area ratio of 2:3 for the remaining peaks suggests an ethyl group, which can be confirmed by noting that the absorption for the CH_2 group are split into a quartet as expected for a set of three equivalent adjacent hydrogens ($n + 1$ rule); the CH_3 absorptions are split into a triplet by the set of two equivalent adjacent hydrogens. The coupling constant is identical for both sets of peaks, as it must be if they are mutually coupled, and has a value close to 7 Hz as expected for hydrogens on carbons joined by a freely rotating single bond (see Table 10.3). The analysis thus far has identified the positions of all hydrogens and eight of the carbons. What is left is a carbon and an oxygen, which could be accounted for as a carbonyl situated between the phenyl ring and the ethyl group. The assigned structure would be propiophenone, which would account for the appearance of the CH_2 group downfield at about $\delta = 3.0$. Confirmation could be achieved chemically (ketone test), by IR (carbonyl stretch), or by ^{13}C NMR as described in the next section.

Other Nuclei. Nuclear magnetic resonance measurements have been most widely made for protons, but some other nuclei also have magnetic moments and NMR spectra. One of the more interesting for structure elucidation is ^{13}C, since, by measurement of NMR spectra of ^{13}C nuclei, it is possible to gain information directly about the carbon skeleton. Until recently a major limitation on ^{13}C spectroscopy was the low natural abundance of ^{13}C atoms, which is only about 1%. The dominant ^{12}C isotope does not have a magnetic moment and gives no NMR signal. Only with the development of advanced digital electronics did it become practical to measure the weak signals from ^{13}C nuclei.[9]

Happily, there is a benefit arising from the low natural abundance of ^{13}C nucleus; there is a low probability that a given ^{13}C nucleus will be adjacent to another ^{13}C nucleus, so that ^{13}C—^{13}C spin–spin coupling is unimportant and the spectra are much simpler than proton spectra. Spin–spin coupling does occur between ^{1}H and ^{13}C nuclei, which makes it possible to identify the number of protons on each carbon atom. By suitable control of the instrument's electronics the ^{1}H—^{13}C spin–spin coupling can be suppressed (to give only one peak for each kind of ^{13}C environment) or left on (to give a simple indication of the number of protons attached to a given carbon). The cost of such a sophisticated instrument is still quite high, but it has become low enough for routine research use.

The ^{13}C chemical shifts are quite large, roughly 20 times those found for the proton attached to the same carbon. Table 10.4 lists ^{13}C chemical shift ranges for

---

[9] Conventional NMR spectrometric measurements are done in a so-called continuous wave (CW) mode. To achieve higher sensitivity ^{13}C NMR uses a Fourier transform (FT) mode in which, in effect, the nuclei are being simultaneously irradiated by all of the frequencies instead of being probed one frequency at a time. To sort out the resulting information requires computer analysis before the spectrum can be plotted.

TABLE 10.4
^{13}C Chemical Shifts

Carbon class	Chemical shift range (TMS = 0), ppm
Alkane	
—CH$_3$	10–30
—CH$_2$	15–55
—CH	25–60
—C	30–40
Alkene	
=CH$_2$	100–120
=CHR and =CR$_2$	110–150
Alkynes	
—C≡C—H	60–70
—C≡C—R	70–90
Aromatic	
Hydrocarbons	120–150
Substituted	90–160
Carbonyl	
Aldehydes	190–220
Ketones	120–220
Carboxylic acids	165–190
Esters	160–180
Amides	150–175

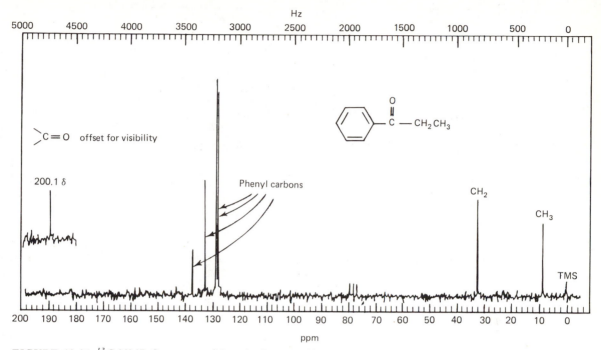

FIGURE 10.21 *^{13}C NMR Spectrum of Propiophenone, Noise Decoupled*

some of the more important classes of carbon atom found in organic molecules. Elaborate rules and correlation charts have been developed that allow prediction of ^{13}C shifts to within a few parts per million. Figure 10.21 is the ^{13}C NMR spectrum of propiophenone.

In Figure 10.21 the different kinds of carbon atoms present stand out clearly; the carbonyl group of propiophenone that had to be assigned indirectly from the ^{1}NMR spectrum appears unambiguously at $\delta = 200.1$ ppm.

Representative Applications of ^{1}H NMR in Structure Elucidation. The NMR spectra of hydrocarbons are quite distinctive. With saturated hydrocarbons, the primary protons absorb near $\delta = 1.0$ ppm (measured downfield from TMS), the secondary protons absorb near 1.25 ppm, and the tertiary protons near 1.5 ppm. Substantial shifts to a lower field (large δ) occur when electronegative groups are present or when the protons fall in the deshielding cone of a nearby aromatic ring. The NMR spectrum of ethylbenzene in Figure 10.22 shows CH$_3$ absorptions centered near $\delta = 1.25$ ppm and CH$_2$ absorptions centered near $\delta = 2.7$ ppm. The CH$_2$ protons are shifted downfield by the nearby aromatic ring. The spin–spin splitting pattern (the CH$_3$ protons are split into a triplet by the adjacent CH$_2$ group; the CH$_2$ protons are split into a quartet by the adjacent CH$_3$ group) confirms this assignment.

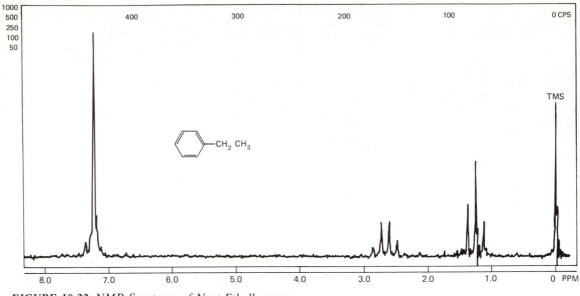

FIGURE 10.22 *NMR Spectrum of Neat Ethylbenzene*

Alkene protons absorb near $\delta = 5.0$ ppm. In Figure 10.23, the NMR spectrum of α-methylstyrene shows a group of peaks for each of the two alkene protons. The fine structure of these peaks comes from the coupling of the alkene protons to each other and to the methyl protons, which absorb at $\delta = 2.1$ ppm.

FIGURE 10.23 *NMR Spectrum of Neat α-Methylstyrene*

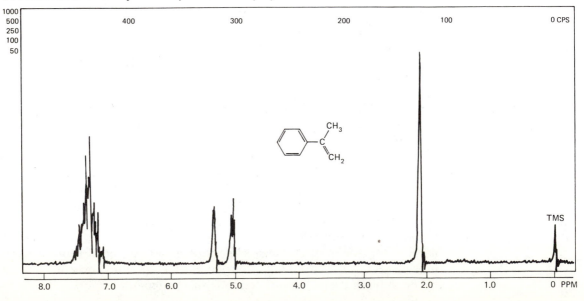

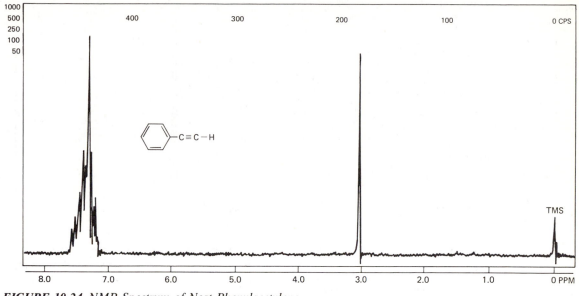

FIGURE 10.24 *NMR Spectrum of Neat Phenylacetylene*

Alkyne protons absorb in the range δ = 2–3 ppm. The NMR spectrum of phenylacetylene, shown in Figure 10.24, contains this distinctive absorption.

Aromatic protons typically absorb near δ = 7.5 ppm. The protons of an aromatic ring are strongly coupled to each other and will appear as a complex set of peaks when they have different chemical shifts (see Figures 10.23 and 10.24). When the aromatic protons have similar chemical shifts (i.e., a large ratio of J/δ), this complexity reduces to a single peak as in Figure 10.22.

In the NMR spectra the protons on carbon atoms adjacent to the carbonyl group of aldehydes and ketones are shifted downfield by about 1 ppm (see the spectrum of acetaldehyde, Figure 10.25). The aldehydic proton occurs at a very low field, near δ = 10 ppm; it is weakly coupled to the protons on the carbon adjacent to the carbonyl.

The position of a hydroxyl proton in an NMR spectrum depends critically on the degree of hydrogen bonding and hence on the choice of solvent. The common range is δ = 3.0–5.5 ppm. If a trace of acid is present, the hydroxyl protons exchange rapidly and appear as a single sharp peak, even though they may be spin–spin coupled to other protons in the molecule. Assignment of an NMR peak to OH can be confirmed by addition of trifluoroacetic acid, which absorbs beyond δ = 10 ppm; by rapid exchange, the OH absorption either disappears or is displaced markedly, whereas the other peaks do not shift significantly.

If there are protons attached to the carbon atom bearing the OH group, they will be shifted downfield by about 2–3 ppm. The NMR spectrum of ethanol containing a trace of acid to sharpen the OH peaks shows these features (see

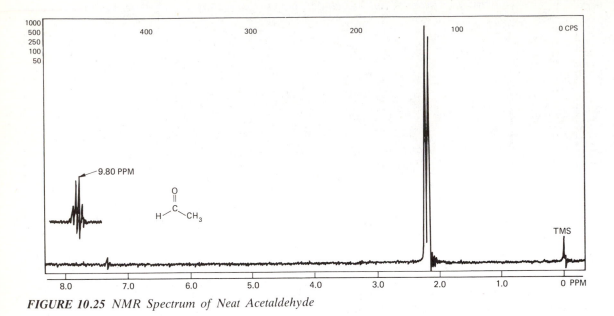

FIGURE 10.25 *NMR Spectrum of Neat Acetaldehyde*

FIGURE 10.26 *NMR Spectrum of Neat Ethanol*

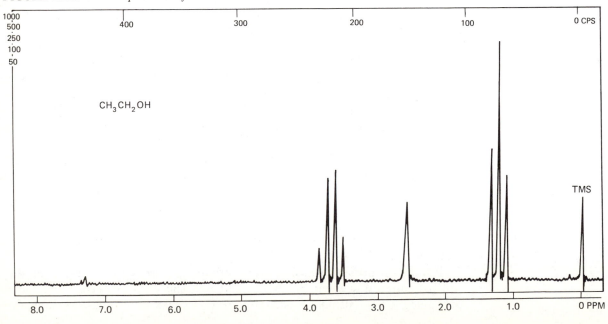

Figure 10.26). This spectrum shows the quartet and triplet peaks characteristic of an ethyl group (compare to Figures 10.20 and 10.22).

NMR Sampling Techniques. Proton NMR spectra are normally determined on 10–20% solutions. After solubility, the most important factor in solvent choice is the presence of interfering solvent absorptions. When it will dissolve the sample, the best solvent is carbon tetrachloride. Another widely used solvent is deuterated chloroform, $CDCl_3$. The deuterium nucleus gives NMR peaks but not in the proton region; most $CDCl_3$ solvent batches contain some $CHCl_3$, which gives a weak peak at $\delta = 7.27$. Because of the possibility for misalignment of the chart paper, it is essential that the NMR sample contain a reference such as 1% TMS. This can be added when the sample is prepared but it is more convenient to use solvents that have been spiked with TMS. If TMS must be added remember that it is extremely volatile (bp 27°) and will evaporate rapidly.

About 0.5 mL of solution is pipetted into a 5-mm NMR tube (TMS is added with a microdropper if it is not already present in the solvent) and the tube capped immediately to prevent evaporation. In drawing the sample into the pipet in preparation for adding it to the NMR tube, it is good practice to twist a small wad of clean surgical cotton around the pipet tip. The cotton filter removes solid impurities from the sample as the pipet is filled; the cotton filter is discarded before the sample is transferred to the NMR tube.

In order to improve the homogeneity of the magnetic field felt by the sample in the NMR machine, the tube is spun about its long axis. This is accomplished by slipping the tube into a tightly fitting plastic "spinner," which spins the tube when the instrument directs a stream of air against it. The positioning of the spinner on the tube is critical, both for even spinning and for proper location of the sample within the NMR instrument. Most installations provide some kind of gauge for proper positioning. One word of caution here: NMR tubes are fragile and the tube must be aligned perfectly as it is placed in the instrument.

The chances are that your laboratory instructor will run your sample, so operation of the instrument will not be described. The principal steps are to fine-tune the machine to give the best-resolved spectrum and adjust the instrument setting to align the TMS peak properly on the chart paper.

10.4 Ultraviolet and Visible Spectroscopy

The electrons of a molecule can be excited to higher energy levels by application of electromagnetic radiation with a frequency corresponding to the energy gap between the ground state and the excited state. Although the processes are identical, a distinction is made for practical reasons between visible spectroscopy (440–800 nm) and ultraviolet spectroscopy (400 nm and below). Ultraviolet spectroscopy and visible spectroscopy are displayed as plots of *absorbance* versus wavelength, where absorbance is defined as

$$\text{Absorbance} = \log_{10} \frac{\text{light intensity entering sample}}{\text{light intensity leaving sample}}$$

It can be shown that, in addition to depending on the electronic structure of the sample, the absorbance is proportional to the concentration of the sample and to the pathlength of the sample through which the light travels (*Beer–Lambert law*). A quantity commonly used to characterize electronic spectra is the *molar extinction coefficient*, ε, defined as

$$\varepsilon = \frac{\text{absorbance}}{\text{concentration (moles/L)} = \text{pathlength (cm)}}$$

The ultraviolet spectrum of methyl benzoate measured in a 1-cm cell is reproduced in Figure 10.27. The value of ε at the absorption maximum at 230 is

FIGURE 10.27
Ultraviolet Absorption Spectrum of Methyl Benzoate

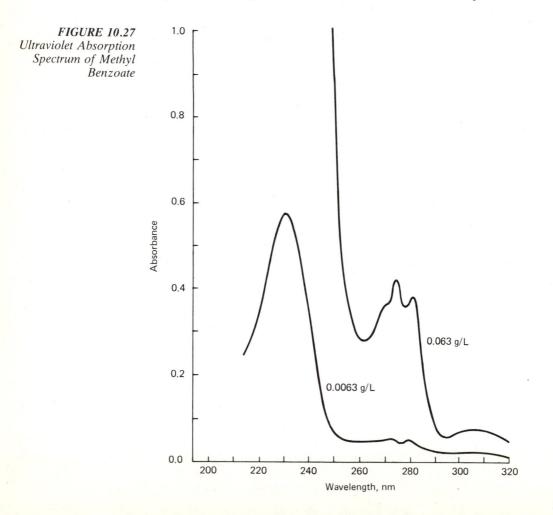

TABLE 10.5
Ultraviolet Absorption
Spectra

Structural unit	Wavelength at maximum absorption, nm	Typical ε
Isolated double bond	180–195	10,000
Isolated carbonyl group	270–290	25
Conjugated diene (*cis*)	near 240	5,000
Conjugated diene (*trans*)	near 220	16,000
Alkylbenzenes	260–280	200
Naphthalene	and near 210	8,000
	314	315
	275	5,625
	220	112,200

12,000; the ε of the smaller peak at 275 nm is 850. Table 10.5 lists several structural units that have characteristic absorptions.

As Table 10.5 suggests, UV–visible absorption is primarily used to detect and identify conjugated π-electron networks. As a general rule, the longer the conjugated chain the longer the wavelength of the absorption maximum. For example, an isolated double bond absorbs at 190 nm, a diene at 220 nm, and the chain of 11 double bonds in β-carotene all the way out in the visible region at 484 nm (in hexane solvent).

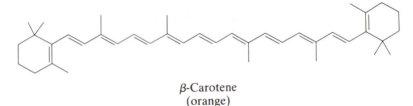

β-Carotene
(orange)

Sometimes the color of the absorbed light is confused with the color of transmitted light. Remember that what makes it through the sample is the light that did *not* get absorbed. The situation is summarized by the data in Table 10.6.

TABLE 10.6
Correlation of
Absorbed and
Transmitted Light

Wavelength, nm	Color absorbed	Color transmitted
Below 400	Ultraviolet	Colorless
400	Violet	Yellow
470	Blue	Orange
500	Blue-green	Red
530	Green	Purple
560	Yellow	Violet
600	Orange	Blue
680	Red	Blue-green
Beyond 800	Infrared	Colorless

Ultraviolet and Visible Spectral Sampling Techniques. Unlike infrared spectroscopy, it is customary in ultraviolet and visible spectroscopy to measure the extinction coefficients of the absorption peaks as well as their positions. Dilute solutions are required, typically 10^{-4} mole/L, which must be prepared by quantitative dilutions using appropriate combinations of transfer pipets and volumetric flasks. Because of the sensitivity of the measurement, considerable care must be exercised to avoid contamination. The most common solvents employed are water, ethanol, acetonitrile, and cyclohexane.

Questions

1. What molecular weights of the parent ion would be observed in a high-resolution mass spectrum for the following molecules?
   (a) C_5H_{12} (b) C_4H_8O (c) $C_3H_8N_2$

2. High-resolution mass-spectrometric analysis of compound A gave a molecular formula $C_9H_{10}O_2$. The infrared spectrum showed strong absorption at 1745 cm^{-1} as well as many other medium-intensity bands. The ultraviolet spectrum showed a maximum at 257 nm, $\varepsilon = 195$. The NMR spectrum consisted of three sharp peaks at $\delta = 7.22$ ppm (area 5), $\delta = 5.00$ ppm (area 2), and $\delta = 1.96$ ppm (area 3). What is the structure of compound A?

3. From a high-resolution mass spectrum of compound B, a molecular formula of $C_6H_{14}O$ could be assigned. The molecule did not absorb in the ultraviolet or visible regions. In the infrared, the strongest absorption above 1400 cm^{-1} occurred at 2900 cm^{-1}. In the NMR, compound B showed a septet at $\delta = 3.62$ ppm (area 1), $J = 7$ Hz, and a doublet at $\delta = 1.10$ (area 6), $J = 7$ Hz. What is the structure of compound B?

11

Isolation and Identification of a Natural Product

The roots of organic chemistry are embedded in the substances derived from natural sources. It was the eventual realization that these "natural products" could also be synthesized from nonliving materials that led to the overthrow of the vital force theory and the creation of the discipline we still call "organic" chemistry.[1]

The experiment described in this chapter is representative of the isolation and characterization approach that was used then and to this day, particularly now in the search for new medicinals. You will be asked to isolate the volatile components of the skin of an orange, which, as it happens, is mostly a single substance. You will then identify this natural product. The experiment is artificial in the sense that instead of an unlimited number of possibilities you will be given a short list of ten candidates based on the known molecular formula of the natural product. It will only take a little imagination to picture yourself isolating a marvelous unknown substance. It is, after all, unknown to you.

11.1 Isolation of Orange Oil

Assemble a steam distillation apparatus as shown in Figure 4.3 using a 500-mL round-bottomed flask as the boiler, allowing enough space for the flask to be heated by a Bunsen burner. The bottom of the flask should rest on a wire gauze supported by an iron ring.

---

[1] The term *organic chemistry* was apparently introduced in 1807 by Berzelius, a firm believer in the vital force theory.

221

||||➤ *CAUTION*   The use of a burner in an organic laboratory always presents a fire hazard. Before the burner is lighted, check to see that no combustible organic solvents are nearby.

Attach a Claisen adapter and on the side arm of the adapter place a distillation head bearing a thermometer; stopper the central arm of the adapter.

Cut or tear the peel of one orange into thin strips, determine the weight to the nearest 0.1 g, and place the material in a blender. Add about 250 mL of water and blend for 15–30 seconds. Transfer the resulting mash to the boiler flask. This is conveniently done by temporarily removing the distillation adapter and slowly pouring the mash through a wide-mouth funnel into the side arm of the Claisen adapter. Reassemble the distillation apparatus and heat as rapidly as possible *without charring the contents or having the foam reach the condenser*. Collect about 100 mL of distillate in a 125-mL flask. Chill the distillate in an ice bath and then extract three times with 10-mL portions of methylene chloride.

||||➤ *CAUTION*   Prolonged exposure to high concentrations of methylene chloride vapors may induce cancer. As with all organic solvents, work in a well-ventilated area when using it.

Dry the extracts with anhydrous sodium sulfate, filter through a filter tube containing a wad of glass wool into a preweighed round-bottomed flask, and remove the methylene chloride by simple distillation. For the identification to be described in the next section it is important that all of the solvent be removed. From the weight of the residue calculate the yield of steam-volatile products.

11.2 Identification of Limonene

The major component of orange oil, a hydrocarbon analyzing for 88.2% carbon and 11.8% hydrogen, is one of those listed Table 11.1. Save a few drops of this product in a tightly closed vial for gas-chromatographic analysis, to be done later in the semester.

(A) Boiling Point Determination

Determine the boiling point of your sample using the micro method described on page 136. You may wish to practice on a compound of known boiling point before you tackle your orange-oil sample. Your boiling point should eliminate more than half of the compounds in the list. A further narrowing of the possibilities can be obtained by determining quantitatively the bromine absorption by a weighed sample of your orange oil.

(B) Quantitative Bromination of Alkenes

The number of double bonds in limonene (and alkenes, in general) can be determined by quantitatively measuring the amount of bromine taken up by the

Hydrocarbon	Structure	Bp, °C
3,7,7-Trimethylbicyclo[4.1.0]-3-heptene	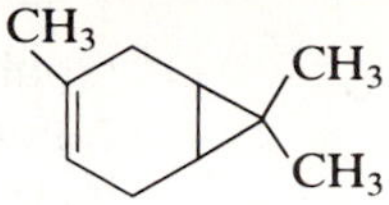	172
3,7,7-Trimethylbicyclo[4.1.0]-2-heptane	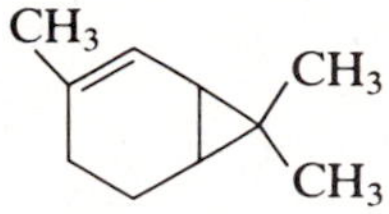	167
4-Isopropenyl-1-methylcyclohexene	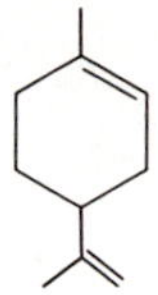	178
4-Isopropylidene-1-methylcyclohexene	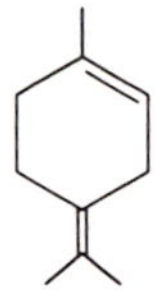	185
(3Z)-3,7-Dimethyl-1,3,6-octatriene		176–178
7,7-Dimethyl-2-methylenebicyclo[2.2.1]heptane		157–159
6,6-Dimethyl-2-methylenebicyclo[3.1.1]heptane	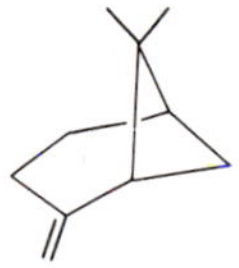	164–166
5-Isopropyl-2-methyl-1,3-cyclohexadiene	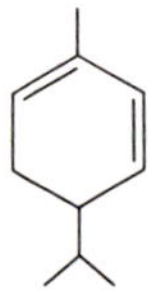	175–176
1-Isopropyl-4-methylenebicyclo[3.1.0]hexane	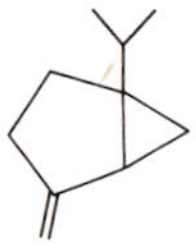	163–165
1-Decen-4-yne		173–174

TABLE 11.1

223

double bonds. Although limonene can be titrated directly with a bromine solution using the persistence of the bromine color as an end-point indicator, it is better to use the indirect titration described below that is both easier to see and much less hazardous.

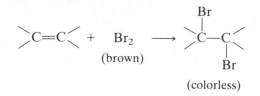

An excess of bromine and limonene are allowed to react; then potassium iodide is added to convert most of the unreacted bromine to iodine and bromide ion. Finally, the mixture of mostly iodine and some bromine is titrated with sodium thiosulfate solution, using the loss of the brown bromine and purple iodine colors as the end-point indicator. The color change at the end point can be enhanced by addition of a starch solution (starch forms a blue-black complex with iodine), but it is not enough of an improvement to be worth the trouble.

Bromine is a hazardous substance; the hazard can be reduced sharply by generating the required bromine solution in the reaction flask by mixing a bromate–bromide solution with dilute acid.

$$5 \, Br^- + BrO_3^- + 6 \, H^+ \longrightarrow 3 \, Br_2 + 3 \, H_2O$$

For example, a solution that is $0.40 \, M$ in potassium bromide and $0.08 \, M$ in potassium bromate, on addition of dilute sulfuric acid, will generate bromine equivalent to having used a $0.24 \, M$ bromine solution. After the reaction with the limonene is finished, sufficient potassium iodide solution is added to convert the unreacted bromine into a mixture of bromine and iodine.

The stoichiometry of the oxidation of thiosulfate ion by the mixed bromine–iodine solution is not clean, with some of it going to the tetrathionate ion, $(S_4O_6^{2-})$, and some going to sulfate (SO_4^{2-}). This ambiguity is sidestepped by standardizing the thiosulfate against a known amount of bromine–iodine solution.

In a stoppered 25-mL Erlenmeyer flask (or better, a 10-mL volumetric flask) weigh out 0.2 g of your orange oil and add 10.0 mL of methylene chloride. Transfer 2.0 mL of the orange-oil solution to a 125-mL Erlenmeyer flask and add 5 mL of an aqueous bromide–bromate solution[2] followed by 5 mL of 10% sulfuric acid. Swirl the flask gently for 5 min to mix the two-phase mixture in order to allow the bromine and limonene solution to react.

---

[2] The bromide–bromate solution is $0.40 \, M$ in potassium bromide and $0.08 \, M$ in potassium bromate.

⫸ *CAUTION*   Pure bromine causes severe, slowly healing burns if it touches the skin. Disposable polyethylene gloves should be worn during the bromination and titration. If any bromine solution is spilled on the skin, wash the area immediately with a large amount of water.

At the end of the 5-min reaction period add 10 mL of 10% potassium iodide solution and, without delay, titrate the mixture with 0.3 *M* sodium thiosulfate solution.[3] The mixture being titrated has two phases; the lower methylene chloride layer contains most of the halogens and the upper aqueous layer contains all of the sodium thiosulfate titrant. The dark purple-brown color will fade as the end point is approached. You should titrate slowly near the end with constant swirling to allow the two phases time to mix and react.

Repeat the titration two more times with fresh 2-mL portions of the orange-oil solution.

Standardize the thiosulfate solution against the brominating solution by repeating in duplicate or triplicate the above procedure, but without the addition of the orange-oil sample.

From the difference between the titer for 5 mL of brominating solution with and without added limonene, calculate the amount of bromine consumed. From this amount and the weight of limonene added, calculate the number of double bonds in limonene.

Sample Calculation for Bromination of Cyclohexene

1. Standardization. With no cyclohexene added, 5.0 mL of the brominating solution (equivalent to 0.24 *M* in bromine) consumed 5.95 mL of thiosulfate solution. The effective normality of the thiosulfate solution was 5.0 mL × 0.24 *M*/5.95 mL = 0.202 *N*.

2. Cyclohexene Bromination. A solution containing 0.20 g of cyclohexene in 10.0 mL of methylene chloride was prepared and 2.0 mL of this solution was brominated with 5.0 mL of the brominating solution. On titration, 3.50 mL of the above sodium thiosulfate solution was needed to reach the end point. The amount of bromine consumed by the added cyclohexene was

$$(5.95 - 3.50) \times 0.202 = 0.495 \text{ meq}$$

The number of moles of cyclohexene (MW = 82) added was

$$(2/10) \times 0.20/82 = 0.488 \text{ meq}$$

The number of double bonds in cyclohexene is thus measured to be

$$0.495/0.488 = 1.02 \text{ double bonds}$$

---

[3] The sodium thiosulfate solution is prepared to be 0.3 *M*. The effective normality as a reducing agent for the bromine–iodine mixture must be measured by repeating the analysis *without* adding the limonene.

(C) Spectroscopic Identification

The combined boiling-point and unsaturation-number data are sufficient to pick out the structure of the hydrocarbon from Table 11.1. However, you could independently identify the compound or at least confirm your assignment by measuring the NMR spectrum. Determine the NMR absorption spectrum of a solution of the hydrocarbon in carbon tetrachloride and compare it to the spectra expected for the hydrocarbons in Table 11.1.

Questions

1. In order for a compound to be steam distilled it must possess two physical properties. What are they?

2. The vapor pressures of the hydrocarbons in Table 11.1 are about 140 mm at 100°. Calculate the ratio of water to hydrocarbon that would be collected if the steam distillation process were 100% efficient. Why might the efficiency of this steam distillation be lower than normal?

3. Write a balanced equation for the bromination of an alkene. Calculate the volume of 0.24 M bromine solution that would react with 0.04 g of hydrocarbon having three double bonds and a molecular weight of 136.

4. Which of the hydrocarbons in Table 11.1 obey the isoprene rule (i.e., can be thought of as *formal* combinations of two isoprene skeletal units)?

III

Preparations and Reactions of Typical Organic Compounds

12 General Remarks

This chapter describes how to calculate the percentage yield of a reaction, how to prepare your notebook in advance of the laboratory, the information to be written in the notebook during the laboratory, and how to submit samples of products. The procedures described here follow closely the current practice at Cornell and represent our view of how to maximize laboratory learning experience. However, since "there are many roads to Rome," your instructor may well modify some of the procedures.

The preparation of typical organic compounds affords a marvelous opportunity to compare your theoretical understanding of organic chemistry obtained from lectures and reading with reality. It is hard at first to accept that those abstract symbols and formulas represent real substances—glistening solids and odoriferous liquids. Also keep in mind that although you are working with particular compounds, the preparations and reactions usually represent general methods that can be applied to entire classes of molecules.

12.1 Preparation Before the Laboratory

For experiments dealing with syntheses and reactions, the advanced preparation of your notebook involves several steps in addition to those listed in Section 1.6 for the exercises on separation and purification of organic compounds. You should review that material now.

The most important addition is a table of physical constants of the substances to be manipulated (see step 5 below). By collecting this information and having it available, you will be able to understand more readily the reasons for the particular procedure and will often be able to overcome independently any small difficulties that may arise in the course of the laboratory work. You should proceed in the following manner.

1. Read the descriptive pages concerning the laboratory operation to be carried out (these are found immediately preceding each experiment). In the notebook, write a title and general statement of the process to be studied.

2. Read the laboratory directions for the entire procedure and note particularly any cautions for handling materials.

   To aid in understanding the reasons for the procedure it is helpful to consult the textbook or lecture notes for a discussion of the particular class of compounds that is to be studied. Consideration should be given to important general principles, such as the law of mass action and the influence of solvents and catalysts on rate of reaction.

3. In your notebook, give a concise statement of the type of reaction that is to be carried out, such as "Conversion of an alcohol to an alkene" or "Oxidation of a secondary alcohol to a ketone." Write balanced equations, using condensed structural formulas, for the main reaction or sequence of reactions involved in converting the starting materials to the final products. Along the reaction arrow indicate the conditions used— temperature, solvent, catalyst (if any), and so forth.

4. Write balanced equations for significant side reactions that may divert an appreciable amount of the starting materials and lead to formulation of by-products that must be removed in the purification of the main product.

   Write balanced equations for any test reactions that are used to test for completion of the reaction, to detect the presence of an impurity, to confirm the identity of the product by conversion to a derivative, and so on.

5. Prepare a table of the physical constants of all organic and inorganic substances that enter into the main and side reactions and are produced in these reactions. The form shown in the accompanying Sample Notebook Page may be used. The physical constants of common organic and inorganic compounds may be found in chemical handbooks.

   Include in the table the weight (in grams) of each reactant and the number of moles, or fraction of a mole, actually used. From these data and the balanced equation for the main reaction, determine which starting material is the limiting factor and calculate the theoretical yield (in grams) based on this reactant. This calculation is described in detail in Section 12.2.

   For each of the reagents used and each of the products produced note any hazard that the material presents. Your instructor may provide this material or ask that you look it up in a book such as the *The Merck Index* (Merck & Co., Inc., Rahway, NJ). The point here is both to

<u>*Experiment 13.3 (A) n-Butyl Bromide*</u>

Conversion of a Primary Alcohol to an Alkyl Bromide

<u>*Main Reactions:*</u>

$$CH_3CH_2CH_2CH_2-OH + HBr \xrightarrow[reflux]{H+} CH_3CH_2CH_2CH_2-Br + H_2O$$

$$NaBr + H_2SO_4 \longrightarrow HBr + NaHSO_4$$

<u>*Side Reactions:*</u>

No important organic side reactions

$$2NaBr + 3H_2SO_4 \text{ (concd)} \longrightarrow Br_2 + SO_2 + 2H_2O + 2NaHSO_4$$

<u>*Test Reactions:*</u> None

<u>*Purification:*</u>

$$\left.\begin{array}{l} C_4H_9-Br, C_4H_9-OH \\ NaHSO_4, H_2SO_4 \\ (Br_2), H_2O \end{array}\right\} \xrightarrow{\text{steam distillation}} \left.\begin{array}{l} C_4H_9-Br \\ C_4H_9-OH \\ (Br_2), H_2O \end{array}\right\} \xrightarrow{80\% H_2SO_4} \left.\begin{array}{l} C_4H_9-Br \\ (H_2O) \\ (H_2SO_4) \end{array}\right\}$$

$$\xrightarrow[\text{(2) } H_2O]{\text{(1) } NaHCO_3} \left.\begin{array}{l} C_4H_9-Br \\ (H_2O) \end{array}\right\} \xrightarrow{CaCl_2} C_4H_9-Br \text{ purified by distillation}$$

<u>*Physical Constants:*</u>

Substance	Mol wt.	Grams used	Moles used	Sp gr. (20°)	Mp	Bp	Solubility (g/100 mL)		
							Water	Ethanol	Ether
$n\text{-}C_4H_9-OH$	74	3.7	0.050	0.810	−80°	117°	9	∞	∞
H_2SO_4	98	7mL	0.13	1.83	11°	d340°	∞	reacts	sol.
$NaBr \cdot 2H_2O$	139	6.2	0.068	—	51°	—	80 cold insol.	slight	insol.
$n\text{-}C_4H_9-Br$	137	6.9	0.050	1.277	−112°	102°	insol.	∞	∞

<u>*Quantities Used:*</u>

$NaBr$ $\dfrac{6.2}{104} = 0.060$ *mole.*

$H_2SO_4 (96\%)$ $\dfrac{7mL \times 1.9g/mL}{98} = 0.13$ *mole*

$n\text{-}C_4H_9-OH$ $\dfrac{3.7}{74} = 0.050$ *mole = limiting factor*

<u>*Theoretical Yield:*</u> based on n-butyl alcohol (0.5 mole) = 0.5 × 137g = 68.5 g n-butyl bromide

<u>*19 March 1987:*</u> Reaction mixture refluxed 2 hr; organic layer had slight brown color. After distillation the organic layer in distillate was colorless. Product was washed and allowed to stand over CaCl₂ in corked flask.

<u>*21 March 1987:*</u> Product filtered into dry 100-mL distillation apparatus and distilled: atmospheric pressure = 745 mm

   Tare of receivers A = 39.5g B = 42.0g.

   Fractions collected: A, up to 99°; B, from 99 to 102°.

   Receiver A, 39.7 g; net weight (39.7 − 39.5) = 0.2 g.

   Receiver B, 46.8 g; net weight (46.8 − 42.0) = 4.8 g

<u>*Percent Yield:*</u> $\dfrac{4.8}{6.9} \times 100\% = 69.6\%$ of theoretical

<u>*Other Methods of Preparation:*</u>

$$3n\text{-}C_4H_9-OH + PBr_3 \longrightarrow 3n\text{-}C_4H_9-Br + H_3PO_3$$

provide the knowledge needed to perform the experiments in a prudent manner and to emphasize that one may not work with organic materials in an innocent, mindless manner.

6. It is helpful also to indicate schematically the successive steps involved in the purification procedure, starting with the substances present in significant amounts in the reaction mixture at the completion of the reaction and showing what substances are removed at each step. The solubility relationships discussed in Sections 6.4 and 9.6 and the table of physical constants (see step 5) are useful in this connection.

7. After you have prepared your notebook in accordance with the foregoing instructions, your instructor may ask to see it for preliminary approval *before you start to perform the experiment*. In certain experiments, you will be asked to arrange the apparatus for the experiment and have it approved by the instructor.

8. In laboratory work it is essential to make efficient use of the time assigned; you are expected to plan your laboratory schedule and to make preliminary preparations before coming to the laboratory.

Since many experiments require that the reactants be refluxed for several hours, you should plan to perform other laboratory work while the operation of refluxing is being carried out.

12.2 Laboratory Directions

The laboratory directions given in this manual are deliberately detailed. In advanced work and research one frequently must follow directions that assume a general knowledge of manipulative technique and of the chemistry involved. As an example, consider the preparation of *n*-butyl bromide described in detail in Section 14.3 (A). Were this preparation to appear in a technical journal, the description might be as follows.

A mixture of 60 mL of water, 130 g (1.3 mole) of concentrated H_2SO_4 (cool), 37 g (0.5 mole) of *n*-butyl alcohol, and 87 g (0.65 mole) of NaBr was refluxed for 2 hr, and then distilled until no more product was collected. The crude distillate was washed with water, cold 80% H_2SO_4 saturated $NaHCO_3$, and finally with water. After drying over $CaCl_2$, the product was distilled; yield 45 g, bp 99–103°.

The journal directions assume the worker realizes that the sodium bromide should be pulverized before addition and that good mixing is essential. Typically, experimental sections of professional journals do not specify the amounts of washing reagents or drying agents. In fact, many journal authors might have condensed the experimental description to read

n-Butyl alcohol (0.5 mole, 37 g) was converted to the bromide by heating with NaBr (0.65 mole) and 70% H_2SO_4 (190 g). The crude bromide was washed, dried, and redistilled; yield 45 g, bp 99–103°.

The reader would have to understand the chemistry well enough to isolate the bromide by distillation from the reaction mixture as well as to use the washes specified in the complete directions in the given order. The ability to fill in experimental detail is an important part of being a good laboratory worker and you are advised, as you carry out the experiments in this manual, to ask yourself at each stage why a certain operation or reagent is used.

12.3 In the Laboratory

Carry out the experiment according to the laboratory directions, and promptly record your observations, in ink, directly in your notebook. Record the quantities of reagents and solvents actually used and the lengths of time taken for operations such as addition and distillation. As you work, compare what you see happening with what you anticipated; record all discrepancies. Always observe and record the boiling point of a liquid preparation, and the melting point of a solid organic preparation, unless instructed not to for reasons of safety. Your instructor may also require that you obtain some spectroscopic proof of purity of your product such as an IR or NMR spectrum.

Record the actual yield and calculate the percentage yield as described in detail in the next section. Record any general observations and conclusions drawn from your experiment.

12.4 Calculation of Yields

The yield (sometimes called the actual yield) is the amount of the purified product actually obtained in the experiment. The theoretical yield (sometimes called the calculated yield) is the amount that could be obtained under theoretically ideal conditions; that is, the main reaction is assumed to proceed to completion without side reactions or mechanical losses, so that the starting materials are entirely converted into the desired product and no material is lost in isolation and purification.

The percentage yield (also called the percent yield) is obtained by comparing the actual yield with the theoretical yield, in the following manner.

$$\text{Percent yield} = \frac{\text{actual yield}}{\text{theoretical yield}} \times 100\%$$

The percentage yield is the measure of the overall efficiency of the preparation since many factors, such as incomplete reactions, side reactions, and mechanical losses, affect the actual yield.

If a preparation involves two reacting substances and the amounts actually used are not in the exact proportions demanded by the equation, it is necessary to determine by calculation which of the reactants is the limiting factor (commonly called the *limiting reagent*) for the calculation of the theoretical yield, as shown by the following example. In this connection the terms *mole* and *moles used* are commonly employed. A *mole* of a compound is equal to the molecular weight expressed in grams. The term *moles used* is employed to express the number of moles or the fraction of a mole of a particular compound actually used in an experiment. The number of moles is equal to the weight of the substance divided by the molecular weight.

Theoretical and Percentage Yields. Suppose that methyl ethyl ether, CH_3—O—C_2H_5, has been prepared by the action of methyl iodide upon sodium ethoxide (the Williamson synthesis of ethers). This preparation is carried out by reacting metallic sodium with ethanol and treating the resulting sodium ethoxide with methyl iodide.

$$C_2H_5\text{—OH} + Na \longrightarrow C_2H_5\text{—ONa} + \tfrac{1}{2} H_2 \qquad (12.1)$$

$$C_2H_5\text{—ONa} + CH_3\text{—I} \longrightarrow C_2H_5\text{—O—}CH_3 + NaI \qquad (12.2)$$

Equations 12.1 and 12.2 may be summarized as follows.

$$C_2H_5\text{—OH} + \quad Na \quad + CH_3\text{—I} \longrightarrow$$
$$\text{1 mole} \qquad \text{1 mole} \qquad \text{1 mole}$$

$$C_2H_5\text{—O—}CH_3 + \quad NaI \quad + \tfrac{1}{2} H_2 \qquad (12.3)$$
$$\text{1 mole} \qquad \text{1 mole} \qquad \tfrac{1}{2}\text{ mole}$$

From this it is evident that the proportions demanded by the equation are 1 mole of C_2H_5—OH : 1 mole of Na : 1 mole of CH_3—I, and the resulting products would be 1 mole of C_2H_5—O—CH_3 : 1 mole of NaI : $\tfrac{1}{2}$ mole of H_2.

Suppose that the following quantities of the reagents are actually used in a laboratory preparation.[1]

$$9.20 \text{ g of absolute ethanol} = \frac{9.20}{46.07} \text{ mole} = 0.200 \text{ mole } C_2H_5\text{—OH}$$

$$0.55 \text{ g of metallic sodium} = \frac{0.55}{22.99} \text{ mole} = 0.0240 \text{ mole Na}$$

$$2.84 \text{ g of methyl iodide} = \frac{2.84}{141.94} \text{ mole} = 0.200 \text{ mole } C_2H_5\text{–OH}$$

---

[1] The number of significant figures shown in the example follow the rules suggested by L. M. Schwartz, *J. Chem. Educ.*, **62**, 693 (1985). The main idea is that the last significant figure shown in the final answer should reflect the uncertainties in the data leading to the answer.

The amounts of the reagents are converted from grams into moles (by dividing by the molecular weights) to compare them with the molar proportions expressed in the equation. The *relative* proportions actually used are thus 10.0 of C_2H_5—OH : 1.20 of Na : 1.00 of CH_3—I. Obviously, ethanol and sodium are used in excess and methyl iodide is the limiting reagent that will determine the theoretical yield.

From equation 12.2 or 12.3, it can be seen that 1 mole of methyl iodide reacting with a sufficient quantity of sodium ethoxide will produce, under ideal conditions, exactly 1 mole of methyl ethyl ether. Consequently, the maximum amount of methyl ethyl ether that could be produced in the above preparation (the theoretical yield) is 0.0200 mole. By multiplying this fraction of a mole by the weight of 1 mole of methyl ethyl ether (60.10 g), the theoretical yield is converted into grams.

$$\text{Theoretical yield} = 0.0200 \times 60.10 \text{ g} = 1.20 \text{ g}$$

If the actual yield of methyl ethyl ether in the above preparation was 0.82 g, the percentage yield would be

$$\text{Percentage yield} = \frac{0.82}{1.20} \times 100 = 68\%$$

Problem. In another preparation of methyl ethyl ether, suppose that the amounts of the reagents actually used were 0.69 g of metallic sodium, 4.60 g of absolute ethanol, and 4.97 g of methyl iodide. Calculate the theoretical yield, in grams. (*Answer*: 1.80 g.)

12.5 Samples and Reports

At the conclusion of the experiment you will be required to submit the final substance prepared (along with samples of intermediates if a sequence of reactions was involved). Each preparation should be placed in an appropriate bottle (wide-mouth bottles for solids and narrow-mouth bottles for liquids) of suitable size, with the experiment number and the name of product, the melting or boiling point, as actually observed, your name, and the actual yield (in grams) and tare of the bottle included on the label, as follows.

Expt 17(A)	Cyclohexene
bp 80–85°	
Mike Clements	
Yield, 1.2 g	Tare, 21.5 g

Some form of report will also be required. At a minimum this could be your notebook[2] with a final report page summarizing the observed and physical properties of the substance and giving the percentage yield. On the other hand, you could be asked to write a formal report that gives all of these data and discusses the chemistry involved. An outline of how to prepare a report is given in Appendix E.

Your instructor will examine your notebook (or report) and your product and may ask questions designed to test your knowledge of the fundamental principles involved in the experiment and your ability to make and apply generalizations of the chemistry. Your grade will depend on the quality and quantity of your product, your laboratory technique, and your notebook and report, as well as your understanding of the chemistry.

---

[2] It is very convenient to use notebooks with tear-out carbon copies. The copies can be submitted after each period, which simplifies notebook checking by the instructor and leaves the notebook in your possession for preparation of the next period's work.

13 Free-Radical Halogenation

13.1 Mechanism of Free-Radical Chlorination

The preparation of chlorinated hydrocarbons is an extremely important industrial reaction because of their wide use as solvents. When a hydrocarbon is heated with chlorine gas to about 120° or the mixture is exposed to light at a lower temperature, the highly exothermic substitution of a chlorine atom for hydrogen occurs.

$$RH + Cl_2 \xrightarrow[\text{or light}]{\text{heat}} R-Cl + HCl \qquad \Delta H \approx -20 \text{ to } -25 \text{ kcal/mole}$$

The mechanism of this reaction has been studied extensively and shown to proceed by way of a free-radical chain process involving chlorine radicals ($Cl\cdot$). Several steps are involved, which can be categorized according to whether they create or destroy free radicals (initiation or termination) or simply transfer the radical character from one atom to another (propagation).

$$\textit{Initiation:} \qquad Cl-Cl \xrightarrow[\text{or light}]{\text{heat}} 2\ Cl\cdot$$

$$\textit{Propagation:} \qquad Cl\cdot + RH \longrightarrow R\cdot + HCl$$
$$R\cdot + Cl-Cl \longrightarrow R-Cl + Cl\cdot$$

$$\textit{Termination:} \qquad Cl\cdot + Cl\cdot \longrightarrow Cl-Cl$$
$$R\cdot + R\cdot \longrightarrow R-R$$
$$R\cdot + Cl\cdot \longrightarrow R-Cl$$

The reaction starts by the heat- or light-induced cleavage of chlorine to form chlorine free radicals. One of these can react with a hydrocarbon molecule (RH) to produce HCl and a different radical ($R\cdot$) (propagation step 1).

The main product-forming step is the reaction of the alkyl radical ($R\cdot$) with molecular chlorine (propagation step 2). In this process a new chlorine radical is produced, which can react with another molecule of hydrocarbon to yield a new alkyl radical, which yields another product molecule and yet another chlorine radical. In principle, the initial formation of a single chlorine atom could lead to the conversion of all of the hydrocarbon and chlorine gas present to chlorinated hydrocarbon, by an unending cycle of these two propagation steps. In practice this does not happen because one or more of the termination steps intervenes. The recombination of two chlorine radicals, for example, breaks the cyclic chain of reactions in which those atoms are involved. If the chain process is to stay alive, the initiation steps must keep up with the termination steps.

In discussing free-radical chain reactions, it is convenient to talk about the "chain length," which is the *average* number of cycles of the propagation steps that occur for each initiation step before the cycle is terminated. The chain length clearly depends on the rate constants of the several steps involved and thus will vary with the reaction conditions. Under favorable conditions the chain length can be 100 to 10,000. If free-radical inhibitors (species that combine quickly with free radicals) are present, the chain length may approach unity or, with particularly effective free-radical scavengers, even zero.

13.2 Chlorination by Means of Sulfuryl Chloride and AIBN

In the teaching laboratory it is undesirable to carry out chlorination with chlorine gas because of the severe hazard it presents. A better alternative is to use sulfuryl chloride (SO_2Cl_2), which can propagate the chain in a similar fashion.

$$R\cdot + Cl\!-\!\overset{\displaystyle O}{\underset{\displaystyle O}{\overset{\|}{\underset{\|}{S}}}}\!-\!Cl \longrightarrow R\!-\!Cl + SO_2Cl\cdot$$

$$SO_2Cl\cdot \longrightarrow SO_2 \text{ (gas)} + Cl\cdot$$

Sulfuryl chloride is both colorless and more stable than chlorine; therefore, neither heat nor light is suitable for initiating the reaction. A widely used industrial substitute is azobisisobutyronitrile (AIBN), which undergoes radical fragmentation when exposed to ultraviolet light ($\sim$340 nm) or heated to about 80° or above.[1] The isobutyronitrile radicals then attack the sulfuryl chloride to generate SO_2 and chlorine radicals.

---

[1] Benzoyl peroxide is also frequently used to initiate free-radical reactions, but great care must be exercised because it can decompose explosively when heated or brought in contact with heavy metal ions. AIBN is considered to be much safer, but even with this chemical there have been E, A reports in the literature of explosions on contact with acetone. The cause of these explosions is unknown.

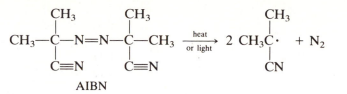

AIBN

13.3 Energetics of Halogenation

Thus far the discussion has centered on chlorination. It is instructive to compare the energies of reaction for the four common halogens in the halogenation of methane.

$$CH_4 + X_2 \rightarrow CH_3X + HX$$

X	ΔH, kcal/mole
F	-102.8
Cl	-24.7
Br	-7.3
I	$+12.7$

These data show that fluorination is so highly exothermic that even carbon–carbon bonds could be cleaved by the explosive release of energy. At best this makes fluorination difficult to control and leads to complex mixtures of carbon skeletons. Fluorocarbons are usually made by indirect methods.

Iodination, by contrast, is endothermic, and thus the favored reaction is the reverse of iodination, reduction by deiodination.

$$CH_3I + HI \longrightarrow CH_4 + I_2$$

This leaves chlorination and bromination as the two accessible halogenations for general laboratory work.

13.4 Selectivity in Halogenations

If each hydrogen atom in a hydrocarbon reacted with a chlorine free radical at the same rate, the proportion of the different halogenated isomers would follow the number of hydrogen atoms of each kind. For example, in 2,3-dimethyl-butane there are 12 primary hydrogens and only 2 tertiary hydrogens. If each hydrogen atom reacted at the same rate, the ratio of primary to tertiary chlorinated products would be 12/2 = 6.

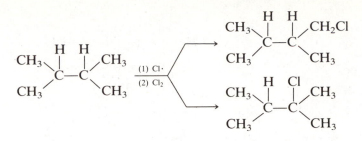

In fact, the tertiary hydrogens react typically about five times as fast (the exact factor depends on the experimental conditions). It follows that the ratio of total rates of attack on the 12 primary and the 2 tertiary hydrogens is

$$\frac{12 \times 1}{2 \times 5} = \frac{12}{10} = 1.2$$

Thus the fraction of primary product is $12/(12 + 10) = 0.55$, a value much less than might have been anticipated from the sixfold preponderance of primary hydrogens.

FIGURE 13.1
Reaction Profiles
for the Reaction
of Cl· with 2,3-
Dimethylbutane

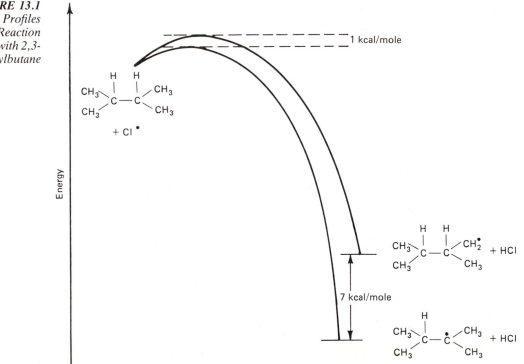

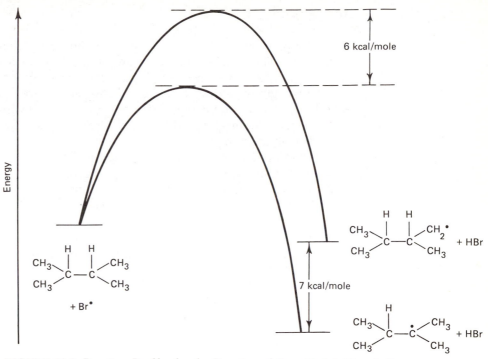

FIGURE 13.2 *Reaction Profiles for the Reaction of Br· with 2,3-Dimethylbutane*

The main reason for the greater reactivity (selectivity) of the tertiary hydrogens lies with the greater stability of the tertiary free radical that is formed in the rate-determining step. However, there is an additional feature to consider. When 2,3-dimethylbutane is brominated, the only product formed in significant amount is the tertiary bromide, corresponding to a tertiary/primary selectivity in excess of 100. Since the same tertiary hydrocarbon radical is being formed, the different results for chlorination and bromination must lie in the relative positions of the two transition states. Free-radical cleavage of a C—H bond by Cl· is a highly exothermic process, and in the transition state relatively little free-radical character on carbon is developed; although the 3° radical is approximately 7 kcal/mole more stable than the 1° radical, only about 1 kcal/mole of the difference shows up in the transition states (Figure 13.1).

In bromination the same 7 kcal/mole difference exists but, because of the weaker HBr bond strength, the reaction profile shown in Figure 13.1 is distorted to give the profile shown in Figure 13.2 by the elevation of the two free radicals on the right in relation to the starting materials on the left. As a consequence, the transition states lie much closer to the right-hand side of the diagram and the difference between them more fully expresses the 7-kcal/mole difference in energy of the 1° and 3° radicals. Bromination is therefore much more selective.

13.5 Substituent Effects

In the previous section we have discussed how the different stabilities of 1°, 2°, and 3° radicals lead to different rates of attack on hydrogen atoms attached to these positions. We have also seen that the selectivity (ratio of rates) depends on whether the transition states lie close to the free radicals being formed (high selectivity as in bromination) or close to the starting hydrocarbon (low selectivity as in chlorination). These differences depend on the differences in exothermicity of the abstraction step.

Another factor to be considered in interpreting the rates of free-radical reactions is the effect of substituents on the stability of transition states. Substituents have little effect on the stability of the starting hydrocarbon or the free radicals formed from it because these species have little polarity. By contrast, the transition state, with a polar partial H—X bond, is quite sensitive to substituents. The magnitude of the substituent effect depends on the polarity of the partial H—X bond, and this depends on the nature of X·.

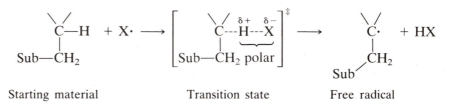

For a chlorine substituent attached to the carbon next to the C—H bond being attacked, the rate of H-atom abstraction is typically diminished by a factor of 2 or 3. When the chlorine substituent is attached directly to the carbon-bearing hydrogen atom, the rate is diminished by a factor of 5 or more.

13.6 Preparations and Reactions

(A) Photochemical Chlorination of 2,3-Dimethylbutane[2]

In this experiment small amounts of hydrogen chloride and sulfur dioxide are produced, so you should work in the hood.

In a 100-mL Erlenmeyer flask place 5 mL (3.4 g, 0.04 mole) of 2,3-dimethylbutane, 1 mL (1.7 g, 0.012 mole) of sulfuryl chloride (SO_2Cl_2), and about 20 mg (0.1 mmole) of azobisisobutyronitrile.[3]

CAUTION   Sulfuryl chloride is extremely corrosive; if any is spilled or spattered on the skin it should be removed immediately by washing thoroughly with water. Even the vapors of sulfuryl chloride are dangerous. All transfers of the material should be carried out in the hood.

---

[2] This experiment is designed to measure the relative rates of attack on 1° and 3° hydrogen atoms. To achieve this measure it is desirable to chlorinate only a small fraction of the hydrocarbon.
[3] Avoid contact of AIBN with acetone or acetone vapors. See footnote 1.

Irradiate the flask for 30 min using a long-wavelength ultraviolet lamp.[4] At the end of the irradiation period, cautiously pour the reaction mixture into a separatory funnel containing 10 mL of methylene chloride and 25 mL of water.

⫸ *CAUTION* Prolonged exposure to high concentrations of methylene chloride vapors may induce cancer. As with all organic solvents, work in a well-ventilated area when using it.

Separate the layers and wash the organic layer with 5% sodium bicarbonate. Wrap a wad of cotton around the tip of a Pasteur pipet and transfer the methylene chloride solution to a dry container for analysis. If the solution is cloudy, it can be dried with magnesium sulfate.

Analyze the mixture by gas–liquid chromatography using a 5-ft 5% DC-200 (silicon oil) column at 50–60°. The approximate relative retention times of the major components are 2,3-dimethylbutane (bp 58°; time = 1.0) and methylene chloride (bp 40°; time = 1.0); 1-chloro-2,3-dimethylbutane (bp 124°; time = 3.5); 2-chloro-2,3-dimethylbutane (bp 112°; time = 4.8).

From the areas of the peaks calculate the relative amounts of the 1° and 3° chloro derivatives (assume that the detector response is the same for each isomer). Calculate the relative reactivities of the 1° and 3° hydrogens (take the 1° hydrogen to be 1.00), taking into account the different numbers of each kind present in the starting material.

Pour the unused methylene chloride solution into the labeled waste jar in the hood.

(B) Substituent Effects in Free-Radical Chlorination[5]

In this experiment small amounts of hydrogen chloride and sulfur dioxide are liberated, so you should work in the hood.

In a 100-mL round-bottomed flask, place 3 mL (2.7 g, 0.028 mole) of 1-chlorobutane, 1 mL (1.7 g, 0.012 mole) of sulfuryl chloride (SO_2Cl_2), and 0.02 g (0.12 mmole) of azobisisobutyronitrile.[3]

⫸ *CAUTION* Sulfuryl chloride is extremely corrosive; if any is spilled or spattered on the skin, it should be removed immediately by washing thoroughly with water. Even the vapors of sulfuryl chloride are dangerous. All transfers of the material should be carried out in the hood.

Attach a water-cooled condenser and an efficient gas trap to remove the HCl and SO_2 gases that are evolved (see Section 8.4). Heat the mixture under gentle reflux on a steam bath for 20 min. Remove the steam bath and after the reaction mixture has cooled somewhat, add another 0.02 g of azobisisobutyronitrile.[3] Resume heating for an additional 10 min to complete the reaction.

---

[4] The azobisisobutyronitrile has an ultraviolet absorption maximum near 340 nm. Sufficient light is emitted by a "grow lamp" used for plants if the flask is held close to it. Germicidal ultraviolet lamps should not be used without a Pyrex filter unless special precautions are taken to protect the eyes from the shorter wavelength radiation.

[5] This experiment is patterned after that described by Reeves, *J. Chem. Ed.*, **48**, 636 (1971).

Cool the reaction mixture to room temperature, pour it into 10 mL of water contained in a separatory funnel, and separate the layers. Wash the organic layer[6] with 5% sodium bicarbonate and then dry it over calcium chloride.

Analyze the mixture by gas–liquid chromatography using a 5-ft 5% carbowax 400 column at about 55–60°. The approximate relative retention times of the various components are 1-chlorobutane (bp 77–78°; time = 1.00); 1,1-dichlorobutane (bp 114–115°; time = 3.0); 1,2-dichlorobutane (bp 121–123°; time = 4.7); 1,3-dichlorobutane (bp 131–133°; time = 6.3); 1,4-dichlorobutane (bp 161–163°; time = 14.7).

From the areas of the peaks calculate the relative amounts of each of the dichlorobutanes (assume that the detector response is the same for each isomer). Calculate the relative reactivities of each hydrogen in 1-chlorobutane (take the 4-hydrogen reactivity to be 1.00) and compare these to the relative hydrogen reactivities of *n*-butane (primary = 1.0; secondary = 3.6 at 80°).

Pour the unused methylene chloride solution into the labeled waste jar in the hood.

Questions

1. It is observed that hydroquinone (*p*-dihydroxybenzene) and catechol (*o*-dihydroxybenzene) inhibit free-radical reactions. Explain.

2. A widely used free-radical initiator is benzoyl peroxide (in spite of its extreme explosive hazard), which decomposes vigorously around 80° to give benzoic acid, benzene, and carbon dioxide in varying amounts depending on the solvent. Explain the origin of these products and the catalytic effect of the peroxide when used to initiate halogenation. What halogenated hydrocarbon will contaminate the product?

3. In the gas phase the relative reactivities of bromine radicals for hydrogen atom abstraction are tertiary:primary = 20,000:1. If 2,3-dimethylbutane were exposed to bromine vapor and light, what amount of primary bromide would be expected to form?

4. Explain in qualitative terms why free-radical bromination is so much more selective than chlorination.

---

[6] The densities of the organic layer and the aqueous washes are similar. The identity of the aqueous layer should be verified at each stage by testing a small portion for water solubility.

14

Conversion of Alcohols to Alkyl Halides

14.1 Preparation of Alkyl Halides

Alkyl halides are prepared by many different methods and by the use of a variety of reagents. Direct halogenation of alkanes to form alkyl chlorides and bromides often leads to mixtures of isomers that are difficult to separate with ordinary laboratory fractionating equipment. Likewise, mixtures are frequently produced by addition of halogen acids to alkenes. The most useful laboratory methods involve the conversion of alcohols to alkyl halides. Fortunately, a large variety of alcohols are available commercially.

Alcohols may be converted to alkyl halides by means of the halogen acids (HCl, HBr, and HI), phosphorus halides (PCl_3, PBr_3, or $P + I_2$), or thionyl chloride ($SOCl_2$).

There are two different mechanisms for nucleophilic substitution, designated as S_N1 (unimolecular) and S_N2 (bimolecular). The S_N1 process involves conversion of the alcohol, via an oxonium-type intermediate (**I**), to the corresponding carbocation (**II**), which reacts rapidly with the nucleophile (Cl^-, Br^-, etc.).

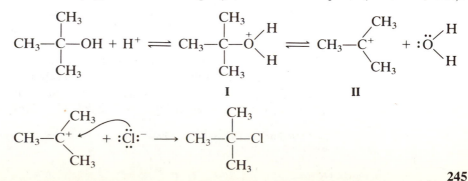

or

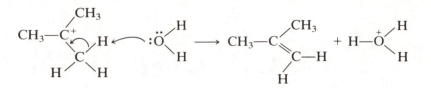

The rate depends only on the oxonium-ion concentration and is independent of the halide-ion concentration (S_N1). This behavior is typical for tertiary alcohols; a side reaction can occur involving elimination of H^+ from the carbocation to form an alkene (El elimination).[1]

In the S_N2 mechanism the protonated alcohol species is approached by the nucleophile from a position directly behind the carbon bearing the protonated hydroxyl group. In terms of the tetrahedral disposition of the groups around the central carbon, this is called "backside attack." In the transition state (**III**) the remaining three substituents have a planar distribution. At a fixed acidity, the rate is second order (S_N2) and depends on the concentration of the attacking nucleophile and the alcohol. This mode is typical for primary alchols. Here also a side reaction leading to an alkene can occur (E2 elimination).

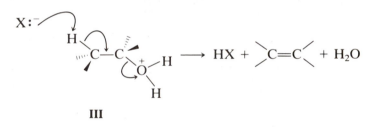

III

The ease of reaction of alcohols toward halogen acids follows the sequence tertiary > secondary > primary, and the tendency to form alkenes, or to undergo rearrangement of the carbon skeleton, follows the same order. Reactivity of the halogen acids toward alcohols declines in the order HI > HBr > HCl. Thus, conditions suitable for the conversion of a particular alcohol to a halide are influenced by the structure of the alcohol and the specific halogen acid to be used.

Tertiary alcohols react readily with any of the three halogen acids, even at 25° and in the absence of catalysts. Secondary alcohols react more slowly; moderate heating and acidic catalysts (50% sulfuric acid, zinc chloride) are used to promote the conversion to bromides and chlorides. Primary alcohols require more vigorous conditions and more active catalysts (65–70% sulfuric acid for conversion to bromides). The Lucas test for differentiating the classes of alcohols (Chapter 9) is based on relative rates of conversion to alkyl chlorides.

[1] For further discussion see Chapters 18 and 19, which deal with formation of alkenes from alcohols.

Primary and secondary alkyl bromides containing two to five carbon atoms are prepared by heating the alcohol with concentrated hydrobromic acid or a hydrobromic–sulfuric acid mixture. A convenient procedure is to use sodium bromide and an excess of strong sulfuric acid to generate the HBr in the reaction flask. Regrettably, this method is not suitable for higher molecular weight bromides because the high concentration of salts present greatly reduces the solubility of the alcohol in the reaction medium. For secondary alcohols, a lower concentration of sulfuric acid (50%) is used; stronger acid is unnecessary and promotes a side reaction (alkene formation). For higher molecular weight alcohols the action of anhydrous hydrogen bromide at 100–120° in the absence of a solvent, or use of phosphorus tribromide, is a satisfactory procedure. Water-soluble tertiary alcohols are converted rapidly to bromides merely by shaking with concentrated aqueous hydrobromic acid.

Hydrochloric acid containing dissolved zinc chloride (a Lewis acid) is used for the conversion of primary and secondary alcohols to alkyl chlorides. Primary alcohols require heating with a saturated solution of zinc chloride in concentrated hydrochloric acid. Thionyl chloride is a useful reagent for the preparation of primary alkyl chlorides.

Alkyl iodides are obtained readily from alcohols by means of strong aqueous hydroiodic acid, but a more economical method is to treat the alcohol with iodine and red phosphorus.

$$6 \ R\!-\!CH_2OH + 3 \ I_2 + 2 \ P \longrightarrow 6 \ R\!-\!CH_2I + 2 \ H_3PO_3$$

Alkyl iodides may be prepared also by reaction of alkyl chlorides or bromides with sodium iodide in acetone solution (Finkelstein reaction; see Chapter 15).

14.2 Reactions of Alkyl Halides

Alkyl halides are relatively reactive molecules and are widely used in laboratory and industrial syntheses. Their reactivity and the mechanism of their reactions vary over a wide range. In some features, they resemble the corresponding alcohols.

Alkyl bromides are particularly useful in laboratory operations. They undergo substitution (replacement) reactions with nucleophilic reagents (alkoxides, phenoxides, cyanide anion, and ammonia) to produce ethers, nitriles, and amines. They alkylate sodium derivatives of malonic and acetoacetic esters, and react with certain metals to form organometallic compounds (such as Grignard reagents).

Reactions of the primary alkyl halides generally follow an S_N2 mechanism, and the relative reactivities of the halides follow the sequence methyl $\gg$ primary $>$ secondary. In many typical reactions the rate for the methyl halide may be 10 to 20 times as fast as the ethyl analog, which will be about 2 times that of the higher primary homologs of straight-chain structures. For primary halides bearing a single group in the α or β position (isobutyl, isopentyl) the rate is about half that of the straight-chain isomer. Greatly enhanced reactivity occurs in the allyl (R—CH=CH—CH$_2$—X) and benzyl (aryl—CH$_2$—X) halides.

Ionic substitution reactions of tertiary halides follow an S_N1 mechanism: the rate-determining step is an ionic cleavage of the carbon–halogen bond to form the carbocation. Secondary halides may react by an S_N1 or an S_N2 mechanism or a hybrid of both. The S_N2 process is favored by powerful nucleophilic reagents (see Section 15.3), high concentration of the reagent, and a weakly polar solvent. The S_N1 mechanism is favored by weakly nucleophilic reagents and low concentrations, and especially by reaction media of high solvating power. Molecular rearrangements often occur in S_N1 replacement reactions and in E1 elimination reactions leading to alkenes but are infrequent in S_N2 reactions and E2 (bimolecular) elimination reactions.

There is a suspicion that many alkyl halides, particularly those that are effective alkylating agents, can cause cancer on prolonged exposure. There is also concern that even unreactive alkyl halides (e.g., carbon tetrachloride) may indirectly induce cancer by disrupting protective fatty layer from some tissues and making them more susceptible to other carcinogens. As noted in Chapter 1, you should always treat organic molecules with respect and avoid unnecessary exposure.

14.3 Preparations

(A) *n*-Butyl Bromide

In a 50-mL round-bottomed flask place 6 mL of water and add in small portions, while cooling the flask in an ice bath, 7 mL (13 g, 0.13 mole) of concentrated sulfuric acid (**Caution**—*corrosive!*). To the cold, diluted acid add 5 mL (3.7 g, 0.05 mole) of *n*-butyl alcohol, while mixing thoroughly with continued cooling. Add 6.2 g (0.06 mole) of sodium bromide crystals (NaBr). Good mixing and thorough cooling are important in these additions to avoid premature reaction.

Add a carborundum boiling chip, attach an upright condenser, and connect the top of the condenser to a gas absorption trap to absorb any hydrogen bromide that evolves during the reaction (see Figure 8.2). Heat the flask *gently* while swirling the contents frequently, until most of the sodium bromide dissolves and the mixture begins to boil.[2] Adjust the heating rate so that the mixture boils *gently* and continue heating at this rate for 30 min.

Discontinue the heating and allow the reaction mixture to cool. Disconnect the condenser and arrange the apparatus for distillation. Add 10 mL of water and a fresh boiling chip, and distill the mixture vigorously into a 15 × 150-mm test tube until about 12–15 mL of distillate has been collected or the boiling point rises above 102°.

---

[2] The sodium bromide will dissolve completely only near the boiling point of the mixture.

With the aid of a Pasteur pipet, remove the upper aqueous layer, add 5 mL of water, and mix the layers well using the squirting technique described in Section 6.5, which consists of adding the water to the crude alkyl bromide and then repeatedly drawing the upper layer into a Pasteur pipet and forcing it back out into the lower layer.[3] Allow the layers to separate, and—*carefully*—remove the upper aqueous layer. To remove unchanged *n*-butyl alcohol, wash the crude alkyl bromide with about 2.6 mL of *ice-cold* 80% sulfuric acid (prepared by *cautiously* adding 2 mL of concentrated acid *to* 0.5 mL of water). The washing is accomplished by the squirting technique used above. Remove the *lower*, slightly yellow sulfuric acid layer carefully and completely. Wash the butyl bromide layer once with a 2-mL portion of water and then twice with 2-mL portions of saturated sodium bicarbonate solution (**Caution**—*foaming!* from residual acid), using the squirting technique as before. Finally, wash the product once again with water. In all of these washings be careful to save the proper layer—the bromide is sometimes on top and sometimes on the bottom, depending on the relative density of the wash solution. It is a good idea to discard nothing until the whole procedure is completed.

To the cloudy *n*-butyl bromide add about 0.5–1.0 g of anhydrous calcium chloride pellets. Cork the test tube, shake it gently to bring the liquid in contact with the drying agent, and set the stoppered tube aside for at least 15 min (longer does no harm).

Transfer the dried product through a filter tube containing a wad of glass wool into a 10-mL distilling flask.[4] Add a boiling chip, attach the one-piece distillation head to the flask and distill. Collect in a weighed 10-mL vial the portion that boils between 99 and 101°. If an appreciable low-boiling fraction is obtained, dry the butyl bromide again and redistill. The yield is 4.5–5.5 g (3.5–4.3 mL).

The NMR spectrum of neat *n*-butyl bromide is shown in Figure 14.1.

(B) Cyclohexyl Chloride

In a 50-mL round-bottomed flask place 15 g (0.11 mole) of granular zinc chloride and 9.0 mL (10.6 g, 0.067 mole) of concentrated hydrochloric acid that has been chilled in an ice bath for 5 min. Shake the mixture (the zinc chloride will not dissolve completely), and then add 5 g (0.05 mole) of cyclohexanol. Attach a reflux condenser, add a boiling chip, and boil the mixture for 40 min. Allow

---

[3] When extracting small volumes of materials, it is desirable to minimize the number of transfers of the product. In the procedure described here the product is collected in a small vessel and remains there as various wash reagents are added, mixed, and removed. With small volumes the transfers are best accomplished with a Pasteur pipet. If total transfer is desired, it would be necessary to rinse the flask and pipet with an appropriate solvent (typically, CH_2Cl_2 or diethyl ether).

[4] The drying agent can be removed completely with minimum loss of material by the following procedure. Place a wad of cotton or glass wool in the constricted neck of a 9×150 mm filter tube, support the tube with the lower end *inside* the receiving flask, and pour the dried product into the tube. If glass wool is used, it is best handled with tweezers or gloves to avoid getting glass fibers stuck in your fingers. Allow the product to drain completely. If any significant amount of product remains in the tube, it can be forced out by applying gentle pressure to the top of the tube with a rubber bulb.

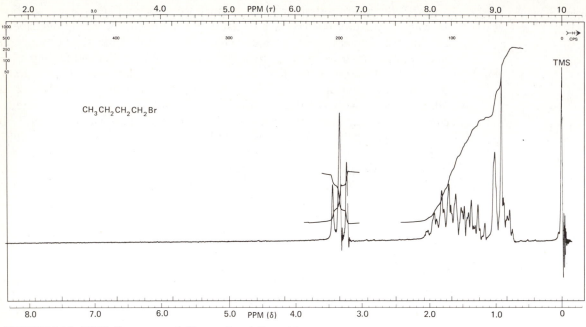

FIGURE 14.1 *NMR Spectrum of Neat n-Butyl Bromide*

the reaction mixture to cool for 5 min, remove the condenser, and then, using a Pasteur pipet, draw off and discard the lower, acid layer.

Add 20 mL of water to the flask, attach a Claisen adapter (Y adapter) to the flask, stopper the central arm, and attach the distilling head to the side arm. Use a 15-mL test tube as the receiver. Heat the mixture of water and chloride so that it boils vigorously and continue the distillation until about 7–8 mL has been collected. Add 2 mL of saturated sodium bicarbonate to the distillate, mix the layers with a Pasteur pipet, using the squirting technique described in Section 6.5, and remove the lower aqueous layer.[3] Wash the product with another 2 mL of saturated sodium bicarbonate (if ordinary water is used, the layers may not separate cleanly because the densities are so close), transfer the chloride into a dry test tube and then dry it over 1 g of anhydrous calcium chloride. After 15 min of drying, transfer the dried chloride into a 25-mL or 50-mL distilling flask (the product tends to foam on distillation), add a boiling chip, and distill. Collect the fraction boiling at 140–143°. The yield is 2.5–3.5 g.

(C) *t*-Butyl Chloride

In a 20 × 150-mm test tube, place 15 mL (17.9 g, 0.18 mole) of concentrated hydrochloric acid (best done in the hood), and add 5 mL (3.9 g, 0.053 mole) of *t*-butyl alcohol with the aid of a Pasteur pipet. Mix the contents of the tube with the same pipet by repeatedly drawing up the liquid and expelling it. Repeat the mixing occasionally over a 15-min period.

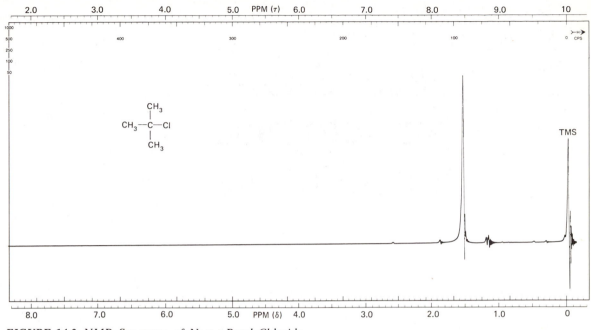

FIGURE 14.2 *NMR Spectrum of Neat t-Butyl Chloride*

Remove the lower aqueous layer with the aid of a Pasteur pipet. Wash the product with 2–3 mL of water (use the squirting technique described in Section 6.5), then with 5% sodium bicarbonate solution (**Caution**—*foaming!*) and again with water.[3] Transfer the alkyl chloride to a dry test tube and dry it with about 1 g of anhydrous calcium chloride pellets.[4] Draw off the dried liquid with a Pasteur pipet, place it in a small distillation apparatus, and add a boiling chip. Distill and collect the fraction boiling at 49–52° in a weighed vial. The yield is 3–4 g. If an appreciable low-boiling fraction is obtained, dry this fraction again with calcium chloride pellets and redistill it.

The ^{1}H NMR spectrum of neat *t*-butyl chloride is shown in Figure 14.2.

Questions
1. In the preparation of *n*-butyl bromide why do you need to keep the reaction cool before the *n*-butyl alcohol is added? What is the equation for the reaction involved?

2. Explain why *n*-butyl alcohol is soluble in sulfuric acid.

3. Potassium hydroxide is sometimes used as a drying agent. Why would it be inappropriate for drying alkyl halides?

4. If during the distillation of the alkyl halides a large amount is obtained, what does this indicate and how can it be corrected?

15

Second-Order Nucleophilic Substitution

15.1 Replacement Reactions

In acetone solution, alkyl bromides react with sodium iodide to produce the alkyl iodide and a precipitate of sodium bromide. This particular reaction, known as the Finkelstein reaction,[1] is but one example of an extremely broad class of organic substitution reactions. Some are named after their discoverers (the Williamson ether synthesis, the Menshutkin reaction, etc.), but today most are known simply by their mechanistic class name—second-order nucleophilic substitutions, or S_N2 for short.

Finkelstein reaction:

$$Na^+I^- + R—BR \xrightarrow{\text{acetone}} I—R + NaBr \downarrow$$

Williamson ether synthesis:

$$CH_3O^- + C_4H_9—I \longrightarrow CH_3O—C_4H_9 + I^-$$

Menshutkin reaction:

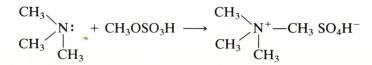

[1] Finkelstein, *Ber*, **43**, 1528 (1910).

15.2 Stereochemistry and Kinetics

In a classic series of papers, Hughes and Ingold[2] worked out the details of nucleophilic displacement reactions. They showed that the entering nucleophile attacks the carbon at the rear of the bond to the leaving group.

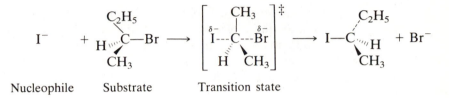

| Nucleophile | Substrate | Transition state | |

The configuration of the groups attached to the central carbon is inverted so that, if the starting material is chiral as a result of the configuration of these groups, there is a corresponding change in optical activity. Inversion is a characteristic feature of S_N2 processes that distinguishes them from S_N1 reaction.

Hughes and Ingold also showed that the rate of the reaction was proportional to the concentration of both the nucleophile and the substrate (the species being attacked).

15.3 Nucleophilicity

In acid–base chemistry basicity is a measure of the affinity of an electron pair of a base for a proton. In the S_N2 reaction the nucleophile shares an electron pair with the carbon atom being attacked. By analogy, nucleophilicity is a measure of the affinity of a base for a carbon atom. Just as there are strong bases (e.g., CH_3O-) and weak bases (NO_3^-), there are good nucleophiles (I^-) and poor ones (CH_3OH). As long as one compares only first-row elements, there is a rough correlation between nucleophilicity and basicity. However, in protic solvents second-and third-row elements are considerably more nucleophilic than predicted by such a correlation. Thus iodide, although much less basic than fluoride, undergoes S_N2 reactions about 100 times faster. The lesser reactivity of first-row elements appears to be related to their stronger solvation in protic solvents. Before these nucleophiles can react some of their strongly stabilizing solvation must be dissociated.

The traditional explanation for the apparently greater nucleophilicity of the second-row elements was in terms of their greater polarizability (ease of distortion of their nonbonding electrons). This explanation is called into question, however, by the observations that in polar, aprotic solvents (e.g.,

---

[2] Ingold, *Structure and Mechanism in Organic Chemistry*, 2nd ed. (Ithaca, NY, and London: Cornell University Press, 1969).

dimethylformamide), where the strength of solvation is much less than in polar, protic solvents, the nucleophilicity of chloride actually is greater than the nucleophilicity of iodide.

15.4 Substrate Structure

In S_N2 reactions there is considerable increase in steric congestion of the backside of the molecule as the nucleophile attacks to form the pentacoordinated transition state. The larger the groups attached to the carbon being attacked, the slower the rate of reaction. The effect of branching at the α-carbon is illustrated in Table 15.1

15.5 Solvent

Many solvents have been used for S_N2 reactions. Methanol and ethanol are readily available and inexpensive; water may be added to increase the solubility of inorganic salts if necessary. Glacial or aqueous acetic acid is also useful. Acetone is a polar, aprotic solvent, which is to say that it has a fairly high dielectric constant but lacks hydroxyl groups.

The choice of solvent can have a profound effect on the rate of S_N2 reactions. The fundamental principle is that polar solvents stabilize ions and, to a lesser degree, dipoles. Dispersed charges are stabilized less than concentrated charges. For example, in the Finkelstein reaction of iodide ion with n-butyl bromide the charge on the iodide ion is concentrated. However, in the transition state the charge is dispersed over both the iodide and the bromide ions. Thus in passing from the starting materials to the transition state there is a net loss of solvation energy, which means that the reaction rate will be smaller in highly polar solvents than in low-polarity solvents. In practice one must not use a solvent of too low polarity (such as a hydrocarbon) or so little of the sodium iodide will dissolve that there will be no reaction.

In the Finkelstein reaction, acetone has a particular advantage as the solvent because it dissolves the starting sodium iodide but not the sodium bromide product, which conveniently precipitates as the reaction proceeds.

As noted above, ionic displacement reactions are usually faster in polar aprotic solvents. In addition to acetone, dimethyl sulfoxide, $(CH_3)_2SO$ (DMSO), dimethylformamide, $(CH_3)_2NCHO$ (DMF), and acetonitrile, CH_3CN, are other examples of effective aprotic solvents.

TABLE 15.1
Effect of Branching at the α-Carbon on S_N2 Reaction Rates

Alkyl group	Relative S_N2 rate
CH_3—X	30
CH_3CH_2—X	1
$(CH_3)_2CH$—X	0.03
$(CH_3)_3C$—X	Immeasurably slow

15.6 Preparation of *n*-Butyl Iodide

In a 50-mL round-bottomed flask place 20 mL of acetone, add 3 g (0.02 mole) of sodium iodide, and 2 mL (2.6 g, 0.018 mole) of *n*-butyl bromide. Add a boiling chip, attach a reflux condenser, and boil the mixture for 30 min.

⫸ *CAUTION*   *n*-Butyl iodide, like most good alkylating agents, is suspected of causing cancer on prolonged exposure. This reaction should be carried out in a hood.

Cool the mixture to room temperature with an ice bath and pour it into a large separatory funnel containing 20 mL of water. Drain off the lower layer of *n*-butyl iodide, wash it with water, and dry it over about 0.5 g of anhydrous calcium chloride.

Transfer the dried *n*-butyl iodide through a filter tube to remove the drying agent and collect it in a distillation apparatus and distill, collecting the fraction 129–131°. The yield is about 2.6 g.

Questions   1. Arrange the compounds of each set in order of reactivity toward an S_N2 displacement reaction.
   (a) 2-chloro-2-methylbutane, 1-chloropropane, 1-chloro-2,2-dimethylpropane, 1-chloro-2-methylpropane
   (b) 2-bromopropane, 2-chloropropane, 2-iodopropane

2. Compare S_N1 and S_N2 reactivities of alkyl halides with regard to
   (a) stereochemical outcomes
   (b) an increase in the nucleophile concentration
   (c) an increase in the alkyl halide concentration
   (d) an increase in the polarity of the solvent
   (e) occurrence of rearrangements

3. Arrange the following nucleophiles in order of increasing reactivity toward ethyl iodide in ethanol as a solvent: $^{131}I^-$ (radioactive iodide); OH^-; H_2O; $CH_3CO_2^-$.

4. What would have been the probable outcome of the *n*-butyl iodide preparation if one had accidentaly substituted isopropyl alcohol in place of the acetone as solvent?

16 Chemical Kinetics: Solvolysis of *t*-Butyl Chloride

16.1 First-Order Kinetics

The measurement of reaction rates (*reaction kinetics*) at different concentrations of reactants provides important clues to the reaction mechanism. For example, the rate of substitution of hydroxide for chloride in the S_N1 reaction of *t*-butyl chloride with potassium hydroxide in aqueous ethanol depends on the concentration of the halide, but is essentially independent of the hydroxide concentration. The molecular interpretation is that the slow, rate-determining step involves the ionization of the halide followed by the fast collapse of the intermediate with either water or hydroxide ion to give *t*-butanol as the product. Collapse with an ethanol molecule leads to *t*-butyl ethyl ether. Because the rates of such reactions depend on the concentration of only *one* species, they are said to follow first-order kinetics.

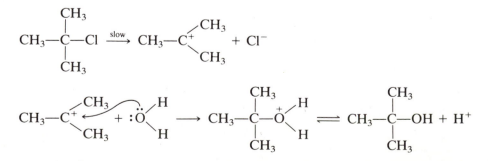

The S_N1 reaction may be contrasted with S_N2 reactions, which involve simultaneous formation of the new carbon–nucleophile bond as the chloride ion departs and thus follow second-order kinetics.

First-order kinetics can be expressed mathematically as

$$\text{Loss of halide per unit time} = \frac{-d[RX]}{dt} = k[RX] \qquad (16.1)$$

where $[RX]$ is the concentration of the halide and k is a proportionality constant known as the *rate constant*. The rate constant varies with the structure of the organic halide, the temperature, the solvent, and the catalyst used. A more practical expression can be obtained by integrating equation 16.1 to obtain

$$\ln \frac{[RX]_0}{[RX]_t} = kt \qquad (16.2)$$

where $[RX]_0$ is the concentration of RX at the start of the reaction, and $[RX]_t$ is the concentration at some later time, t. Note particularly that ln is the *natural* logarithm (base e). After some mathematical manipulation, equation 16.2 can be rewritten as

$$\log_{10} [RX]_t = -\frac{kt}{2.303} + \log_{10} [RX]_0 \qquad (16.3)$$

where now the more familiar base 10 logarithm is used. The form of equation 16.3 is of a straight line

$$y = mx + b$$

If the reaction rate is first order, a plot of each $\log_{10} [RX]_t$ value (calculated from measurement of RX at various times) against the time of measurement will give a straight line of slope $-k/2.303$.

In practice any means of measuring the halide concentration will do and spectroscopy, chromatography, or classical analytical chemistry might be used. Because of the properties of logarithms, the actual concentration of RX is not required and any quantity proportional to it is acceptable. Thus if gas chromatography is used, the heights of the RX peaks at different times can be substituted in equation 16.3 and the same value of k will result. Under some circumstances it is more convenient to measure the concentration of a reaction product than a reactant. From the stoichiometry the two can be related and the integrated rate equation expressed in terms of the product concentrations.

$$RX + H_2O \longrightarrow ROH + HX$$

$$[HX]_t = [RX]_0 - [RX]_t$$

At very large times (approximated as t equal to infinity, ∞), $[HX]_\infty = [RX]_0$, since there is no RX remaining. On this basis equation 16.3 becomes

$$\log_{10}([HX]_\infty - [HX]_t) = -\frac{kt}{2.303} + \log_{10}[HX]_\infty \qquad (16.4)$$

In equation 16.4 it is permissible to use any quantity proportional to the HX concentration.

The units of k are time^{-1}. A quantity frequently used to characterize a first-order rate process is the "half-life," the time required for half of the reactant to react. From equation 16.2, this is seen to be

$$\text{Half-life} = t_{1/2} = \frac{\ln 2}{k} = \frac{0.69}{k}$$

Conversely, if the time for half of the reaction to occur can be estimated, the value of k can be estimated without making a graph.

It is interesting to note that $t_{1/2}$ does not depend on the concentration of any reactant. If a reaction follows first-order kinetics, as do S_N1 reactions, the time required for completion does not change if solvent is removed or added. This lack of concentration dependence is *not* true of reactions following other kinetic orders.

16.2 Laboratory Practice

A successful kinetics experiment requires both an understanding of the operations to be performed and *advance* preparation of many solutions and reagents, as well as collection of essential apparatus. Kinetics experiments once started must be completed without interruption. To avoid mistakes and "lost points" a table should be prepared in which the laboratory data can be recorded as the experiment progresses.

In the following experiment the S_N1 reaction of *t*-butyl chloride with a 50 : 50 mixture of water and 2-propanol will be examined. In this *solvolysis* of the chloride it is convenient to follow the reaction progress by titrating small samples of the reaction mixture of determine the amount of hydrochloric acid developed. Since it takes some time to perform a titration, it is necessary to stop (*quench*) the reaction before the sample is titrated. S_N1 reaction rates are very sensitive to the amount of water present in the solvent, going slower the smaller the amount of water. A convenient technique for quenching the *t*-butyl chloride solvolysis is to add the aliquot to an equal volume of acetone.

Reaction rates are also quite dependent on temperature, typically doubling for each rise of 10°. In precise kinetic work the reacting solution is placed in a constant temperature bath that holds the temperature fluctuation to ±0.05°. In this experiment the solution is kept on the bench top, and if it is not in a drafty location, it will probably hold its temperature to $\pm 1°$ during the experiment, which corresponds to about $\pm 10\%$ variation in the rate constant. If greater temperature control is desired, the reaction flask can be placed in a large beaker of water.

The time at which each sample is quenched must be recorded, so that the times from the start of the reaction can be calculated for use in equation 16.4. It is desirable to titrate several samples from each of the first two half-lives. There is no point in measuring time intervals with greater precision than is obtained with the titrations. In the following example the half-life is about 50 min, so that recording the times to the nearest half-minute (about 1%) is adequate.

If the course of the reaction is being followed by measuring one of the products, equation 16.4 is to be used. This requires knowing the product concentration at "infinite" time. Since the starting material diminishes by 50% for each half-life, the "infinity" value can be approximated with adequate precision by titration of a sample after 8–10 half-lives, which corresponds to 99.5–99.9% reaction. Another strategy for obtaining the required "infinity" is to withdraw a sample and accelerate the rate of reaction either by raising the temperature or changing the solvent. With S_N1 reactions, which are very sensitive to the polarity of the solvent, a common technique is to place a sample in an equal volume of water. This causes about a 10-fold increase in rate and permits the "infinity" to be measured after one half-life corresponding to the original reaction conditions (1 : 1 H_2O–2-propanol).

16.3 Measurement of the S_N1 Reaction Rate of t-Butyl Chloride

In a cork-stoppered 250-mL Erlenmeyer flask prepare 100 mL of a 1 : 1 (by volume) solution of 2-propanol and water; after mixing it well, allow it to equilibrate to the laboratory temperature while the next step is carried out.[1]

Obtain 150 mL of approximately 0.04 N standardized aqueous sodium hydroxide in a 250-mL Erlenmeyer flask fitted with a cork or rubber stopper. Record the concentration of the base in your notebook. Set up a 25- or 50-mL buret, and fill it with base. Obtain a dropping bottle of bromothymol blue solution and a white background card to enhance the visibility of the green end point. Secure a stopwatch or clock that will record up to 2 hr of elapsed time with a resolution of at least 0.5 min.

In a 100-mL volumetric flask weigh *accurately* a sample (about 1 g) of t-butyl chloride and add the water–2-propanol solvent to the mark. Stopper the flask, mix the contents well, and note the time and laboratory temperature. The flask should remain stoppered at all times except when a sample is being withdrawn.

The first sample should be taken about 10 min after the reaction is started.

---

[1] Another suitable solvent is 1:1 (by volume) acetone and water. The samples should be withdrawn at 10-min intervals.

Subsequent samples should be withdrawn at about 20, 35, 50, 75, and 100 min after the start of the reaction. For each sample a 10-mL aliquot is drawn into a 10-mL pipet using a pipet bulb (*not by mouth!*) and transferred to a 125-mL Erlenmeyer flask containing about 10–15 mL of ordinary acetone as a reaction quench. Record the time of addition in the notebook. Three drops of bromothymol blue solution are added and the solution titrated to a green end point that persists for about 20 s. Before the next sample is withdrawn, the pipet must be cleaned by rinsing it with a *little* acetone and dried by drawing air through it by connecting it to a rubber hose leading to the water pump.

To determine the "infinity" titer, a 10-mL sample is added to 10 mL of water. The additional water causes about a 10-fold increase in reaction rate so that the sample is ready to titrate after about 1 hr. If sufficient sample is available it is desirable to have duplicate or triplicate "infinity" values so that the error can be reduced by taking an average value.

Prepare a plot of $\log_{10}$(titer at infinity − titer at time *t*) versus the time of sampling as in Figure 16.1. Draw the best straight line through the points and

FIGURE 16.1
Plot of Kinetic Data

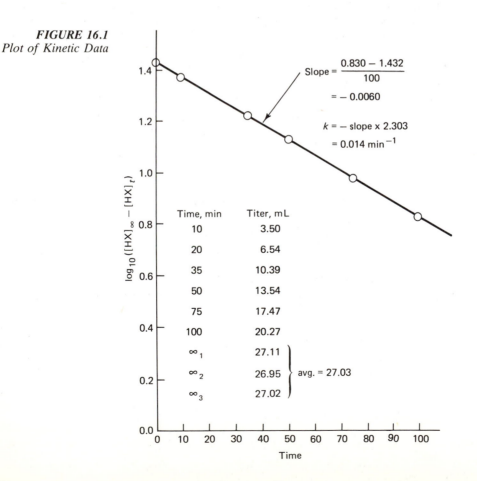

from the slope calculate the value of the rate constant. Compare the measured infinity value with the quantity calculated from the concentrations of the chloride, the base, and the volume of the pipet.

   If a calculator or a computer with a program for fitting a straight-line function is available, evaluate the slope (k) and compare the result with the slope obtained from the graph.

Questions

1. Show how equation 16.2 can be transformed into equation 16.3.
2. Derive equation 16.4.
3. In the solvolysis of *t*-butyl chloride some di-*t*-butyl ether is formed. Does this affect the application of equation 16.4?
4. Explain why addition of water to the solvent in the solvolysis of *t*-butyl chloride causes an increase in reaction rate.

17 Alkenes: E1 and E2 Reactions

17.1 Sources of Alkenes

Alkenes are often obtained from alcohols or alkyl halides by elimination reactions. Alcohols undergo elimination of water by heating with sulfuric or phosphoric acid,[1] or by passing the alcohol vapor over alumina or silica catalysts at high temperatures. Alkyl halides undergo loss of halogen acid (dehydrohalogenation) by heating with a solution of potassium hydroxide in ethanol.

Alkenes are produced industrially in enormous quantities by the pyrolysis (*cracking*) of alkanes at 400–600°, by passage over metal-oxide catalysts. The large alkane molecules undergo rupture of carbon–carbon bonds to form a mixture of smaller alkanes and alkenes. Catalytic dehydrogenation of alkanes is also an important industrial method for producing alkenes. Alkanes may also be converted to aromatic hydrocarbons (arenes); thus, *n*-heptane is changed stepwise to methylcyclohexane and finally to toluene, and *n*-hexane furnishes benzene.

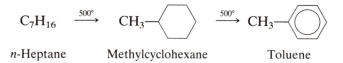

$$C_7H_{16} \xrightarrow{500°} CH_3- \bigcirc \xrightarrow{500°} CH_3- \bigcirc$$

n-Heptane   Methylcyclohexane   Toluene

The ease of dehydration of alcohols follows the sequence: tertiary > secondary ≫ primary. *t*-Butyl alcohol is converted rapidly to isobutylene

[1] Other dehydrating agents (iodine, oxalic acid, potassium bisulfate) are illustrated in Chapter 18.

(2-methylpropene) by 40–50% sulfuric acid at 85°; *sec*-butyl alcohol requires 60–65% acid at 100° for alkene formation; and *n*-butyl alcohol 75–80% acid at 135–140°. Phosphoric acid causes less oxidative degradation then strong sulfuric acid, but the rate of reaction is slower and higher temperatures are required. In the present experiment, the dehydration of secondary alcohols by 65% sulfuric acid at 95–110° is illustrated.

17.2 Carbocation Rearrangements

The mechanism of dehydration of alcohols varies in detail with the reagent and the structural type of the alcohol. For primary alcohols the mechanism is complex: at the high temperatures required for alkene formation with strong acids (170° for ethanol), the alcohol is in equilibrium with the corresponding dialkyl ether. Attack of the protonated alcohol or ether (oxonium ions) by the acid anion can occur by way of a one-step concerted process (E2 mechanism) in which the proton and water molecule are lost simultaneously. Primary carbocations are relatively unstable, and under most reaction conditions do not form.

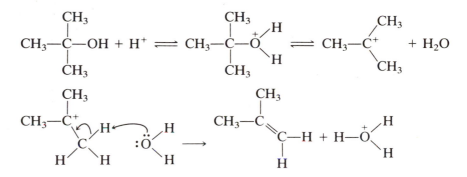

With tertiary and most secondary alcohols, protonation occurs more readily and the intermediate oxonium structure leads to a carbocation that loses a proton to form the alkene (E1 mechanism).

$$(CH_3)_3C-OH \xrightarrow{H^+} (CH_3)_3C-\overset{+}{O}H_2 \longrightarrow (CH_3)_3\overset{+}{C} \longrightarrow (CH_3)_2C=CH_2$$

If two different alkenes can be formed, there is usually a selectivity in the mode of elimination and one of the isomers predominates. For example, 2-pentanol yields mainly 2-pentene and very little 1-pentene. Likewise 4-methyl-2-pentanol furnishes only about 5% of 4-methyl-1-pentene.

$$CH_3CH_2CH_2-\overset{\underset{\displaystyle OH}{|}}{C}H-CH_3 \longrightarrow CH_3CH_2CH_2-\overset{+}{C}H-CH_3 \longrightarrow$$

$$CH_3CH_2CH=CH-CH_3$$

Sometimes the double bond is formed at a position removed from the carbon atom that was bonded to the hydroxyl group and occasionally the carbon skeleton itself is altered during the reaction. The formation of 2-pentene from 1-pentanol may occur through rapid protonation of 1-pentene as it is formed, under the reaction conditions, to give the more stable secondary carbocation. Alternatively, the secondary carbocation might be formed directly from the protonated alcohol, as the water molecule departs, by rapid migration of hydrogen with its bonding electrons (hydride shift).

Similar migration of a methyl group with its bonding electrons (methide shift) is observed in the dehydration of neopentyl alcohol ((CH_3)$_3$C—CH_3—OH). Shift of the methyl group accompanies formation of the primary neopentyl carbocation and converts it into the more stable tertiary pentyl carbocation, which leads to the formation of mostly 2-methyl-2-butene.

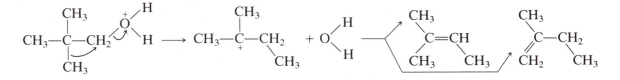

17.3 Dimerization of Isobutylene (2-Methylpropene)

Addition of a carbocation to an alkene furnishes a new carbocation and if the process continues, leads stepwise to alkene polymers of increasing molecular weight (dimers, trimers, and higher polymers). Usually small amounts of such dimers and trimers are formed in laboratory preparations of alkenes from alcohols.

In the presence of 60–65% sulfuric acid under mild conditions, t-butyl alcohol undergoes dehydration to isobutylene, which is converted mainly to octenes rather than high polymers (see Section 17.5(B)).

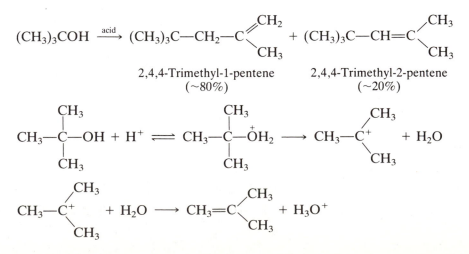

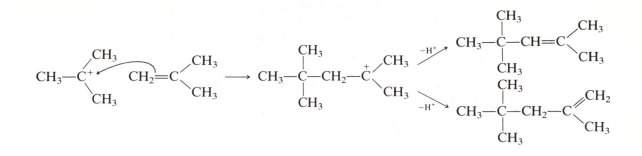

17.4 Reactions of Alkenes

The double bond of alkenes is easily oxidized by a number of reagents, including aqueous potassium permanganate. The organic product formed depends on the reaction conditions with *cis* hydroxylation being favored by mild conditions and cleavage to a dicarboxylic acid under more vigorous conditions. The inorganic product under neutral or basic conditions is manganese dioxide; the color change from the initial purple permanganate ion to the brown manganese dioxide forms the basis of the Baeyer test for unsaturation (Chapter 9).

$$3 \; \text{>C=C<} + 2 \; MnO_4^- + 4 \; H_2O \longrightarrow 3 \; \text{>C—C<} + 2 \; MnO_2 + 2 \; OH^-$$
$$\qquad\qquad\qquad\qquad\qquad\qquad\qquad\qquad OH \quad OH$$

$$3 \; \text{>C—C<} + 4 \; MnO_4^- \longrightarrow 3 \; \text{>C=O} \quad O=C< + 4 \; MnO_2 + 2 \; OH^-$$
$$\quad\;\; OH \quad OH \qquad\qquad\qquad\quad O^- \qquad O^-$$

17.5 Preparations

(A) Cyclohexene

Arrange a distillation assembly similar to that shown in Figure 2.10 using a 25-mL round-bottomed flask and an unpacked short fractioning column. Fit the lower end of the condenser into a 15-mL test tube cooled in an ice–water bath.

Place 5 mL of water in a 50-mL Erlenmeyer flask and add *carefully*, while swirling, 4.0 mL (8.0 g, 0.08 mole) of concentrated sulfuric acid (**Caution**— *corrosive!*). Cool the diluted acid (65% by weight H_2SO_4)[2] to 20–25° and add slowly 3.2 mL (3.0 g, 0.03 mole) of cyclohexanol (practical or technical grade).

---

[2] In place of sulfuric acid, 6 mL of 85% phosphoric acid may be used, but the reaction is slower. There is less discoloration.

Add a boiling chip and heat the reaction flask *gently* so that cyclohexene and water distill through the column. Continue the distillation until the liquid remaining in the distilling flask turns black and begins to evolve white vapors. About 3 mL of distillate will be collected.

◢▶ *CAUTION*   Cyclohexene is a volatile flammable liquid. Take care to minimize fire hazards and loss by evaporation.

The distillate will form two layers. With the aid of a Pasteur pipet remove *most* of the lower aqueous layer; wash the remaining upper cyclohexene layer with about 1 mL of 10% sodium carbonate solution to neutralize traces of acid (use the "squirting" technique described in Section 6.5; about a dozen squirts should mix the layers adequately). Remove *most* of the basic wash layer and then wash the residue with about 1 mL of water. Remove as much water as possible and dry the cyclohexene layer with about 0.5 g of anhydrous calcium chloride for 10–20 min with occasional swirling.

Decant the dried liquid into a 10-mL distilling flask, add a boiling chip, and distill carefully. Collect the material boiling at 77–83° in a weighed flask or bottle. If there is an appreciable fraction boiling below 77°, combine the fractions, dry them again, and redistill. The yield is 0.7–1.0 g.

If the reagents are available, carry out the bromine and potassium permanganate tests for unsaturation described in Chapter 9 or, if assigned, the oxidation of cyclohexene to adipic acid described below.

(B) Oxidation of Cyclohexene to Adipic Acid

In a 250-mL Erlenmeyer flask prepare a solution of 3.16 g (0.020 mole) of potassium permanganate in 100 mL of water. The permanganate dissolves slowly and occasional stirring is required to effect solution. Add the permanganate solution in 10-mL portions over a period of 10 min to a solution of 0.62 mL (0.62 g, 0.0076 mole) of cyclohexene in 5 mL of acetone contained in a second 250-mL Erlenmeyer flask. Between additions swirl the flask to mix the reagents. After the additions are complete, heat the mixture in a 55–60° water bath for 30 min.

Remove the flask from the hot water bath, add 1 g of sodium bisulfite to the dark reaction mixture, and chill the flask in an ice bath. Filter the reaction mixture on a large Büchner funnel into a clean filter flask; rinse the flask with about 25 mL of water and pour the rinse over the brown precipitate in the funnel. Transfer the clear filtrate to a 400–600-mL beaker and acidify it to about pH 2 with concentrated hydrochloric acid (follow the acidification with pH paper).

Reduce the volume of the filtrate to about 10 mL by boiling it in the hood. Chill the concentrated solution in an ice bath and collect the white precipitate of adipic acid on a Hirsch funnel. Draw air through the funnel to dry the product. The yield is 0.4–0.5 g, mp 149–152° (153–154° after recrystallization from water).

(C) 2,4,4-Trimethyl-1- and -2-pentenes (Diisobutylenes)

In this experiment you will be using *t*-butyl alcohol, which melts at 25–25.5°. If your laboratory is much warmer than 25°, you can conveniently measure out *t*-butyl alcohol as a liquid, but if the temperature is close to 25°, or below, the alcohol should be weighed even though it is a liquid in the bottle.

In a 50-mL round-bottomed flask, place 7 mL water and add cautiously, with swirling, 7 mL (12 g, 0.12 mole) of concentrated sulfuric acid (**Caution—** *corrosive!*). Cool the diluted acid to about 50° and add slowly 6.3 mL (5 g, 0.07 mole) of *t*-butyl alcohol. Attach at once a reflux condenser and boil the material gently for 30 min.

Cool the reaction mixture to room temperature, transfer it to a separatory funnel, and carefully draw off the aqueous acid layer. Wash the hydrocarbon layer with water to remove traces of acid and dry it with about 0.5 g of anhydrous calcium chloride. Decant the dry liquid into a small flask arranged for distillation and distill, collecting the fraction boiling at 100–108°. The yield is 1.8–2.0 g (2.7 mL). The recorded boiling points of the octenes are 2,4,4-trimethyl-1-pentene, 101–102°; 2,4,4-trimethyl-2-pentene, 104°.

If the reagents are available, carry out the bromine and potassium permanganate tests for unsaturation described in Chapter 9.

Gas chromatography on a SE-30 column is an excellent method for determining the ratio of the mixed octenes. For details of this method see Chapter 7.

Questions

1. Using the curved arrow formalism show the electron movement in the formation of cyclohexene from protonated cyclohexanol.

2. What alkene will be the main product when each of the following alcohols is dehydrated with acid?
   (a) 2-methyl-2-butanol
   (b) 3-methyl-2-butanol
   (c) 3-methyl-2-butanol

3. What isomeric hexenes (propylene dimers) can be formed by addition of the 2-propyl carbocation to propylene and subsequent loss of a proton ? How could the structure of these hexenes be established (by either chemical or spectrometric methods)?

4. Give equations to show the stepwise reactions to form diisobutylene from *t*-butanol.

5. Write a balanced equation for the permanganate oxidation of cyclohexene to adipic acid and manganese dioxide.

6. In the oxidation of cyclohexene by potassium permanganate the reaction mixture at the end is quite basic even though no base is added. Where does the base come from?

7. Write a reaction for the synthesis of cyclopentanone from adipic acid.

18

Alkenes: A Multiple-Step Synthesis

18.1 The Road from *n*-Butyl Alcohol to 2-Methylhexenes

The sequence of reactions undertaken in this experiment involves four steps and illustrates a typical situation in synthetic organic chemistry. Chemical transformations and isolation procedures always take place with some loss of material, and this reduces, sometimes quite drastically, the yield of product.

$$C_4H_9\text{—}OH + HBr \longrightarrow C_4H_9\text{—}Br + H_2O \qquad (18.1)$$

$$C_4H_9\text{—}Br + Mg \longrightarrow C_4H_9\text{—}MgBr \qquad (18.2)$$

$$C_4H_9MgBr + CH_3\overset{\overset{\displaystyle O}{\|}}{\text{—}C}CH_3 \longrightarrow C_4H_9\underset{\underset{\displaystyle OMgBr}{|}}{\text{—}C}(CH_3)_2 \qquad (18.3)$$

$$C_4H_9\underset{\underset{\displaystyle OMgBr}{|}}{\text{—}C}(CH_3)_2 + HBr \longrightarrow C_4H_9\underset{\underset{\displaystyle OH}{|}}{\text{—}C}(CH_3)_2 \qquad (18.4)$$

$$C_4H_9\underset{\underset{\displaystyle OH}{|}}{\text{—}C}(CH_3)_2 \xrightarrow{\text{acid}} \underset{\text{(alkene mixture)}}{C_7H_{14}} \qquad (18.5)$$

With four steps, each with an average yield of 80%, 1 mole of starting material will furnish 0.4 mole of finished product. This figure falls rapidly as more steps are involved; with the same average yield of 80% six steps will give

0.26 mole and eight steps, 0.17 mole. Lower average yields lead to more drastic losses; for an assumed yield of 70%, eight steps give only 0.057 mole of the end product. This emphasizes the need for good yields of the intermediate products and the desirability of short, direct syntheses without unnecessary steps.

In the present sequence, the first step is the conversion of _n_-butyl alcohol to _n_-butyl bromide, as described in Section 14.3 (A). The following step, formation of an organomagnesium halide (Grignard reagent), requires that _all_ of the reactants be strictly anhydrous. Conversion of the resulting tertiary alcohol, 2-methyl-2-hexanol (**I**; R = C_3H_7), to the 2-methylhexenes affords an example of the partitioning of a tertiary carbocation between isomeric alkenes during an E1-type elimination of water.[1]

For the final dehydration step, one of several different reagents may be used. This will give you an opportunity to see how a specific reagent can affect the distribution of isomers in the resulting alkenes.

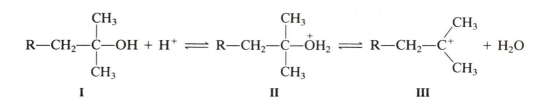

The carbocation intermediate (**III**) can lose one of the six hydrogens from the adjacent methyl groups to give the alkene **IV** (2-methyl-1-hexene) or lose a hydrogen from the adjacent CH_2 group to give the alkene **V** (2-methyl-2-hexene).

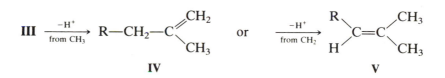

If there were no selectivity, one would expect **IV** and **V** to be formed in the ratio 3 of **IV** to 1 of **V**. The more highly substituted alkene **V** is the more stable product and at equilibrium would be present in the ratio of about 1 of **IV** to 10 of **V**. It will be of interest to compare the results with different dehydrating agents on the distribution of isomers **IV** and **V** and the yields. If a free carbocation is involved in the process, all of the catalysts should give the same result. But if the catalyst is intimately associated with the elimination step, then it may influence the relative amounts of **IV** and **V**.

---

[1] See also Section 17.3.

18.2 Grignard Synthesis of an Alcohol[2]

Organomagnesium halides, Grignard reagents, are among the most versatile synthetic intermediates for laboratory work. They are formed by simple, direct reaction of magnesium metal with alkyl or aryl halides (usually bromides) in the presence of a solvent such as ether or tetrahydrofuran. There is some uncertainty about the structure and the detailed mechanism of reaction of Grignard reagents. They seem to exist as coordination compounds in a complex equilibrium involving R_2Mg, $MgBr_2$, and $RMgBr$, but for convenience, they are simply designated $RMgBr$. The relative amounts of each form depend on the reaction conditions and may change during the reaction.

Unless the reactants, solvent, and apparatus are dried carefully and the magnesium is pure and relatively free of oxide coating, the reaction does not start readily. Addition of a small crystal of iodine aids in inducing reaction, probably by exposing a small fresh surface of the metal. A few drops of 1,2-dibromoethane (**Caution**—*cancer suspect agent!*) also may be used to start the reaction. Alkyl and aryl bromides are the preferred halides in most cases. With chlorides (except relatively reactive ones) the reaction is more difficult to start; with iodides there is greater tendency to favor a side reaction—coupling at the metal surface to form the hydrocarbon R—R (the Wurtz reaction).

The most important uses of the Grignard reagent involve two types of reaction, in both of which the alkyl or aryl group of R—MgX is transferred *with its bonding electrons* to a carbon atom of the reactant.

1. Addition to the carbonyl function of an aldehyde, ketone, ester, amide, acid halide, or carbon dioxide (or the cyano group of a nitrile).[3]

$$(CH_3)_2C{=}O \xrightarrow{RMgX} (CH_3)_2\underset{\underset{R}{|}}{C}{-}OMgX \xrightarrow[H^+]{H_2O} (CH_3)_2\underset{\underset{R}{|}}{C}{-}OH$$

$$C_6H_5{-}C{\equiv}N \xrightarrow{RMgX} C_6H_5{-}\underset{\underset{R}{|}}{C}{=}N{-}MgX \xrightarrow[H^+]{H_2O} C_6H_5{-}\underset{\underset{R}{|}}{C}{=}O$$

2. Replacement of the alkoxyl groups of esters and acetal and of the halogen atom of a reactive organic halide (also ring opening of alkylene oxides).

---

[2] Section 35.2 gives another example of the Grignard reaction in which an ester, methyl benzoate, is allowed to react with 2 moles of phenylmagnesium bromide to form triphenylmethanol.

[3] The relative reactivity of various functional groups toward phenylmagnesium bromide follows roughly the order: —CH=O, —CO—CH₃, —N=C=O, —CO—F, —CO—C₆H₅, —CO—Cl, —CO—Br, —CO₂Et, —C≡N. See Entemann and Johnson, *J. Am.. Chem. Soc.*, **55**, 2900 (1933).

$$C_6H_5-CO-OCH_3 \xrightarrow{\text{2 RMgX}} C_6H_5-\underset{\underset{R}{|}}{\overset{\overset{R}{|}}{C}}-OMgX \xrightarrow[H^+]{H_2O} C_6H_5-\underset{\underset{R}{|}}{\overset{\overset{R}{|}}{C}}-OH$$

$$\underset{\diagdown O \diagup}{CH_2-CH_2} \xrightarrow{\text{RMgX}} R-CH_2CH_2-OMgX \xrightarrow[H^+]{H_2O} R-CH_2CH_2-OH$$

The halomagnesium complex produced in the reaction is usually hydrolyzed by cold dilute mineral acid, to liberate the organic product. For acid-sensitive compounds strong aqueous ammonium chloride solution may be used.

Compounds containing active hydrogen (water, alcohols, ammonia and amines, acetylenes, phenols, acids) convert a Grignard reagent to the parent hydrocarbon R—H. Halogens, oxygen, and atmospheric carbon dioxide also react with R—MgX. Halides of boron, tin, mercury, and many other metals react with R—MgX to produce organic derivatives of these less reactive elements.

18.3 Preparation of 2-Methyl-1-hexene and 2-Methyl-2-hexene

(A) *n*-Butyl Bromide

Follow the procedure given in Experiment 14 (A). If your yield of bromide is less than the amount specified in the next step, adjust the quantities of the other materials to correspond.

(B) *n*-Butylmagnesium Bromide

In this experiment it is essential that all of the apparatus be clean and *dry* since traces of impurities or water can markedly reduce the yield (for example, do not handle the magnesium turnings with your fingers). If a drying oven is available, it is helpful to dry the components of the apparatus for about 20 min or longer before it is assembled.

Assemble an apparatus like that shown in Figure 8.1b with reflux condenser, separatory funnel for addition, and a 50-mL reaction flask. Prepare a bath of ice and water to permit rapid cooling if the reaction should become too vigorous. During all of the operations that follow make certain that there are no flames *anywhere nearby* that could ignite ether vapor.

CAUTION Ether is extremely volatile and highly flammable. Extreme care must be taken to avoid flames and electrical sparks.

In the reaction flask place 0.80 g (0.033 mole) of magnesium turnings (oven-dried magnesium works better) and add a *small* crystal of iodine. Place about 2.5 mL of *anhydrous* diethyl ether[4] in a small dry test tube and stopper the tube with a cork. In a dry 25-mL Erlenmeyer flask place 3.6–3.7 mL (4.5–4.6 g, 0.033 mole) of dry *n*-butyl bromide and add 10 mL of *anhydrous* diethyl ether. Pour the ether–bromide solution into the separatory funnel and allow 1–1.5 mL to flow onto the magnesium in the flask. Under favorable conditions, the reaction will start within 5 min without heat, accompanied by vigorous boiling of the ether and disappearance of the iodine color. As soon as this occurs, introduce the 2.5-mL portion of dry ether previously set aside directly through the top of the condenser to moderate the vigor of the reaction.

If the reaction does not start promptly, warm the flask gently in a bath of tepid water, but be prepared to cool the reaction quickly with the ice-water bath if the reaction starts suddenly. If the warming does not initiate reaction within 5 min, the yield will be low and it is best to start over with more carefully dried equipment and reagents. For the success of the experiment it is *absolutely essential* that the reaction begin before the remainder of the butyl bromide solution is added to the magnesium.

When the initial vigorous reaction has moderated, allow the remainder of the butyl bromide solution to flow dropwise into the flask at a rate such that the ether refluxes gently without external heating (about 10 min). Swirl the flask frequently to avoid a buildup of local concentrations of Grignard reagent. After all of the butyl bromide has been introduced, reflux the mixture gently (*no flames!*) for 20–30 min. Do not heat so vigorously that the ether vapor escapes through the condenser. At this point most of the magnesium will have reacted. The volume of solution should not be less than 15 mL; if it is, add more dry ether to bring the volume to about 15–20 mL.

(C) 2-Methyl-2-hexanol

Before adding acetone to the Grignard solution, cool the reaction flask in an ice–salt mixture to as low a temperature as possible.

In a small, dry Erlenmeyer flask mix 2.6 mL of dry acetone[5] with 4 mL of anhydrous ether. Transfer the solution to the separatory funnel and allow it to drop *very slowly* into the cooled Grignard solution, while swirling the solution to ensure good mixing and effective cooling. Each drop of the solution reacts vigorously, producing a hissing sound and forming a white precipitate that usually redissolves when the solution is swirled. After all the acetone has been added, remove the cooling bath and allow the reaction mixture to stand at room temperature with occasional shaking for 20 min or longer.

At this point one no longer need be concerned about keeping water away from the reaction mixture but remember that the ether fire hazard is still as great as ever.

---

[4] The ether used at this stage *must* be anhydrous.
[5] Analytical Reagent grade acetone is suitable.

Decant the reaction mixture from the residual magnesium chips slowly onto a mixture of chipped ice and dilute sulfuric acid (prepared by adding 1.1 mL of concentrated acid to 10 mL of water, and adding about 10 mL of chipped ice). Rinse the reaction flask with a little of the dilute sulfuric acid and a little ether,[6] and add these washings to the main product. Transfer the mixture to a separatory funnel and separate the two layers; *save both layers*. Extract the aqueous layer with two 5-mL portions of ether and combine the ether extracts with the ether layer from the first separation. The *aqueous* layer may now be discarded.

Wash the ether layer with 2–3 mL of cold water, then with 2–3 mL of saturated sodium bicarbonate, and once again with water. Separate the layers carefully and dry the ethereal layer over anhydrous magnesium sulfate. Pass the dried solution through a filter tube containing a small wad of glass wool[7] to remove the drying agent and let it drain into a small distilling flask. Remove most of the ether by distillation using a steam or hot-water bath.

Replace the bath with an electrical heating device (*no flames or sparks allowed when ether vapor is present in the lab*) and continue the distillation. Collect any impure material boiling above 75° and below 135°. The product distills over the range 135–144°. The typical yield is 1.2–2.0 g (1.4–2.4 mL). The reported boiling point of 2-methyl-2-hexanol is 141–142°.

Reserve 1.0 mL (0.8 g) of the product for conversion to alkenes, and hand in the remainder in a small, labeled vial.

(D) 2-Methylhexenes

Arrange a 10-mL round-bottomed flask for simple distillation (Figure 2.10), using as a receiver a centrifuge cone or a small test tube that can be cooled in an ice bath.[8] In the reaction flask, place a dehydration catalyst chosen from the following list or a special catalyst provided by your instructor.

1. Iodine: Use a small crystal; only a little is needed. The distillate is likely to have a red color from traces of iodine, which will fade in several hours; it can be removed quickly with a little aqueous bisulfite.
2. Oxalic acid: Use 0.5 g finely ground; the solid should dissolve on heating. Oxalic acid is quite toxic!
3. Potassium bisulfate ($KHSO_4$): Use 0.5 g finely ground; the solid may not dissolve on heating.
4. Phosphoric acid: Use 0.5 mL of 85% H_3PO_4.

[6] From here on, the ether used may be ordinary, wet diethyl ether.
[7] You can use a small fluted filter to remove the drying agent but the losses from evaporation are greater.
[8] Alternatively, the dehydration can be carried out in the inexpensive Hickman still described in Section 2.3. The advantage of this still is that it can be used with very small samples; its disadvantage is that the distillation temperatures cannot be monitored.

Place 1.0 mL (0.8 g, 0.007 mole) of 2-methyl-2-hexanol in the 10-mL flask, mix it well with the catalyst, and heat the material *slowly*. Distill the resulting alkenes *carefully*, not allowing the distillate temperature to rise above 90°. Rapid distillation or overheating will result in incomplete reaction by forcing unreacted alcohol into the distillate.

◀▶ *CAUTION*   The 2-methylhexenes are volatile and flammable.

Wash the cloudy distillate with 1 mL of saturated sodium bicarbonate solution (squirting technique), draw off the lower aqueous layer, and dry the remaining alkene layer with a little anhydrous magnesium sulfate. Pipet the alkenes into a tared vial and store the sample, tightly capped, until it can be analyzed by gas chromatography or NMR spectrometry. Record the weight and yield of the alkenes; the recorded bp of pure 2-methyl-1-hexene is 91°; of 2-methyl-2-hexene, 94.5°.

(E) Chromatographic Analysis of the 2-Methylhexenes

Analyze your alkene sample by gas chromatography using a polar silcon oil column such as DC-710 (see Chapter 7; directions for preparing the column are given in the Appendix). Although a 5% solution of the alkene mixture in methylene chloride gives usable results, a 5% solution in diethyl ether (**Caution—** *fire hazard!*) avoids solvent inteference. To confirm the identity of the components, it is highly desirable to have pure samples of 2-methyl-2-hexene and 2-methyl-1-hexene available for comparing the individual retention times (at 25° on DC-210 the relative retention times are about 11.6 and 13.0 for the 1- and 2-isomers). In calculating the ratio of isomers in your mixture, assume a molar response factor of 1 : 1. Compare your observed ratio with the expected "statistical" ratio calculated by assuming equal rate of attack on each proton.

(F) NMR Analysis of the 2-Methylhexenes

If an NMR instrument is available, the ratio of isomers can be determined by comparing the mixture NMR spectrum with the spectra of the pure components shown in Figure 18.1. The protons bound to the sp^3-hybridized carbons give rise to complex, but different patterns between 0.7 and 2.3 ppm. In a mixture of the two alkenes, these peaks will be superimposed, which prevents accurate differentiation. However, the protons bound to the sp^2-hybridized carbons give rise to distinct absorptions: a singlet at 4.6 ppm for the 1-alkene, and a triplet at 5.1 ppm for the 2-alkene. With a mixture of these two, these peaks are easily distinguished and can be integrated to give, after correction for the relative number of protons giving rise to the peaks, the ratio of 2-methyl-1-hexene and 2-methyl-2-hexene in the sample.

Questions   1. In the preparation of *n*-butylmagnesium bromide, what reaction is minimized by the dilution with additional ether? Why is the *n*-butyl bromide added dropwise?

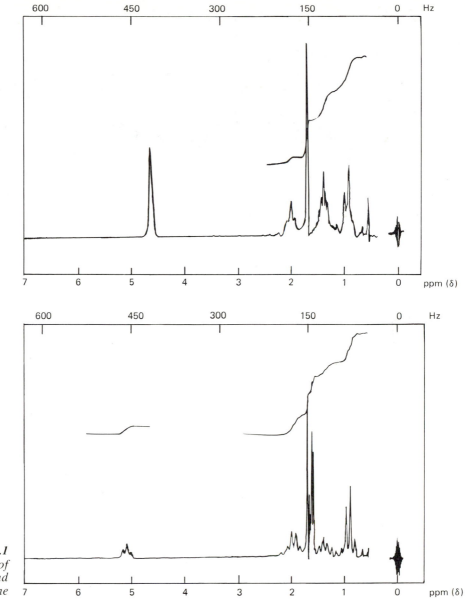

FIGURE 18.1
NMR Spectra of
2-Methyl-1-hexene and
2-Methyl-2-hexene

2. If diethyl ether were not available for the Grignard reaction, explain why each of the following would/would not be a suitable substitute solvent.
   (a) petroleum ether   **(b)** tetrahydrofuran
   (c) ethanol   **(d)** acetonitrile

3. The flammability of a solvent is determined in part by its volatility and in part by its rate of reaction with oxygen. What intermediate is formed in the reaction of ether with oxygen that causes it to react so rapidly?

19 Hydration of Alkenes and Alkynes

19.1 Hydration of Double Bonds

An important industrial reaction is the hydration of alkenes to produce alcohols. For example, *t*-butyl alcohol, a valuable antiknock fuel additive, is prepared by the hydration of isobutylene with 60–65% aqueous sulfuric acid as described in Chapter 17.

$$(CH_3)_2C\!=\!CH_2 + H^+ \rightleftharpoons (CH_3)_3C^+ \xrightarrow{H_2O}$$
$$(CH_3)_3C\overset{+}{O}H_2 \rightleftharpoons (CH_3)_3COH + H^+$$

The reaction proceeds through a *t*-butyl cation and is the reverse of the acid-catalyzed dehydration of alcohols. The direction of the reaction is determined by the reaction conditions. Hydration is favored by low temperatures and an aqueous medium.

In the isobutylene hydration, two carbocation intermediates might conceivably be formed, giving rise to two different alcohols.

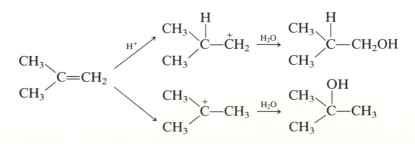

In practice, the tertiary cation is so much more stable than the primary cation that only the tertiary alcohol is observed. In general, the position of the hydroxyl group is the site of the more stable carbocation, that is, the reaction follows Markovnikov's rule.

One consequence of the carbocation mechanism for hydration is that in some circumstances the intermediate carbocation can rearrange to an altered carbon skeleton before it reacts with water to give rise to the alcohol. Such rearrangements are more likely when the rearranged cation is more stable than the initial cation. For example, when 3,3-dimethyl-1-butene is hydrated with sulfuric acid and water, the major product is the rearranged alcohol 2,3-dimethyl-2-butanol. The initial secondary cation rearranges to the more stable tertiary cation which then reacts. Apparently the rate of rearrangement of the methyl group is faster than the rate of attack by water on the secondary cation.

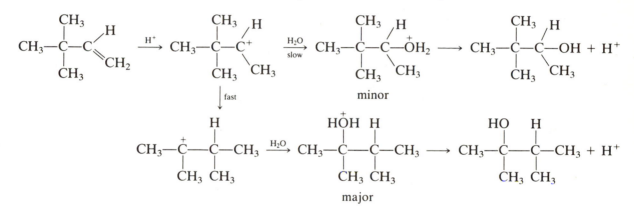

19.2 Oxymercuration–Demercuration of Alkenes

A major improvement on the acid-catalyzed hydration of alkenes is provided by the oxymercuration–demercuration procedure developed by Brown and Geohegan.[1] Their standard procedure utilizes a 1:1 water–tetrahydrofuran solution of mercuric acetate, which reacts with most alkenes to give a mercury-containing intermediate that yields an alcohol on reduction with sodium borohydride.

$$R-CH=CH_2 \xrightarrow[H_2O/THF]{Hg(OAc)_2} \overset{OH}{\underset{H}{\overset{R}{\diagup}}}C-CH_2-HgOAc \xrightarrow[NaOH]{\overset{H^-}{from}\ NaBH_4} \overset{OH}{\underset{H}{\overset{R}{\diagup}}}C-CH_3 + Hg + \bar{O}Ac$$

The alcohol is formed in excellent yields (typically 90–100%) and follows the Markovnikov substitution pattern. The advantage of the Brown procedure over

[1] Brown and Geohegan, Jr., *J. Org. Chem.*, **35**, 1844 (1970).

acid-catalyzed hydration is that it is much faster and generally gives the unre-arranged alcohol.

The oxymercuration–demercuration of 1-hexene to give 2-hexanol is de-scribed in detail in Section 19.4 (A).[2] In this standard procedure, tetrahydrofu-ran is used as the solvent. All chemists should be aware that tetrahydrofuran tends to form peroxides on exposure to air, and that these peroxides have led to a number of laboratory explosions. As supplied by the manufacturers, tetrahy-drofuran is stabilized by adding small amounts of antioxidant, which prevents buildup of peroxides from short air exposure. However, on purification the antioxidant may be removed or destroyed and the tetrahydrofuran becomes susceptible to peroxide formation with even brief exposure to air. As a general rule, only freshly purchased, stabilized tetrahydrofuran should be used.

19.3 Hydration of Alkynes

The triple bond of alkynes can also be hydrated with aqueous acid. The intermediate vinyl alcohols are unstable and rearrange immediately to give a carbonyl compound.

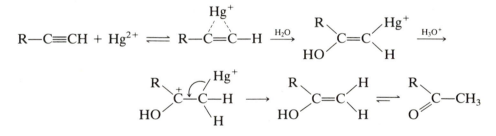

The initial addition follows Markovnikov's rule so that with terminal acetylenes a methyl ketone rather than an aldehyde is formed. The only exception is acet-ylene ($HC\equiv CH$), which as a consequence of its unique structure must yield acetaldehyde. In practice, better yields are obtained if a mercuric-ion catalyst is added. In this respect the hydration of alkynes differs sharply from that of alkenes, which require that a full equivalent of mercuric ion be used.

19.4 Reactions and Preparations

(A) Oxymercuration–Demercuration of 1-Hexene

In a 50-mL round-bottomed flask, dissolve 1.3 g (0.0040 mole) of mercuric acetate, $Hg\,(OCOH_3)_2$, in 5 mL of water.

---

[2] The preparation of 2-hexanol by acid-catalyzed hydration of 1-hexene with 85% sulfuric acid at 100° has been described by McKee and Kauffman, *J. Chem. Educ.*, **59**, 695 (1982).

▐▶ *CAUTION*   Mercuric acetate is very poisonous. It is advisable to wear disposable gloves during this experiment and to wash your hands thoroughly after you are finished.

Add 5 mL of tetrahydrofuran and 0.5 mL (0.34 g, 0.0040 mole) of 1-hexene.

▐▶ *CAUTION*   Tetrahydrofuran is volatile and extremely flammable. Only stabilized material should be used.

A yellow precipitate forms immediately. Swirl the flask to dissolve the precipitate. Attach a water-cooled condenser, add a boiling chip, and add through the condenser 2 mL of 6 M aqueous sodium hydroxide followed by 4 mL of a 0.5-M solution of sodium borohydride (0.002 mole) in 3 M aqueous sodium hydroxide.[3] Boil the mixture for 15 min on a steam bath.

Cool the mixture to room temperature and allow the mercury droplets to coalesce and settle (this may require waiting until the next laboratory period). Carefully decant the upper liquid layers from the mercury into a small separatory funnel. Discard the mercury into the container situated in the hood. (Do not pour the mercury into the sink.)

Add 10 mL of methylene chloride to the water–tetrahydrofuran mixture contained in the separatory funnel and separate the lower organic layer. Dry the organic layer over magnesium sulfate.[4]

▐▶ *CAUTION*   Prolonged exposure to high concentrations of methylene chloride vapors may induce cancer. As with all organic solvents work in a well-ventilated area when using it.

Remove the drying agent by filtration and analyze the organic solution by gas chromatography (silicon oil column prepared as described in the Appendix) for the relative concentrations of the 2-hexanol (Markovnikov) and 1-hexanol (anti-Markovnikov) products. A reference solution of the alcohols in methylene chloride can be used to determine their relative retention times.

(B) 2-Heptanone by Hydration of 1-Heptyne

In a 100-mL round-bottomed flask place 4 mL of water and add to it *slowly* and cautiously 2 mL of concentrated sulfuric acid (**Caution—***corrosive*!). To the warm acid solution add 0.1 g (0.46 mmole) of red mercuric oxide.

▐▶ *CAUTION*   Mercuric oxide is very poisonous. It is advisable to wear disposable gloves during this experiment and to wash your hands thoroughly after you are finished.

---

[3] Sodium borohydride hydrolyzes slowly in aqueous sodium hydroxide to give sodium borate and hydrogen gas. The solution should be freshly prepared.

[4] A quicker way to dry the solution is to first remove any dispersed water droplets with anhydrous sodium sulfate and then, after filtration or careful decantation, distill off about half of the methylene chloride.

Add a boiling chip, attach a reflux condenser, and add through the condenser 10 mL of methanol followed by 3 mL (2.2 g, 0.022 mole) of 1-heptyne.[5] The mercury complex of the alkyne will precipitate, and an exothermic reaction will begin. Swirl the mixture to promote smooth reaction and after the initial boiling subsides, heat the reaction mixture until it boils, and continue heating for 30 min.

Cool the reaction mixture to room temperature and add 20 mL of water. Rearrange the apparatus for simple distillation and distill the mixture until about 25 mL of distillate has been collected. Add approximately 3 g of sodium chloride to salt out the product.[6] Stir the mixture to hasten solution of the salt and then decant the mixture into a separatory funnel, leaving behind any undissolved salt. Separate the layers, extract the lower aqueous layer twice with 4-mL portions of dichloromethane, and combine the original organic layer with the extracts.

⮕ CAUTION   Prolonged exposure to high concentrations of methylene chloride vapors may induce cancer. As with all organic solvents, work in a well-ventilated area when using it.

The residue from the distillation, which contains the mercuric salts, should be poured carefully into the mercury waste container in the hood. (Do not pour this severely poisonous solution into the sink. Wash your hands carefully after this operation.)

Dry the combined organic layers over anhydrous magnesium sulfate, filter the sample into a 25-mL round-bottomed flask, and distill off the solvent and any unreacted 1-heptyne (bp 35–100°). Continue the distillation and collect the product, bp 148–150°, in a preweighed container. The yield of 2-heptanone is about 1.6–1.8 g. Place the low-boiling distillate, which contains methylene chloride, in the marked waste container.

2-Heptanone has a penetrating fruity odor and is said to be responsible for the "peppery" odor of Roquefort cheese. It is used commercially as a constituent of artificial carnation essence.

Small samples of 2-heptanone can be converted to the semicarbazone and 2,4-dinitrophenylhydrazone as described in Chapter 9.

If IR or NMR instruments are available, determine the spectra of the starting 1-heptyne and the 2-heptanone product. Identify the characteristic spectral features.

Questions   1. Explain why hydration of double bonds is favored by low temperature and an aqueous medium. (Hint: Consider the equilibrium expression for the reaction.)

2. What is Markovnikov's rule and what is its mechanistic basis?

---

[5] 1-Heptyne is available inexpensively from Heico Division, Whittaker Corporation, Delaware Water Gap, PA.

[6] See Section 6.3 for a discussion of "salting out."

3. Why does the oxymercuration–demercuration reaction on 1-hexene yield 2-hexanol rather than 1-hexanol? Why does 1-hexyne yield 2-hexanone rather than hexanaldehyde?

4. Give the equations for the formation of 2-heptanone from 1-heptyne by mercuric-ion-catalyzed hydration.

5. What are the expected characteristic NMR and IR absorptions for 1-heptyne and 2-heptanone?

20 Glaser–Eglinton–Hayes Acetylene Coupling

20.1 Introduction

A major challenge in the design of an organic synthesis is the construction of the carbon skeleton. For this reason any method for creating new C—C bonds is of interest, particularly if it leaves reactive sites that can subsequently be converted into the desired functional groups.

The usual methods for forming new C—C bonds add one or two carbon atoms at a time. When appropriate, it is far more efficient to couple the two halves of the molecule in a single step. More than a century ago, Glaser discovered that it was possible to couple oxidatively two acetylene molecules using air and a basic solution of cuprous ion as oxidant.

$$C_6H_5-C\equiv C-H \xrightarrow[\text{air}]{Cu^+,\ NH_4OH} C_6H_5-C\equiv C-C\equiv C-C_6H_5$$

In 1956 Eglinton and Galbraith discovered that catalytic quantities of cupric ion (Cu^{2+}) could be used but reported that the reaction was slow. In 1962 Hayes found that superior catalysis was obtained with tertiary amine complexes of cuprous (Cu^+) salts. Under these conditions, even at room temperature, the reaction proceeded almost as rapidly as the oxygen could be added. For example, the acetylene 1-ethynylcyclohexanol had a reaction half-life (the time for 50% reaction) of 13 min and gave a 93% yield of the coupled acetylene product.

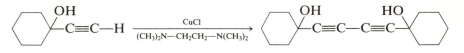

An interesting application of the acetylene coupling reaction was the synthesis of cyclic hydrocarbons containing very large rings.

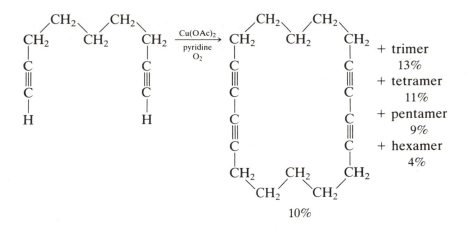

Catalytic reduction gave the saturated macrocyclic rings including the hexamer, which has a ring of 54 CH_2 groups.

20.2 Mechanism of Acetylene Coupling

The mechanism of the Glaser–Eglinton–Hayes acetylene coupling reaction has been examined and, although the details are still obscure, at least the broad features can be understood. Both the Cu^+ and Cu^{2+} ions appear to be required. The Cu^{2+} is the actual oxidizing agent, but Cu^+ appears to be required to form the copper acetylide, $R—C\equiv C—Cu$. It is proposed that a dimer of the copper(I) salt is formed, which is then oxidized by copper(II) to the coupled diacetylene product. The required copper(II) is generated by air oxidation of copper(I).

$$2\ R—C\equiv C—H + 2\ Cu^+(amine)_2 \rightleftharpoons 2\ H^+ + \begin{array}{c} amine \\ \downarrow \\ R—C\equiv C—Cu\leftarrow amine \\ \vdots\qquad\vdots \\ amine\rightarrow Cu—C\equiv C—R \\ \uparrow \\ amine \end{array} \xrightarrow{2\ Cu^{2+}\ (amine)_2}$$

$$R—C\equiv C—C\equiv C—R + 4\ Cu^+(amine)_2$$

In this mechanism the amine serves both as a base to consume the acid generated in the formation of the complex and as a solubilizing agent to prevent the precipitation of the acetylide salt.

20.3 Preparation

Oxidative Coupling of 1-Ethynylcyclohexanol

Assemble the apparatus shown in Figure 20.1 using a 250-mL filter flask and a 100-mL round-bottomed flask. The purpose of the filter-flask bubbler is to saturate the air with acetone to minimize evaporation of the reaction solvent in the round-bottomed flask. Place about 30 mL of acetone in the filter flask and 12 mL of acetone in the round-bottomed flask. Add 7 drops (0.07 g, 0.5 mmole) of tetramethylethylenediamine and 0.06 g (0.6 mmole) of cuprous chloride (the

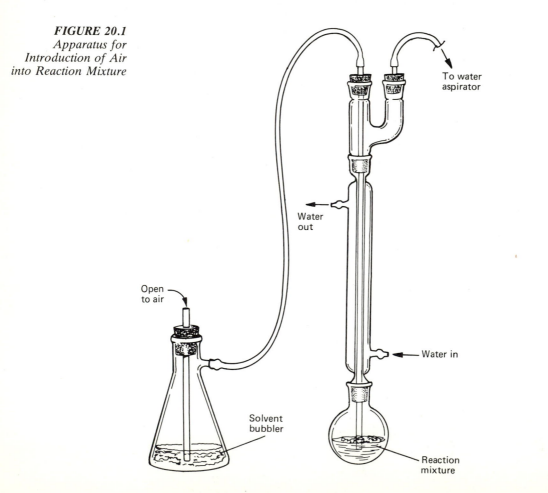

FIGURE 20.1
Apparatus for Introduction of Air into Reaction Mixture

To water aspirator

Water out

Water in

Open to air

Solvent bubbler

Reaction mixture

preparation is described below). Swirl the flask until most of these reagents have dissolved and then add 1 g (8.0 mmole) of 1-ethynylcyclohexanol. Immerse the two flasks in pans of water heated to about 40°, turn on the water aspirator, and for 20 min draw air through the bubbler and through the reaction mixture.

At the end of the reaction period, disconnect the filter-flask bubbler, turn off the water flow to the condenser, and continue to draw air through the round-bottomed flask until the acetone has evaporated. Add 4 mL of water and 0.2 mL of concentrated hydrochloric acid to the residue. The cupric salts will go into solution and leave the product as a solid. Collect the solid on a Büchner funnel, wash with 10-mL portions of water, and draw air through the funnel until the product is dry. The yield of crude product is about 0.7 g and is quite pure. A purer material can be obtained by recrystallization from 3 : 1 hexanes; ethyl acetate, mp 174–177°.

Preparation of Cuprous Chloride Solution. In a 25-mL round-bottomed flask, prepare a solution of 1.2 g (0.004 mole) of powdered copper sulfate crystals ($CuSO_4 \cdot 5H_2O$) and 0.38 g of sodium chloride in 5 mL of hot water. In an Erlenmeyer flask prepare a solution of 0.9 g of solid sodium hydroxide in 1.4 mL of water. Transfer 0.3 mL of the hydroxide solution to a small beaker and add 0.28 g sodium bisulfite. Add this solution with swirling to the hot copper sulfate solution. Chill the mixture in an ice bath and collect the precipitated cuprous chloride on a Hirsch funnel. Wash the solid with two small portions (2–3 mL each) of water and then with acetone. Transfer the solid to a filter paper and allow it to dry for about 5 min. It is best to use the material immediately, but it can be stored for several days under acetone.

Questions 1. Give the principal products for the reaction of 1-ethynylcyclohexanol and its Glaser–Eglinton–Hayes coupling product with the following reagents.
   (a) H_2/Pt
   (b) aqueous $AgNO_3$
   (c) H_2 (1 equiv)/Pd–BaSO$_4$–quinoline
   (d) excess Br_2/CCl_4, 0°

2. The starting material for this preparation, 1-ethynylcyclohexanol, is an industrial product. Suggest how it is made.

3. Vinylacetylene is synthesized industrially by the dimerization of acetylene with a mixture of cuprous chloride, ammonium chloride, and HCl (the Nieuwland enyne synthesis). Do you think that the mechanism is the same as the Glaser–Eglinton–Hayes reaction? If not, what might the mechanism be?

21 Oxidation of Alcohols to Ketones and Their Transformation to Amides

21.1 Chromic Acid Oxidation of Alcohols

Chromic acid mixture (dichromate salts and 40–50% sulfuric acid) oxidizes a primary alcohol stepwise to the aldehyde and the corresponding carboxylic acid (see Section 9.7 (A)). Unfortunately, it is difficult to stop the oxidation at the aldehyde stage and special reagents or reaction conditions must be used for a satisfactory preparation.

$$Na_2Cr_2O_7 + 2\,H_2SO_4 \rightleftharpoons H_2Cr_2O_7 + 2\,NaHSO_4$$

$$H_2Cr_2O_7 + H_2O \rightleftharpoons 2\,H_2CrO_4$$

$$3\,R{-}CH_2OH + 2\,H_2CrO_4 + 3\,H_2SO_4 \longrightarrow 3\,R{-}CH{=}O + Cr_2(SO_4)_3 + 8\,H_2O$$

(orange-red)　　　　　　　　　　　(green)

Oxidation of secondary alcohols to ketones by means of chromic acid mixture is generally a satisfactory method of preparing ketones, because the latter do not undergo further oxidation so easily. The oxidation is exothermic and the temperature must be controlled to avoid a violent reaction. For water-insoluble compounds, chromic acid ($CrO_3 + H_2O$) in an acetone–sulfuric acid medium (Jones' reagent) or in glacial acetic (Fieser's reagent) may be used. Oxidation with chromic acid in acetone solution is rapid and quite selective; usually a double bond is not attacked. During the oxidation the orange-red chromic acid is converted to the green chromium (3+) ion, which is the basis of the chromic acid oxidation test for alcohols (see Section 9.7 (A) and Chapter 38).

Under ordinary conditions tertiary alcohols are relatively stable to cold chromic acid, but under more vigorous conditions tertiary alcohols, and ketones, may be degraded by cleavage of carbon–carbon bonds.

Chromic acid oxidation of primary and secondary alcohols occurs through formation of the chromate ester. The next step is the slowest in the reaction and involves removal of a proton from the adjacent carbon atom, forming the ketone and H_2CrO_3 (Cr^{4+}). Disproportionation of this acid gives CrO_3 and a salt of Cr_2O_3 (Cr^{3+}).

$$R-CH_2OH + CrO_3 \longrightarrow \underset{\text{Chromate ester}}{R-CH_2O-CrO_3H} \longrightarrow H_2CrO_3 + R-CH{=}O$$

$$R-CH{=}O \xrightarrow[\text{slow}]{CrO_3} R-CO_2H$$

$$R-CH{=}O \underset{H_2O}{\rightleftharpoons} \underset{\underset{H}{\overset{|}{O}}}{\overset{|}{R-CH-OH}} \quad \uparrow CrO_3 \text{ fast}$$

$$3\, H_2CrO_3 \xrightarrow{H_2SO_4} CrO_3 + Cr_2(SO_4)_3$$

$$3\, H_2CrO_3 + 3\, H_2SO_4 \longrightarrow CrO_3 + Cr_2(SO_4)_3 + 6\, H_2O$$

Replacement of the pertinent hydrogen by deuterium causes the oxidation rate of 2-propanol to decline to about one-sixth the normal rate (kinetic isotope effect).

21.2 Oxidation of Alcohols with Sodium Hypochlorite

Chromic acid oxidations work well but there is considerable ecological concern about the disposal of chromium wastes. There is evidence that factory workers exposed to fine particles of chromium salts have a higher than normal incidence of cancer. An alternative oxidation procedure is to use aqueous hypochlorous acid.

$$\overset{|}{\underset{|}{H-C-OH}} + HOCl \rightleftharpoons \overset{|}{\underset{|}{H-C-OCl}} + H_2O$$

$$H_2O + \overset{|}{\underset{|}{H-C-OCl}} \rightleftharpoons H_3O^+ + C{=}O + Cl^-$$

$$H_3O^+ + Cl^- \rightleftharpoons H_2O + HCl$$

A convenient source of hypochlorous acid is liquid "swimming pool chlorine," which is a 12.5% solution, by weight, of aqueous sodium hypochlorite. This material is prepared commercially by bubbling chlorine gas into aqueous sodium hydroxide, where it reacts rapidly to develop a mobile equilibrium between chlorine and hypochlorite.

$$Cl_2 + OH^- \rightleftharpoons OCl^- + Cl^-$$

Addition of acid to this basic solution converts the hypochlorite ion into hypochlorous acid. Analogous inorganic chemistry, discussed in Chapter 9, involves the equilibrium between iodine and hydroxide to form hypoiodite and iodide in the iodoform test (page 149). Because the equilibrium is established rapidly, "swimming pool chlorine" is a good source of both chlorine and hypochlorite.

21.3 Transformation of the Carbonyl Group

One of the most useful features of the carbonyl group is the rich chemistry it affords. The next several chapters of this book illustrate a few of the many reactions carbonyl groups undergo. In this experiment we will explore the two-step transformation of a carbonyl group into an amide, illustrated here by the conversion of cyclohexanone into the cyclic amide, ε-caprolactam, which is used for the commercial production of nylon-6. The first step involves the conversion of cyclohexanone into cyclohexanone oxime (the chemistry of oxime formation is described in greater detail in Chapters 9 and 22). This oxime when heated in polyphosphoric acid undergoes the Beckmann rearrangement to produce the ε-caprolactam. The same transformation applied to 2-pentanone would yield a mixture of N-propylacetamide and N-ethylpropanamide.

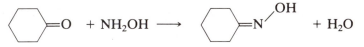

Cyclohexanone   Cyclohexanone oxime

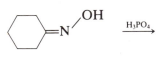

Cyclohexanone oxime   ε-Caprolactam

21.4 Preparations

(A) Methyl n-Propyl Ketone (2-Pentanone)

In a 100-mL round-bottomed flask place 10 mL of water and add carefully, with cooling, 2.2 mL (4 g, 0.04 mole) of concentrated sulfuric acid (**Caution—**
corrosive!). To the cold, diluted acid add 6.5 mL (5.3 g, 0.06 mole) of 2-pentanol,[1] and swirl the flask to obtain good mixing. To prepare the chromic

[1] Diethyl ketone (3-pentanone) may be prepared from 3-pentanol by the same method; in the preliminary distillation of the product the azeotrope, containing 14% water, distills at 82.9°.

acid oxidizing solution, dissolve 6 g (0.02 mole) of sodium dichromate dihydrate in 10 mL of water, add carefully 2.2 mL (4 g, 0.04 mole) of concentrated sulfuric acid (**Caution**—*corrosive!*), and cool the solution to room temperature.

⟱⟱⟩ *CAUTION*   Dust containing chromium salts is suspected of causing cancer. Do not grind the sodium dichromate to hasten its solution.

Introduce the oxidizing solution into the pentanol solution in small portions, swirling the solution between additions, and observe the temperature. By intermittent cooling in a pan of water, as needed, maintain the internal temperature at 30–50°. If the temperature is kept too low, the oxidizing agent may accumulate in the solution and react suddenly with great vigor. When all of the oxidizing agent has been added, and the temperature no longer rises spontaneously, attach a water-cooled condenser to the flask and heat the flask on a steam bath for 20 min.

Allow the flask to cool for a few minutes, remove the condenser, fit the flask with a short fractioning column (it need not be packed), and attach a distillation head. Add 30 mL of water and a boiling chip, and steam distill the mixture (steam distillation is discussed in Chapter 4) until a test portion of the distillate is essentially free of oil droplets. Do not collect an excessive amount of aqueous distillate; the ketone is appreciably soluble in water and a larger portion will be lost in the solution. The azeotrope of 2-pentanone and water, containing 20% of water, distills at 83.3°; stop the distillation when the temperature of the distilling vapor has risen to 98–99°.

After the spent oxidizing solution has cooled to room temperature, transfer it into the labeled waste jar situated in the hood.

To the distillate add 0.1 g of solid sodium carbonate to neutralize any acid, and salt out[2] the dissolved ketone by adding 0.5 g of sodium chloride for each 2 mL of water present. Draw off the aqueous layer and transfer the ketone to a small Erlenmeyer flask. Add 0.5–1 g of anhydrous magnesium sulfate and shake well. If an aqueous layer is formed, draw this off and add a fresh portion of the drying agent. Filter (or decant) the dried liquid into a small distilling apparatus, add a boiling chip, and distill (include a short Vigreux column if one is available). If the crude product has been well dried, it will boil in the range of 97–102°; otherwise it will boil low. The yield is 2.2–3 g. the product should be colorless.[3] The product may be checked for contaminating starting material by either IR or GC (3% Carbowax, prepared as described in the Appendix, works well).

If much of the product boils below 97°, combine the distillation fractions, dry them, and redistill. The low-boiling product can be used directly for preparation of the semicarbazone 2,4-dinitrophenylhydrazone and other derivatives given in Chapter 9.

---

[2] For a discussion of salting out, see Section 6.3.

[3] Yellow discoloration of the product may arise from the presence of a little of the intensely yellow diketone, 2,3-pentanedione. This impurity is alkali-sensitive and can be removed by adding about 0.1 g of crushed sodium hydroxide and redistilling.

(B) 2-Pentanone by Hypochlorite Oxidation

This oxidation of 2-pentanol to 2-pentanone uses "swimming pool chlorine," which is a 12.5% by weight aqueous sodium hypochlorite solution. Because of the equilibrium between aqueous hypochlorite and chlorine gas this reaction must be carried out in the hood or other well-ventilated area.

In a 125-mL Erlenmeyer flask, dissolve 6.5 mL (5.3 g, 0.06 mole) of 2-pentanol in 15 mL of glacial acetic acid. Cool the flask in an ice–salt bath to about $-5°$ and add dropwise by means of an addition funnel 45 mL of fresh concd aqueous sodium hypochlorite solution (about 5.6 g, 0.09 mole of NaOCl). Swirl the mixture during the addition to ensure thorough mixing and observe its temperature. The mixture rapidly becomes warm; the temperature should be maintained between 30 and 35° by controlling the addition rate and the degree of cooling. The greenish color of the hypochlorite solution will be discharged as it reacts, but a pale yellow color will persist near the end of the addition.[4] Swirl the mixture occasionally for 15 min and then add saturated sodium bisulfite solution until the yellow color disappears (about 4 mL).

Pour the mixture into a 250-mL round-bottomed flask, add an additional 25 mL of water, and attach a short unpacked fractionating column (or a Claisen head). To the top of the column (or Claisen head) attach a distillation head, add a boiling chip, and steam distill the mixture into a 100-mL receiver until about 30 mL of distillate, consisting of water, acetic acid, and an upper layer of cyclohexanone, has been collected.

Neutralize the acetic acid by adding slowly, with stirring, solid sodium carbonate (5–6 g). Then saturate the aqueous layer with 10 g of sodium chloride to salt out any dissolved cyclohexanone.[2] Decant the mixture into a separatory funnel, separate and save the 2-pentanone layer, and extract the remaining aqueous layer with 10 mL of methylene chloride, or pentane (**Caution**—*flammable!*).

IIII➤ CAUTION Prolonged exposure to high concentrations of methylene chloride vapors may induce cancer. As with all organic solvents, work in a well-ventilated area when using it.

Combine the solvent extract with the 2-pentanone layer and dry it with 0.5–1.0 g of anhydrous magnesium sulfate after removing any water layer that may have separated. Filter the dried solution (a filter tube works best) into a 25-mL distilling flask, attach a distillation head (and a short Vigreux column if one is available), and distill off the solvent and collect the 2-pentanone fraction boiling at 97–102°. The yield is 2.2–3 g. The product can be checked for contaminating starting material by either IR or GC (3% Carbowax, prepared as described in the Appendix, works well).

---

[4] If the "swimming pool chlorine" is old, there may not be sufficient hypochlorite (as indicated by the yellow color of the chlorine in equilibrium with it) to complete the oxidation. Add more hypochlorite solution (no more than 3 mL) until a persistent pale yellow color develops.

If much of the product boils below 97°, combine the distillation fractions, dry them, and redistill. The low-boiling product can be used directly for preparation of the semicarbazone, 2,4-dinitrophenylhydrazone, and other derivatives given in Chapter 9.

(C) Cyclohexanone by Chromic Acid Oxidation

In a 100-mL beaker, dissolve 4.2 g (0.014 mole) of sodium dichromate dihydrate in 25 mL of water. Add carefully 3.4 mL (6 g, 0.06 mole) of concentrated sulfuric acid (**Caution**—*corrosive!*), stirring the mixture, and cool the deep orange-red solution to 30°. Place 4.2 mL (4 g, 0.04 mole) of cyclohexanol and 12 mL of water in a 250-mL Erlenmeyer flask and to it add the dichromate solution in one portion. Swirl the mixture to ensure thorough mixing and observe its temperature. The mixture rapidly becomes warm; when the temperature reaches 55°, cool the flask in a basin of cold water, or under the tap, and regulate the amount of cooling only as long as necessary to maintain this temperature; when the temperature of the mixture no longer rises above 60° on removal of external cooling, heat the flask on a steam bath for 15 min.

Pour the mixture into a 100-mL round-bottomed flask, add an additional 25 mL of water, and attach a short, unpacked fractionating column (or a Claisen head). To the top of the column (or Claisen head) attach a distillation head, add a boiling chip, and distill the mixture until about 10 mL of distillate, consisting of water and an upper layer of cyclohexanone, has been collected.

Saturate the aqueous layer with 3 g of salt to salt out the dissolved cyclohexanone.[2] Separate the cyclohexanone layer, and extract the aqueous layer with 5 mL of methylene chloride, or pentane (**Caution**—*flammable!*).

IIII➤ *CAUTION*   Prolonged exposure to high concentrations of methylene chloride vapors may induce cancer. As with all organic solvents, work in a well-ventilated area when using it.

Combine the solvent extract with the cyclohexanone layer and dry it with 0.5–1.0 g of anhydrous magnesium sulfate. Filter the dried solution (a filter tube works best) into a 25-mL distilling flask, attach a distillation head, and distill off solvent and collect the cyclohexanone fraction boiling at 150–160° (mainly 153–157°). The yield is 2.2–3 g.

Conversion of cyclohexanone to the oxime is described in part (E) below.

(D) Cyclohexanone by Hypochlorite Oxidation

This oxidation of cyclohexanol to cyclohexanone uses "swimming pool chloride," which is a 12.5% by weight aqueous sodium hypochlorite solution. Because of the equilibrium between aqueous hypochlorite and chlorine gas this reaction must be carried out in the hood or other well-ventilated area.

In a 125-mL Erlenmeyer flask, dissolve 4.2 mL (4 g, 0.04 mole) of cyclohexanol in 10 mL of glacial acetic acid. Cool the flask in an ice–salt bath to about

−5° and add dropwise by means of an addition funnel 30 mL of fresh concd aqueous sodium hypochlorite solution (about 4.5 g, 0.06 mole of NaOCl). Swirl the mixture during the addition to ensure thorough mixing and observe its temperature. The mixture rapidly becomes warm; the temperature should be maintained between 30–35° by controlling the addition rate and the degree of cooling. The greenish color of the hypochlorite solution will be discharged as it reacts, but a pale yellow color will persist near the end of the addition.[4] Swirl the mixture occasionally for 15 min and then add saturated sodium bisulfite solution until the yellow color disappears.

Pour the mixture into a 100-mL round-bottomed flask, add an additional 25 mL of water, and attach a short unpacked fractionating column (or a Claisen head). To the top of the column (or Claisen head) attach a distillation head, add a boiling chip, and distill the mixture into a 100-mL receiver until about 25 mL of distillate, consisting of water, acetic acid, and an upper layer of cyclohexanone, has been collected.

Neutralize the acetic acid by adding slowly, with stirring, solid sodium carbonate (2–3 g). Then saturate the aqueous layer with 3 g of salt to salt out any dissolved cyclohexanone.[2] Decant the mixture into a separatory funnel, separate the cyclohexanone layer, and extract the aqueous layer with 5 mL of methylene chloride, or pentane (**Caution**—*flammable!*).

⫸ *CAUTION*   Prolonged exposure to high concentrations of methylene chloride vapors may induce cancer. As with all organic solvents, work in a well-ventilated area when using it.

Combine the solvent extract with the cyclohexanone layer and dry it with 0.5–1.0 g of anhydrous magnesium sulfate. Filter the dried solution (a filter tube works best) into a 25-mL distilling flask, attach a distillation head, and distill off the solvent and collect the cyclohexanone fraction boiling at 150–160° (mainly 153–157°). The yield is 2.2–3 g.

Conversion of cyclohexanone to the oxime is described in part (E) below.

(E) Cyclohexanone Oxime

In a 100-mL beaker dissolve 1.8 g (0.025 mole) of hydroxylamine hydrochloride in 25 mL of water and add 2.6 mL (2.5 g, 0.025 mole) of cyclohexanone. To this add dropwise, with stirring, a solution of 1.4 g (0.013 mole) of sodium carbonate in 20 mL of water. As the sodium carbonate is added, the hydrochloride salt is converted to unprotonated hydroxylamine and the carbonate produces sodium chloride and carbon dioxide. Stir the mixture occasionally for another 10 min and then chill it in an ice bath. Collect the precipitated cyclohexanone oxime on a Büchner or Hirsch funnel and wash it with a few milliliters of cold water. Press out as much water as possible and draw air through the funnel until the product is dry. The yield is about 1.6 g, mp 86–89°. The product is sufficiently pure to use in the Beckmann rearrangement described below.

(F) Beckmann Rearrangement of Cyclohexanone Oxime to Caprolactam

This experiment should be conducted in a hood. In a 100-mL beaker place about 10 mL (21 g) of syrupy polyphosphoric acid (**Caution**—*corrosive!*) and add 1.4 g (0.014 mole) of cyclohexanone oxime and stir the mixture with a thermometer. Heat the beaker on a hot plate that has been set such that the temperature of the mixture rises to 130° over a 15-min period (about 7° per min). Stir the mixture occasionally as the temperature rises. When the temperature of the mixture reaches 130°, turn off the hot plate and place the hot beaker on a transite board to cool (*use beaker tongs!*). When the temperature of the mixture falls to 100°, pour it cautiously into a mixture of 30 mL of water and a similar volume of ice contained in a 250-mL beaker. Rinse the 100-mL beaker with 5 mL of water and add the rinse to the ice–water mixture.

Add solid sodium carbonate to the mixture until the pH is about 6. Extract the aqueous layer with two 20-mL portions of methylene chloride; wash the combined extracts with 10 mL of water and then dry the extract with anhydrous magnesium sulfate.

⫸ *CAUTION*   Prolonged exposure to high concentrations of methylene chloride vapors may induce cancer. As with all organic solvents work in a well-ventilated area when using it.

Filter the solution through a filter tube to remove the drying agent and concentrate the solution to a volume of 1–2 mL in the hood. Chill the residue to obtain crystals of the ε-capralactam. Collect the solid on a Hirsch funnel and recrystallize it from about 10 ml of mixed hexanes. The yield is 0.7–0.8 g, mp 64–66°.

Questions   1. Compare the behavior of 2-methyl-2-butanol, 3-methyl-2-butanol, and 2,2-dimethylpropanol toward mild oxidation with chromic acid solution or hypochlorite.

2. How many moles of potassium dichromate are required to oxidize 1 mole of 2-pentanol?

3. Write balanced equations for the oxidation of cyclohexanol with chromic acid solution and with sodium hypochlorite solution.

4. In the oxidation of cyclohexanol, why does the cyclohexanone product (bp 155°) distill even though the thermometer never reads above 120°?

5. Write equations showing the action of the following reagents on 2-pentanone.
   (a) hydrogen cyanide   (b) ethylmagnesium bromide
   (c) iodine–potassium iodide + alkali (iodoform test)

6. When ε-caprolactam is refluxed with 6 N HCl for 1 hr, a water-soluble product, $C_6H_{14}O_2NCl$, is produced. Neutralization of the mixture gives a compound with the molecular formula $C_6H_{14}O_2N$. Draw the structures for these compounds and write equations for their production.

7. Nylon-6 is produced by the polymerization of ε-caprolactam when the lactam is heated with a *catalytic amount* of water. What is the structure of the polymer and the mechanism for its formation? Why is only a catalytic amount of water used? (*Hint*: consider Question 6.)

8. Neat cyclohexanone has a strong IR absorption band at 1715 cm^{-1}; cyclohexanone oxime absorbs at 1660 cm^{-1}; and ε-caprolactam has a broad absorption at 1660 cm^{-1}. What is the origin of these absorptions?

22 Addition Reactions of Aldehydes and Ketones

22.1 Carbonyl Addition Reactions

In Chapter 9 a number of carbonyl addition reactions were described that are useful in the chemical identification of an unknown aldehyde or ketone. The reactions can be summarized according to the equation

$$R_2C{=}O + NH_2{-}X \rightleftharpoons R_2C{=}N{\diagdown}_X + H_2O$$

Here X stands for OH (with oxime formation), for $NH{-}C_6H_3(NO_2)_2$ (with dinitrophenylhydrazone formation), and for $NH{-}CO{-}NH_2$ (with semicarbazone formation). In Chapter 9 the reaction conditions were chosen so that, for most unknowns, the equilibrium was achieved and the product was favored. However, in typical laboratory experiments, most organic reactions are performed under conditions that do not achieve equilibrium between the products and the starting materials. This means that the sole or major product is due to kinetic control. Under equilibrium conditions the principal final product may be different from that resulting from kinetic control.

A study of equilibria and rates[1] in the reaction of aldehydes and ketones with semicarbazide has led to examples that illustrate the effect of these factors.

$$R_2C{=}O + NH_2NH{-}CO{-}NH_2 \rightleftharpoons R_2C{=}NNH{-}CO{-}NH_2 + H_2O$$

---

[1] Conant and Bartlett, *J. Am. Chem. Soc.*, **54**, 2881 (1932).

In this reaction two opposing forces are at work. Increased acidity below about pH 4.9 decreases the amount of free semicarbazide through salt formation, and this reduces the rate because of the low nucleophilic activity of the cation. On

$$
\underset{\text{reactive}}{NH_2NHCNH_2} + H^+ \rightleftharpoons \underset{\text{unreactive}}{\overset{+}{N}H_3NHCNH_2}
$$

the other hand, high acidity favors partial protonation of carbonyl group, which enhances its electrophilic activity (and rate of reaction). High acidity also increases the rate of dehydration of the carbinolamine intermediate. The combination of these opposing factors leads to a range of pHs over which the reaction proceeds readily. Above and below this pH range, the rate falls off sharply. (See Questions 1 and 2 at the end of this chapter.)

$$
R_2C{=}O + H^+ \rightleftharpoons [R_2C{=}OH^+ \leftrightarrow R_2C^+{-}OH]
$$

The optimum conditions for interaction of carbonyl compounds with reagents to form derivatives such as semicarbazones, oximes, and arylhydrazones involve buffered solutions. Sodium acetate or phosphate buffers are often used.

In the experiment described in Section 22.3 (A), the semicarbazones of a mixture of 2-furaldehyde and cyclohexanone will be prepared under both kinetically controlled and equilibrium conditions. One of these carbonyl compounds forms a semicarbazone much faster than the other, but at equilibrium the second semicarbazone is formed in greater amount. Both situations will be examined and the dominant product identified.

2-Furaldehyde   Cyclohexanone

22.2 Reactions of Carbonyl Compounds

Equilibria and Rates in Carbonyl Reactions; Formation of 2-Furaldehyde and Cyclohexanone Semicarbazones

In a 50-mL flask dissolve 0.5 g of semicarbazide hydrochloride and 1 g of potassium monohydrogen phosphate (K_2HPO_4) in 13 mL of water. Mix together in a small flask 0.5 mL (0.48 g) of cyclohexanone, 0.50 mL (0.54 g) of freshly distilled 2-furaldehyde, and 2.5 mL of 95% ethanol. Place half of the solution in each of two small test tubes.

In a 50-mL Erlenmeyer flask place 6 mL of the semicarbazide solution and cool it to 0–5° in an ice bath. In the same cooling bath chill one of the test tubes containing the furaldehyde–cyclohexanone solution. When it is thoroughly chilled, empty the contents of the test tube into the semicarbazide solution and mix them well. Crystals will form quickly; replace the flask in the cooling bath for 5 min, then filter the crystals with suction and wash them with about 1–2 mL of cold water. After drying them thoroughly, weigh them and determine the melting point. Which semicarbazone is this?

The recorded melting points of the semicarbazones are 2-furaldehyde, 202°; cyclohexanone, 166°.

Place the remaining 6 mL of semicarbazide solution in a 50-mL flask and heat it on a steam bath to 85°. Add the 2-furaldehyde–cyclohexanone solution from the second test tube and swirl the solution. Heat the flask for 15 min longer on the steam bath and then cool the solution to room temperature. Finally, chill the reaction mixture in an ice bath for a few minutes to complete crystallization of the product. Collect the crystals with suction, and wash them with a little cold water. Dry the product thoroughly, record the weight, and take the melting point. Which semicarbazone is this?

Account for the results that you have observed.

Questions

1. In considering the effects of pH on rate, it is convenient to plot log [S] or log [SH$^+$] versus pH, where [S] and [SH$^+$] are the concentrations of species S in its unprotonated and protonated forms. Such plots are called "pH profiles."
   (a) What is the pH profile of a species S that is half-protonated at pH 7?
   (b) What is the pH profile for SH$^+$?

2. (a) If the rate of reaction between A and BH$^+$ is proportional to the product of their concentrations (rate = k[A] [BH$^+$]), show that log (rate) = constant + log [A] + log [BH$^+$].
   (b) What is the pH profile (Question 1) of the rate if A is half-protonated at pH 7 and B is half-protonated at pH 3?

3. Explain why the yield of 2-furaldehyde semicarbazone obtained from a mixture of 2-furaldehyde and cyclohexanone changes with temperature.

23 Reduction of the Carbonyl Group

23.1 Introduction

An important reaction of aldehydes and ketones is their reduction to an alcohol, and many reagent–reaction condition combinations have been developed to carry out this process cleanly. For small-scale syntheses sodium borohydride ($NaBH_4$) is a particularly convenient reducing agent. It has extremely high reducing capacity—1 mole can reduce 4 moles of a ketone.

$$4\ (C_6H_5)_2C{=}O + Na^+\ {}^-BH_4 \longrightarrow Na^+\ {}^-B[OCH(C_6H_5)_2]_4 \xrightarrow{H_2O}$$
$$4\ (C_6H_5)_2CH{-}OH$$

Since sodium borohydride decomposes at an appreciable rate in water[1] or methanol, it is desirable to carry out the reduction in ethanol or 2-propanol. In these media at room temperature sodium borohydride reduces aldehydes and ketones to the corresponding primary and secondary alcohols. It is quite selective and does not reduce nitriles, nitro compounds, carboxylic acids, esters, or lactones. Lithium aluminum hydride ($LiAlH_4$) is a stronger reagent and will reduce esters, lactones, and amides, but must be used in aprotic media.[2]

In aprotic solvents such as dioxane and 1,2-dimethoxyethane (glyme), sodium borohydride will reduce acid chlorides to alcohols. Secondary and

---

[1] The decomposition is slowed markedly by addition of alkali.
[2] Surveys of the uses of these reducing agents are found in Fieser and Fieser, *Reagents for Organic Synthesis*, Vol. 1 (New York; Wiley, 1967), pp. 1049–1055 (for $NaBH_4$) and 581–600 (for $LiAlH_4$). See also, for $LiBH_4$, W. G. Brown, *Organic Reactions*, **6**, 649 (1951).

tertiary halides are reduced to the hydrocarbon at 50° by 4 M solutions of sodium borohydride in a 65 volume percent solution of diglyme and 1 M aqueous sodium hydroxide. Sodium borohydride is used in the Brown procedure for hydration of double bonds (see Chapter 19). Other uses of sodium borohydride are replacement of the diazonium group by hydrogen, and reduction of ozonides to the corresponding alcohols.

23.2 Reduction by Sodium Borohydride: Diphenylmethanol (Benzhydrol)

In a 25-mL round-bottomed flask place 0.55 g (0.003 mole) of diphenyl ketone (benzophenone) and add a slurry of 0.06 g (0.0015 mole, a large excess) of sodium borohydride in 3 mL of 2-propanol. Add a boiling chip and reflux the mixture for 30 min on a steam bath. Allow the solution to cool; no harm is done if it stands overnight or longer.

➠ *CAUTION*   Sodium borohydride is strongly caustic. Handle it carefully and do not permit it to touch the skin.

To decompose the borate ester complex, add 3 mL of 10% aqueous sodium hydroxide and swirl the reaction mixture vigorously until the precipitate has dissolved completely. Break up any resistant lumps carefully with the aid of 3 mL of water. Extract the diphenylmethanol by shaking it with two successive 5-mL portions of methylene chloride (CH_2Cl_2).

➠ *CAUTION*   Prolonged exposure to high concentrations of methylene chloride vapors may induce cancer. As with all organic solvents work in a well-ventilated area when using it.

Combine the extracts, transfer them to a distilling apparatus, and carefully distill off the methylene chloride (traces of water in the methylene chloride will steam-distill). On cooling and standing the residue will crystallize to give a nearly quantitative yield of almost pure diphenylmethanol, mp 68–69°. If the product melts below this range, it can be recrystallized from 60% water–methanol.

The infrared spectra of benzophenone and diphenylmethanol are shown in Figures 23.1 and 23.2.

Reactions. Diphenylmethanol has a reactive hydroxyl group: rupture of the C—OH bond is facilitated by resonance stabilization by the two phenyl groups of the incipient carbocation in the transition state. Diphenylmethanol is easily converted into bis-diphenylmethyl ether merely by boiling with dilute mineral acids and reacts readily with hydrogen chloride to give diphenylchloromethane. Diphenylmethanol may be characterized by conversion to the acetate (mp 41–42°) or benzoate (mp 88°).

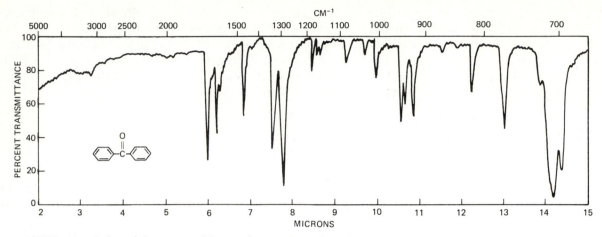

FIGURE 23.1 *Infrared Spectrum of Benzophenone (Nujol Mull)*

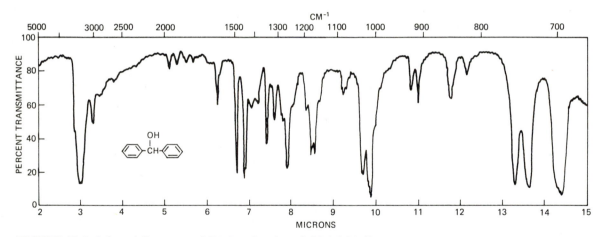

FIGURE 23.2 *Infrared Spectrum of Diphenylmethanol (Nujol Mull)*

Questions
1. Examine the IR spectrum given in Figures 23.1 and 23.2 and comment on how they reveal the change in functional groups (see Chapter 10 for a discussion of IR spectra).

2. Predict where the methine (tertiary) hydrogen of the diphenylmethanol would appear in the NMR (see Chapter 10 for a discussion of NMR spectra).

3. Sodium borohydride is a less-powerful reducing agent than lithium aluminum hydride. In what other way do these two reagents differ?

4. Devise a synthesis for the following compounds.
   (a) 4-methyldiphenylmethanol, starting from benzene and toluene
   (b) 3,3'-dibromodiphenylmethanol, starting from bromobenzene

24 A Modified Wittig Synthesis

24.1 The Wittig Reaction

In the Wittig reaction,[1] aldehydes and ketones are converted into alkenes by reaction with an alkylidene phosphorane (a phosphorus ylid,[2] **II**), generally in high yield. The requisite ylids can be obtained by the action of strongly basic reagents, such as phenyllithium or sodium hydride, upon appropriate quaternary phosphonium halides (**I**).

$$[(C_6H_5)_3\overset{+}{P}{-}CH_2R]\bar{X} \xrightarrow{\ C_6H_5Li\ } [(C_6H_5)_3P{=}CH{-}R \leftrightarrow (C_6H_5)_3\overset{+}{P}{-}\overset{-}{C}H{-}R]$$

$$\textbf{I} \qquad\qquad\qquad\qquad \textbf{IIa} \qquad\qquad\qquad \textbf{IIb}$$

$$\textbf{II} + (C_6H_5)_2C{=}O \longrightarrow C_6H_5{-}CH{=}C(C_6H_5)_2 + (C_6H_5)_3P{=}O$$

Since the ylids are unstable, they are usually generated in the reaction mixture in the presence of the carbonyl compound. Advantages of the Wittig synthesis are that carbon–carbon double-bond formation occurs without skeletal rearrange-

---

[1] Wittig and Geisler, *Ann.*, **580**, 44 (1953); Maercker, *Organic Reactions*, **14**, 270 (1965).

[2] An ylid is a species with formal positive and negative charges on adjacent covalently bound atoms that also have full octets.

ment, and acid-sensitive alkenes can be prepared because the reaction occurs under basic conditions.

If the organic halide (R—CH$_2$X) used for the formation of the quaternary phosphonium halide (**I**) is highly reactive, such as C$_6$H$_5$—CH$_2$—Cl, a simpler procedure may be used. An ester of phosphorous acid is converted by the Arbusov reaction[3] to a phosphonic ester (**III**), which reacts with carbonyl compounds, in the presence of a base, in the same way as the Wittig intermediate.

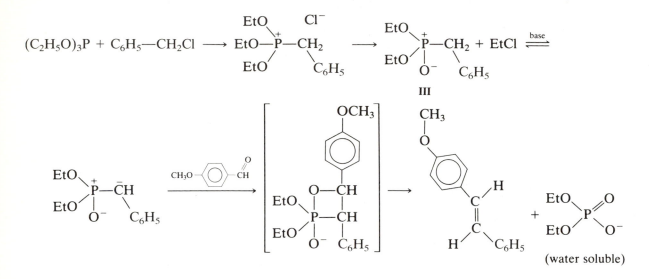

In this reaction a proton is abstracted from the phosphonic ester (**III**) to give an anion that adds to the aldehyde. A transitory four-membered ring (containing the phosphorus and the original aldehyde oxygen) is formed, which undergoes carbon–oxygen and carbon–phosphorus bond cleavage to give the alkene and diethyl phosphate.

From benzyl-type halides and substituted benzaldehydes, unsymmetrical stilbenes can be synthesized,[4] and with benzophenones, triarylethylene derivatives are formed. α,β-Unsaturated aldehydes such as crotonaldehyde and cinnamaldehyde furnish derivatives of 1,4-butadiene. The reaction with aldehydes generally produces the *trans* stereochemistry.

$$C_6H_5—CH{=}CH—CH{=}O + C_6H_5—CH_2—PO(OC_2H_5)_2 \xrightarrow{\text{NaOEt}}$$
$$C_6H_5—CH{=}CH—CH{=}CH—C_6H_5$$

[3] Kosolapoff, *Organic Reactions*, **6**, 276 (1951).
[4] Seus and Wilson, *J. Org. Chem.*, **26**, 5243 (1961); Wadsworth and Emmons, *J. Am. Chem. Soc.*, **83**, 1733 (1961).

24.2 Preparation of *p*-Methoxystilbene[5]

(A) Diethyl Benzylphosphonate

In a 25-mL round-bottomed flask place 1.8 mL (1.66 g, 0.01 mole) of triethyl phosphite and 1.2 mL (1.26 g, 0.01 mole) of benzyl chloride (**Caution—** *lachrymator!*).

⫸ *CAUTION*   Avoid contact of phosphorus compounds with the skin. Wash off any spilled material thoroughly with soap and water.

Add a boiling chip, attach a condenser, and heat the mixture gently for 1 hr. When the temperature reaches 130–140°, evolution of ethyl chloride (bp + 12°) begins. The internal temperature continues to rise and attains about 190° by the end of the hour. Allow the product to cool, remove the condenser, and add 2 mL of dimethylformamide with swirling to dissolve the phosphonate ester.

(B) *p*-Methoxystilbene

To the cooled phosphonate ester solution prepared above, add 0.56 g (0.0104 mole) of fresh sodium methoxide.[6]

⫸ *CAUTION*   Handle sodium methoxide carefully. Any material spilled on the hands should be washed off promptly with a large quantity of water.

Swirl the mixture, and add drop by drop a solution of 1.2 mL (1.36 g, 0.01 mole) of *p*-methoxybenzaldehyde in 8 mL of dimethylformamide with intermittent cooling in an ice bath so that the temperature of the reaction mixture is maintained between 30 and 40°. Allow the reaction mixture to stand overnight or longer.

Pour the reaction mixture into about 10 mL of water, while stirring it, and collect the product on a suction filter. After washing thoroughly with water, recrystallize the product from ethanol. The recorded melting point of *p*-methoxystilbene is 136°. The yield is 1.2–1.4 g.

---

[5] The specific example of the modified Wittig synthesis given here may be varied at the second step by using *p*-chlorobenzaldehyde (0.01 mole, 1.2 g) to give *trans*-4-chlorostilbene, mp 129°. Another example is the use of cinnamaldehyde (0.01 mole, 1.3 g) to produce 1,4-diphenyl-1,3-butadiene [see Fieser and Williamson, *Organic Experiments*, 4th ed. (Lexington, MA: Heath, 1979)].

[6] Commercial sodium methoxide gives erratic results unless fresh reagent is available. Sodium methoxide sufficient for twenty five preparations can be prepared by the procedure of Cason described in *Organic Syntheses*, Collective Volume IV, 651 (1963). To 130 mL of anhydrous methanol contained in a 250-mL round-bottomed flask equipped with an upright condenser add through the condenser tube 6.0 g of clean sodium cut in small pieces. To keep the reaction under control one piece of sodium should be allowed to react completely before another is added. After all of the sodium has reacted, the excess methanol is removed from distillation, first at atmospheric pressure and then under an aspirator vacuum using a heating bath maintained at 150°. The resulting free-flowing sodium methoxide can be stored in a desiccator for several weeks.

Questions   **1.** What is the product of the reaction of diethyl benzyl phosphonate with
(a) furfuraldehyde, C_4H_3O—CHO   (b) acetone, CH_3COCH_3

2. Predict the NMR spectrum of *trans-p*-methoxystilbene. How would the *cis-p*-methoxystilbene differ?

3. Indicate the difference in structure between a phosphonate ester, a phosphite ester, and a phosphate ester. (Use your organic lecture text.)

4. When bromine is added to *trans-p*-methoxystilbene, *trans* addition dominates. Give the R,S designations for the two components of the racemic mixture.

25 The Cannizzaro Reaction

25.1 Reactions of Aromatic Aldehydes

Aromatic aldehydes, like aliphatic aldehydes, undergo addition reactions of the carbonyl group leading to cyanohydrins, acetals, oximes, phenylhydrazones, and similar derivatives (see Chapters 9 and 22). They also undergo reactions such as the Cannizzaro reaction and the benzoin condensation (Chapter 48), which require the absence of acidic hydrogens on the carbon adjacent to the aldehyde group (α-hydrogens).

In the presence of strong alkalis, benzaldehyde (like formaldehyde) undergoes disproportionation to form the corresponding primary alcohol and a salt of the carboxylic acid: the Cannizzaro reaction.[1]

$$C_6H_5-\overset{\overset{\textstyle O}{\|}}{C}-H + C_6H_5-\overset{\overset{\textstyle O}{\|}}{C}-H \xrightarrow[H_2O]{KOH} C_6H_5-C\overset{\textstyle O}{\underset{\textstyle O^- K^+}{}} + C_6H_5-CH_2OH$$

The mechanism involves addition of hydroxide ion to the carbonyl group of one benzaldehyde molecule and transfer of a hydride anion from the adduct to a second molecule of benzaldehyde. Finally, the resulting benzoic acid is transformed to the benzoate anion and the benzyl alcoholate gives benzyl alcohol.

[1] For a discussion of the Cannizzaro reaction see Geissman, *Organic Reactions*, 2, 94 (1944).

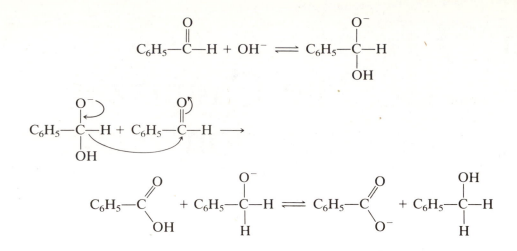

If the reaction is carried out under anhydrous conditions with the sodium derivative of benzyl alcohol ($NaOCH_2C_6H_5$) as the base, the products are the ester, benzyl benzoate, and another molecule of sodium benzyl alcoholate. Because the initial alcoholate is regenerated, only catalytic amounts are required. Aluminum alkoxides in catalytic amounts, under anhydrous conditions, convert aromatic and aliphatic aldehydes to esters (the Tishchenko reaction).

25.2 Preparations and Reactions

(A) Benzyl Alcohol[2]

In a small beaker dissolve 0.054 mole (3.6 g of 85% pure solid) of solid potassium hydroxide in 3.6 mL of water and cool the solution to about 25°. Place 4 mL (4.2 g, 0.04 mole) of benzaldehyde in a 50-mL Erlenmeyer flask and to it add the potassium hydroxide solution. Cork the flask firmly and shake the mixture thoroughly until an emulsion is formed. Allow the mixture to stand for 24 hr or longer. At the end of this period, the odor of benzaldehyde should no longer be detectable.

To the mixture add just enough water to dissolve the precipitate of potassium benzoate. Shake the mixture thoroughly to facilitate solution of the precipitate. Extract the alkaline solution with three or four 4-mL portions of methylene chloride to remove the benzyl alcohol and traces of any unconverted benzaldehyde. Combine the methylene chloride extracts for isolation of benzyl alcohol and reserve the aqueous solution to obtain the benzoic acid.

---

[2] In planning the laboratory schedule, it should be observed that this experiment requires materials to be mixed and allowed to stand for 24 hr or longer.

▌▶ *CAUTION* Prolonged exposure to high concentrations of methylene chloride vapors may induce cancer. As with all organic solvents, work in a well-ventilated area when using it.

Concentrate the methylene chloride solution of benzyl alcohol by distillation using a steam bath into a cooled receiver, until the volume of the residual liquid has been reduced to 3–4 mL. Cool the residue, transfer it to a large test tube (using a little methylene chloride to rinse the distilling flask), and extract it using the squirting technique (see Section 6.5; use about a dozen "squirts" for each washing reagent) with two 1-mL portions of 20% aqueous sodium bisulfite to remove any benzaldehyde. Wash the methylene chloride solution finally with two 2-mL portions of water and dry it with 0.5–1 g of anhydrous magnesium sulfate. Filter the solution into a small distilling flask and distill off the methylene chloride. Change the receiver and continue the distillation to collect the benzyl alcohol. Collect the material boiling at 200–206°. The yield is 1–1.2 g.

Reactions. Benzyl alcohol may be characterized by reaction with *p*-nitrobenzoyl chloride (*fresh!*), in the presence of pyridine, to obtain the crystalline *p*-nitrobenzoic ester as described in Chapter 9. Alternatively, the IR or NMR spectrum of the alcohol will readily reveal the presence of benzaldehyde impurity.

$$C_6H_5-N=C=O + C_6H_5-CH_2OH \longrightarrow C_6H_5-NH-CO-OCH_2C_6H_5$$
$$(\text{mp } 78°)$$

$$O_2N-C_6H_4-CO-Cl + C_6H_5-CH_2OH \xrightarrow{C_5H_5N} O_2N-C_6H_4-CO-OCH_2C_6H_5$$
$$(\text{mp } 85°)$$

(B) Benzoic Acid

To isolate the benzoic acid, pour the aqueous solution of potassium benzoate (from which the benzyl alcohol has been extracted) into a vigorously stirred mixture of 8 mL of concentrated hydrochloric acid, 8 mL of water, and 8–10 g of chipped ice. Test the mixture with indicator paper to make sure that it is strongly acidic. Collect the benzoic acid with suction and wash it once with cold water. Crystallize the product from hot water, collect the crystals, and allow them to dry thoroughly. The yield is about 1.5 g.

Aromatic and aliphatic carboxylic acids generally are characterized by conversion to crystalline amides. Benzoic acid may be converted to benzamide (mp 130°) or benzanilide (mp 160°). For this purpose the acid usually is converted by means of thionyl chloride to the acid chloride, which is treated with ammonia or an arylamine to obtain the desired amide (Chapter 9).

A valuable aid in the identification of an unknown organic acid is the determination of its equivalent weight (neutralization equivalent) by titration with a standard base. This method may be used also to check the purity of a sample of a known acid. The procedure is described in Chapter 9.

Questions **1.** Which of the following aldehydes will undergo the Cannizzaro reaction? For those that do not react this way, what product will form?
   (a) 2-methylpropanal   **(b)** 2,2-dimethylpropanal
   (c) 1-methylcyclobutanecarbaldehyde

2. Sketch the expected NMR spectra of benzaldehyde, benzyl alcohol, and benzoic acid.

3. If one started with benzaldehyde in which the aldehydic proton was replaced by deuterium, what products would be formed in the Cannizzaro reaction?

4. Give the mechanism for the reaction of benzaldehyde with catalytic amounts of aluminum oxide to form benzyl benzoate (Tishenko reaction).

5. Write equations for the preparation of benzaldehyde from
   (a) benzene   **(b)** toluene   **(c)** benzoic acid

6. Write equations for the reaction of benzaldehyde with the following reagents.
   (a) methanol (+ hydrogen chloride catalyst)
   (b) semicarbazide
   (c) *p*-tolylmagnesium bromide, followed by dilute acid
   (d) sodium cyanide and ammonium chloride, followed by hydrolysis (the Strecker reaction)
   (e) aluminum isopropoxide (the Meerwein–Pondorff reaction)

26 Esters

26.1 Esterification and Saponification

Esters may be prepared by direct esterification of an acid with an alcohol in the presence of an acid catalyst (sulfuric acid, hydrogen chloride) and by alcoholysis of acid chlorides, acid anhydrides, and nitriles. Occasionally they are prepared by heating the metallic salt of a carboxylic acid with an alkyl halide or other alkylating agent.

Direct esterification is an acid-catalyzed nucleophilic addition of an alcohol to the carboxylic acid group. The reaction occurs through the mechanism illustrated on page 310 with acetic acid and ethanol: (1) protonation of the carboxyl group, (2) addition of the alcohol and transfer of a proton to one of the hydroxyl groups, (3) elimination of water and deprotonation. It has been demonstrated that an oxygen atom of the carboxyl group is eliminated as water and the oxygen atom of the alcohol is retained in the ester. Since the reaction is reversible and the equilibrium constant is usually not too different from unity, the equilibrium must be driven to the right to obtain good conversion to the ester. The use of an excess of one of the starting materials, removal of one of the products, or a combination of both serves this purpose.

The composition of the equilibrium mixture is given approximately by the mass law, shown in equation 26.1, where K is the equilibrium constant for esterification and the symbols [ester], [water], etc., refer to concentrations expressed in moles per liter or as mole fractions.

$$K_E = \frac{[\text{ester}][\text{water}]}{[\text{acid}][\text{alcohol}]} \qquad (26.1)$$

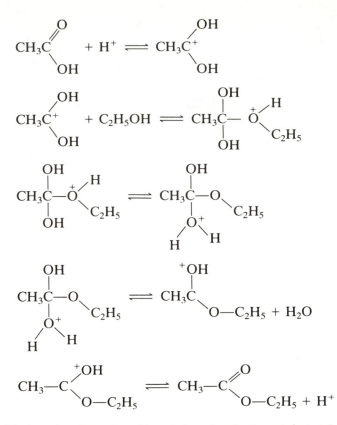

Starting with 1 mole of acetic acid and 1 mole of ethanol (a total of 2 moles), the equilibrium mixture is found experimentally to contain 0.66 mole of ethyl acetate (and an equimolar amount of water). Thus, the mole fractions of ester and water are 0.66/2, and the mole fractions of unesterified acid and alcohol are 0.34/2. Putting these equilibrium concentrations into the mass-law expression yields a K value of 3.77 for this particular estrification.

$$K_E = \frac{(0.33)(0.33)}{(0.17)(0.17)} = 3.77 \tag{26.2}$$

Inspection of the equilibrium expression shows that the use of an excess of the alcohol (or an excess of the organic acid) will increase the amount of ester formed. Calculations based upon the K value of 3.77 indicate that the use of 2 moles of ethanol to 1 mole of acetic acid will bring about an 80% conversion of the acid of ethyl acetate, and 3 moles of ethanol will effect almost 90% conversion to ester. The choice of reactant to be used in excess will depend upon factors such as availability, cost, and ease of removal of excess reactant from the product.

Under ideal conditions, the composition of an equilibrium mixture is not affected by the presence or absence of a catalyst, but experiments have shown that the observed K values may increase as much as twofold if a relatively large amount of the acid catalyst is used. In these situations, the "catalyst" changes the environment within the system and partly removes through its hydration, the water formed in the reaction.

Driving an esterification to completion by removal of the water formed in the reaction is a common practice, especially in larger-scale preparations. One method is to add benzene or a similar hydrocarbon and distill out a ternary azeotropic mixture, benzene–alcohol–water.

The rate of reaction is influenced significantly by the structure of the alcohol and the acid, and steric factors play an important part. Increasing the number of bulky substituents in the α- or β-position of the acid markedly reduces the rate constant for esterification. The reaction rates for two series of acids are given in the following sequences.

$$H{-}CO_2H > CH_3{-}CO_2H > (CH_3)_2CH{-}CO_2H \gg (CH_3)_3C{-}CO_2H$$

$$C_2H_5{-}CH_2{-}CO_2H > (CH_3)_3C{-}C\hat{O}_2H > (CH_3)_3C{-}CH_2{-}CO_2H > (C_2H_5)_3C{-}CO_2H$$
$$\quad\;\;(0.51) \qquad\qquad\;\; (0.037) \qquad\qquad\quad (0.023) \qquad\qquad\quad\;\; (0.00016)$$

Specific rates of esterification with methanol at 40°, relative to acetic acid, are shown in the second series (in parentheses). Because of the rate problem, esters of sterically hindered acids are prepared by methods other than direct esterification: conversion of the acid to the acid chloride, followed by treatment with an alcohol; or reaction of a salt of the acid with an alkyl halide, in the presence of a secondary amine as catalyst.

Acid-catalyzed esterification is a practical method for the preparation of esters of primary and secondary alcohols with typical organic acids but is not useful for tertiary alcohols. They react very slowly and their equilibrium constants are low; K_E for t-butanol is about 0.005, compared with approximately 2 for 2-propanol and 3-pentanol and 4 for 1-propanol and 1-butanol.

Acid chlorides and anhydrides react rapidly with primary and secondary alcohols to give the corresponding esters. In the absence of a base, acid chlorides convert tertiary alcohols into alkyl chlorides; but in the presence of a tertiary amine (pyridine, triethylamine), tertiary alcohols furnish esters. Acid anhydrides are less reactive than acid chlorides but react with most alcohols upon heating. Acetylations with acetic anhydride are promoted by acid (sulfuric acid, zinc chloride) and base catalysts (sodium acetate, tertiary amines).

Hydrolysis of an ester is the reverse of esterification. With an acid catalyst, even in the presence of a large amount of water, an appreciable amount of the ester may be present in the equilibrium mixture. Hydrolysis by strong alkalis, saponification, is more rapid, and is more effective because hydroxyl ion reacts with the organic acid and drives the reaction to completion.

$$CH_3{-}CO_2C_2H_5 + OH^- \longrightarrow [CH_3{-}CO_2]^- + C_2H_5{-}OH$$

Saponification affords a means of establishing the structure of an unknown ester, through identification of the resulting alcohol and organic acid.

Since many esters are insoluble in water, solutions of potassium hydroxide or sodium hydroxide in 85–90% aqueous methanol or ethanol are used frequently for saponifications. A high-boiling solvent such as diethylene glycol (HO—CH$_2$CH—O—CH$_2$CH$_2$—OH, bp 245°) is advantageous for the saponification of esters of high molecular weight, and esters of sterically hindered acids or tertiary alcohols. Compounds of the latter group are very resistant to saponification.

26.2 Preparations and Reactions

(A) n-Butyl Acetate: Esterification of Acetic Acid[1]

In a 25-mL round-bottomed flask provided with a water-cooled reflux condenser, mix thoroughly 3.0 mL (2.2 g, 0.03 mole) of 1-butanol, 5.1 mL (5.4 g, 0.09 mole) of glacial acetic acid, and 0.4 mL (0.72 g) of concentrated sulfuric acid (**Caution**—*corrosive!*). Add a carborundum boiling chip, heat the mixture to boiling, and continue heating under reflux for 1 hr. Cool the contents of the flask in an ice bath and decant it from the boiling chip into a separatory funnel containing 5 mL of water. Rinse the flask with another 5 mL of water and add the rinse to the funnel. Shake the funnel gently (vigorous shaking may form an emulsion) and after the layers have had time to separate, remove the lower aqueous layer. Wash the organic layer with 5 mL of saturated aqueous sodium carbonate. Swirl the funnel for several minutes to allow most of the carbon dioxide to escape, then cap the funnel, shake, and separate the aqueous layer. Transfer the organic layer into an Erlenmeyer flask and dry it with a small amount of anhydrous magnesium sulfate.[2] Transfer the dried material into a filter tube (Figure 5.6) and allow it to drain into a small distilling flask. In order to minimize losses from ester clinging to the walls of the Erlenmeyer flask and the drying agent, rinse the flask with 2 mL of pentane and pass the rinse through the funnel into the distilling flask. Add a boiling chip to the flask, attach a distilling head, and distill the product slowly. Collect the fraction boiling at 119–125°. The yield is 1.9–2.4 g.

Take the IR spectrum of your product and compare it to the spectrum of n-butyl acetate shown in Figure 10.8. Examine your spectrum for traces of alcohol and acetic acid.

---

[1] Other esters such as n-propyl acetate or ethyl propionate may be prepared from the appropriate alcohol and acid combination by using proportionate molar quantities of reactants. A twofold excess of acetic acid is used to help drive the equilibrium reaction toward complete conversion of the alcohol into ester. The excess acid is easily removed by base extraction.

[2] Anhydrous calcium chloride is not suitable because it forms addition complexes with esters.

(B) Saponification of *n*-Butyl Acetate

In a 50-mL round-bottomed flask provided with a reflux condenser, place 1 g of *n*-butyl acetate. To this add 5 mL of 10% aqueous sodium hydroxide and about 10 mL of water. Add a boiling chip and boil until the odor of *n*-butyl acetate can no longer be detected (about 2 hr). Arrange the apparatus for distillation, distill off about 6 mL of liquid, and examine the distillate. Is it homogenous? Add enough ordinary salt (sodium chloride) to saturate the liquid (about 2 g will be required), shake it thoroughly, and allow it to stand undisturbed for a short while. Note the results.

(C) Methyl Benzoate[3]

In a 25-mL round-bottomed flask place 2.5 g (0.021 mole) of benzoic acid and 8 mL (6.4 g, 0.2 mole) of methanol. Carefully pour 0.6 mL of concentrated sulfuric acid (**Caution**—*corrosive!*) down the wall of the flask and swirl the flask to obtain good mixing. Add a boiling chip, attach an upright condenser, and reflux the mixture for 1 hr. Cool the solution to room temperature and decant it from the boiling chip into a separatory funnel containing about 5 mL of water and 5 mL of methylene chloride (dichloromethane).

⫸ *CAUTION*   Prolonged exposure to high concentrations of methylene chloride vapors may induce cancer. As with all organic solvents, work in a well-ventilated area when using it.

Rinse the reaction flask with 2–3 mL of methylene chloride and pour this into the separatory funnel. Shake the mixture and separate the organic liquid from the aqueous layer, which contains sulfuric acid and methanol. Wash the organic liquid with 5 mL of water and then with 5 mL of 5% aqueous sodium carbonate (**Caution**—*foaming!*) to remove unesterified benzoic acid. Swirl the mixture gently at first without a stopper, then insert the stopper and shake more vigorously; invert the separatory funnel and release the internal pressure by opening the stopcock. Separate the layers carefully (save the aqueous layer until you are certain that you have your product).

   Extract the organic layer with a second 5-mL portion of sodium carbonate solution and separate the layers carefully (the aqueous layers may be discarded). Finally, wash the organic layer with water, separate the layers, and place the organic layer in a dry Erlenmeyer flask. Add a small amount of anhydrous

---

[3] An interesting variation of this esterification experiment is to use a mixture of methanol and isopropyl alcohol in a 1 : 1 molar ratio and to determine by gas chromatography the composition of the esters formed. It should be noted that the boiling point of isopropyl benzoate is about 18° higher than methyl benzoate, so that the distillation is carried far enough to obtain the isopropyl ester.

   An independent value for the composition of the mixed esters may be obtained by determination of the saponification equivalent.

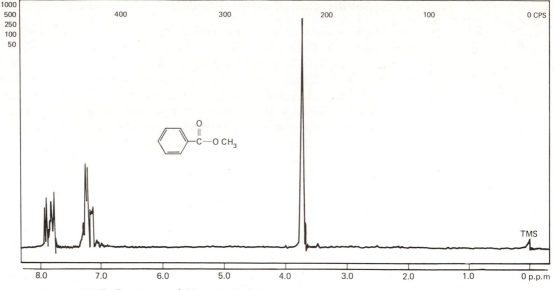

FIGURE 26.1 *NMR Spectrum of Neat Methyl Benzoate*

magnesium sulfate,[2] swirl the mixture thoroughly and allow it to stand for at least 20 min.

Filter the liquid through a filter tube into a dry flask, attach a distillation head, add a boiling chip, and distill off the methylene chloride from a water bath or steam bath. Collect the recovered solvent in a receiver cooled in an ice bath. When no more solvent distills, turn off the flow of cooling water in the jacket[4] and allow the water to drain. Add a boiling chip, and distill the product into a small, weighed round-bottomed flask that is cooled in an ice bath. Collect the fraction boiling above 190° (mainly in the range 192–196°). The yield is 1.8–2.0 g.

The saponification equivalent of the product may be determined by the procedure described in Section 9.8(G), using a 0.25-g sample. Nitration of methyl benzoate furnishes a crystalline derivative, methyl *m*-nitrobenzoate, mp 78° (see Chapter 30).

Obtain IR and NMR spectra of your product. The NMR spectrum of neat methyl benzoate is shown in Figure 26.1.

---

[4] For substances that boil above 160–170° an air-cooled condenser is preferred since the hot vapors might crack a water-cooled condenser. It would be desirable to draw air through the condenser by connecting one of the inlet–outlet tubes to an aspirator. Alternatively, the receiver can serve as the condenser by being mated to the distillation head and cooled in an ice bath.

Questions 1. What is the purpose of adding sulfuric acid in the preparation of *n*-butyl acetate? Would any of the ester be formed in the absence of sulfuric acid?

2. What procedures may be used to drive esterifications toward completion?

3. Assuming a *K* value of 4, calculate the percentage conversion of butyl alcohol to ester (at equilibrium) with the molar ratio of reactants used in this preparation.

4. Assuming a *K* value of 3, calculate the percentage conversion of benzoic acid to methyl benzoate (at equilibrium) with the molar ratio or reactants used in this preparation.

5. In the preparation of butyl acetate (or methyl benzoate) what two materials are removed by extraction with base?

6. Suggest a method for preparing *t*-butyl benzoate from benzoic acid and *t*-butyl alcohol.

7. Contrast the behavior of primary, secondary, and tertiary alcohols in esterification with a carboxylic acid and in reaction with hydrobromic acid. Explain.

8. In what way would the saponification equivalent of a sample of *n*-butyl acetate be affected by the presence of the following impurities?
   (**a**) water   (**b**) butyl alcohol   (**c**) acetic acid

27 Ionization of Carboxylic Acids

27.1 Introduction

Carboxylic acids contain the functional group $-\overset{\displaystyle O}{\overset{\|}{C}}-OH$, which ionizes to release a proton according to the equilibrium of equation 27.1.

$$R-\overset{\displaystyle O}{\overset{\|}{C}}-OH \rightleftharpoons R-\overset{\displaystyle O}{\overset{\|}{C}}-O^- + H^+ \tag{27.1}$$

The acid dissociation constant, K_a, for this equilibrium is written in terms of the concentrations[1] of the species involved.

$$K_a = \frac{[RCO_2^-][H^+]}{[RCO_2H]} \tag{27.2}$$

In water, typical carboxylic acids have K_a values of about 10^{-4} to 10^{-5}, which means that they are much weaker than acids such as sulfuric or hydrochloric but much stronger than very weak acids such as the alcohols ($K_a \approx 10^{-15}$) or the exceedingly weak acids such as the hydrocarbons ($K_a \approx 10^{-50}$).

For convenience in expressing the wide range of observed acidities, it is common practice to substitute pK_a values, defined by analogy to $pH = -\log_{10}[H^+]$ as the negative base 10 logarithm of the K_a.

---

[1] The value of K_a, defined in terms of concentrations, varies slightly with the concentration of the carboxylic acid or on addition of salts. The so-called thermodynamic constant substitutes *activities* for concentrations and is truly constant, but the estimation of activities is complex, and organic chemists normally employ the simpler concentration expression.

$$pKa = -\log_{10} K_a$$

In these terms the pK_a values of typical carboxylic acids in water are 4–5, alcohols are about 15, and hydrocarbons are about 50. The important point is that larger pK_a values correspond to *weaker* acids.

It is convenient to take the negative logarithm of both sides of equation 27.2 to give equation 27.3, which in the biological literature is called the Henderson–Hasselbalch equation.

$$pK_a = -\log_{10}\left(\frac{[RCO_2^-]}{[RCO_2H]}\right) + pH$$

or

$$pH = pK_a + \log_{10}\left(\frac{[RCO_2^-]}{[RCO_2H]}\right) \tag{27.3}$$

Equation 27.3 shows that when the concentration of a carboxylic acid equals the concentration of the carboxylate anion, the logarithmic term is zero and the pH of the solution is the pK_a of the acid.

27.2 Inductive Effects

When electron-withdrawing substituents are attached to the R group of a carboxylic acid, the acid strength increases (larger K_a, smaller pK_a). For example, the pK_a of acetic acid in water is 4.74, whereas the pK_a of α-chloroacetic acid is 2.86.

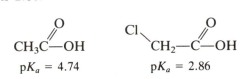

The increase in acidity by 1.88 pK_a units is usually attributed to the electrostatic stabilization by the polar C— Cl bond of the negative charge formed on the anion during ionization.

Actually, the stabilizing interaction (inductive effect) is quite complex and the full explanation requires consideration of the solvent not shown in the equilibrium represented by equation 27.1. In the gas phase, where there is no solvent, inductive effects can be greater by a factor of 10. When the solvent is changed from water to one with more hydrocarbon character, which is poorer at solvating the carboxylate anion, the pK_as become much larger.

The acid-stabilizing inductive effect of a substituent falls off markedly with increasing distance from the anion centers. This phenomenon is exemplified by the chloropropionic acids.

$$
\begin{array}{ccc}
 & \underset{|}{\text{Cl}} & \\
\text{CH}_3\text{CH}_2\text{CO}_2\text{H} & \text{CH}_3\text{CHCO}_2\text{H} & \text{ClCH}_2\text{CH}_2\text{CO}_2\text{H} \\
pK_a = 4.87 & pK_a = 2.83 & pK_a = 3.83
\end{array}
$$

The falloff can be understood as a consequence of the inverse distance dependence of the electrostatic interaction of charges.

When more than one substituent is included, the inductive effect is greater but not strictly additive. This is exemplified by the chloroacetic acids.

$$
\begin{array}{cccc}
\text{CH}_3\text{CO}_2\text{H} & \text{ClCH}_2\text{CO}_2\text{H} & \text{Cl}_2\text{CHCO}_2\text{H} & \text{Cl}_3\text{CO}_2\text{H} \\
pK_a = 4.7\dot{4} & pK_a = 2.86 & pK_a = 1.26 & pK_a = 0.64
\end{array}
$$

27.3 Analysis of pH/Titer Data for pK

From a set of pH versus basic titer data for the titration of a carboxylic acid, there are several ways to analyze the data to determine the pK of the acid. The method chosen depends on the precision desired.

In all cases, one starts by plotting the pH as a function of the total volume of base added, as shown in Figure 27.1. The steep rise at the right of the pH curve is the region of neutralization of the acid; the volume of base required to produce the steep rise is the equivalence point of the acid. From this volume and the known normality of the base, one can calculate the equivalents of acid present in the original solution. This quantity can be compared with the amount of acid placed in the starting solution and is a useful check on the purity of the acid. The pK of the acid is (approximately) the pH of the solution at the base volume where half of the acid has been neutralized. This result follows directly from equation 27.3. At the half-neutralization point, the concentration of the remaining undissociated acid, $[\text{HA}]$, is almost exactly equal to the concentration of the anion, $[\text{A}^-]$, so that $[\text{A}^-]/[\text{HA}] = 1$ and $K_a = [\text{H}^+]$, and thus pK_a = pH.

One problem with this simple approach is that it puts undue weight on a single pH measurement. If the measurement does not happen to fall exactly at the half-neutralization point, additional error is introduced. These problems can be overcome by applying equation 27.4 to all of the pH measurements.

$$
\left(\frac{[\text{A}^-]}{[\text{HA}]}\right)_x \approx \frac{(\text{fraction of acid neutralized})_x}{1 - (\text{fraction of acid neutralized})_x} = \frac{V_x}{V_{eq} - V_x} \qquad (27.4)
$$

The required ratios of $[\text{A}^-]/[\text{HA}]$ can be well approximated from the fraction of

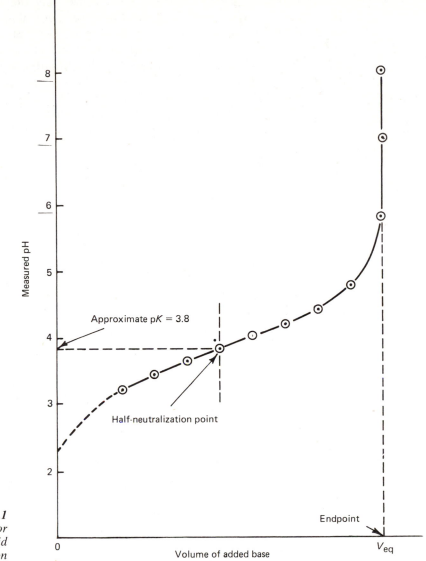

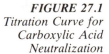

FIGURE 27.1
Titration Curve for
Carboxylic Acid
Neutralization

acid neutralized, which in turn is derived from the volume of added base, V_x, divided by the volume of base at the equivalence point V_{eq}. On insertion of this expression into the Henderson–Hasselbalch equation, the pK evaluated at point x becomes

$$pK = pH_x - \log\left(\frac{V_x}{V_{eq} - V_x}\right) \qquad (27.5)$$

In applying equation 27.5 it is best to use only the measurements lying between 20 and 80% neutralization. For lower percentages of neutralization, equation 27.5 is slightly inaccurate because it does not include a correction for the ionization of water; at higher percentages of neutralization, the term $V_{eq} - V_x$ approaches zero and the resulting uncertainty can produce large errors in the calculated pK.

The pK of the acid is taken as the average of the several estimates obtained from applying equation 27.5 to the measurements. Even with error-free measurements, there will be some discrepancy among the different measurements because equation 27.5 does not include any corrections for the variation in concentration of ions that occurs during titration. However, if the measurements are carried out as described in the next section, the discrepancy from this source should be less than ± 0.003 pK unit.

27.4 Measurement of the pK of a Carboxylic Acid

In a 100-mL volumetric flask, weigh out a 1-g sample of a carboxylic acid. (**Caution**—*many carboxylic acids are quite corrosive!*) Fill the flask about half full with distilled water and swirl the flask until the acid is dissolved; then fill it to the mark with distilled water. Transfer the solution as completely as possible to a 400-mL beaker.

Standardize your pH meter according to the directions provided with the meter, wash the electrodes with distilled water from a wash bottle, and gently dry them with a tissue. Clamp the electrodes so that they dip about 2–3 cm into the acid solution contained in the 400-mL beaker. Read and record the pH.

Fill a 50-mL buret about two-thirds full with 0.5 M NaOH solution. Read the volume and be sure to record the normality of the base, Add about 1.0 mL of base and stir the solution gently with a stirring rod. Read and record the buret volume and the pH. Repeat the addition, mixing, reading, and recording operations until the difference between successive increments becomes greater than 0.25 pH unit. When this point is reached, decrease the volume of added base to about 0.2 mL, and continue the titration until the pH rises above 9.

Clean the electrodes with distilled water from a wash bottle and leave them clamped in a beaker of fresh water with the tips dipping 2–3 cm below the surface. The pH electrodes should always be stored vertically and never allowed to dry out.

Analyze the volume/pH data as described in Section 27.3 and determine the pK of the acid. If a least-squares program is used in the analysis, determine the standard deviation of the pK as well.

Questions
1. Calculate the percent dissociation of monochloroacetic acid in the following solutions at 25°.
   (a) 0.1 M (b) 0.01 M (c) 0.001 M
   What is the pH of each of these solutions?
2. Which is the stronger base at 25°, $ClCH_2CO_2^-$ or $CH_3CO_2^-$?
3. The pK_as of very strong carboxylic acids such as trichloroacetic acid are difficult to determine on a 0.01 M solution. Why?

28 Side-Chain Oxidation of Aromatic Compounds

28.1 Oxidation of Side Chains

Oxidation of methyl groups and other side chains is an important method of preparing aromatic carboxylic acids. Side-chain oxidation is also useful in the identification of aromatic hydrocarbons and their derivatives because the acids are crystalline solids that can be purified and characterized readily.

Practically all side chains with a carbon atom directly attached to the benzene ring may be oxidized to carboxylic acid groups. Thus, the xylenes are oxidized to the corresponding phthalic acids, C_6H_4 $(CO_2H)_2$. Since the oxidation of an alkyl side chain is initiated by attack of a C—H group adjacent to the ring, a t-butyl side chain is extremely resistant. Common reagents for oxidizing side chains in aromatic compounds are chromic acid, nitric acid, and potassium permanganate. The first two reagents are used under strongly acidic conditions, whereas the permanganate is used under neutral–basic conditions. The choice is usually made on the basis of other structural features present in the molecule.

The presence of halogen, nitro, and sulfonic acid groups on the benzene ring does not interfere with the oxidation of the alkyl group, but if hydroxyl or amino groups are present, most oxidizing agents will attack these groups first and eventually destroy the ring completely. On the other hand, if these groups are protected by conversion to alkoxy or acetamido groups, then alkyl groups can be satisfactorily oxidized to carboxylic acids (see the synthesis of PABA in Chapter 47).

In order to balance the equations for oxidation it is convenient to use the method of half-cells. The half-cell for chromic acid (expressed in terms of dichromate ion) is

$$Cr_2O_7^{2-} + 14\ H^+ + 6\ e^- \longrightarrow 2\ Cr^{3+} + 7\ H_2O$$

The half-cell for permanganate ion to give manganese dioxide (neutral–basic conditions) is

$$MnO_4^- + 2\ H_2O + 3\ e^- \longrightarrow MnO_2 + 4\ OH^-$$

The half-cells for side-chain oxidations include

$$Ar\text{—}CH_3 + 2\ H_2O \longrightarrow Ar\text{—}CO_2H + 6\ e^- + 6\ H^+$$

$$Ar\text{—}CH_2CH_3 + 4\ H_2O \longrightarrow Ar\text{—}CO_2H + CO_2 + 12\ e^- + 12\ H^+$$

$$Ar\text{—}CH(CH_3)_2 + 6\ H_2O \longrightarrow Ar\text{—}CO_2H + 2\ CO_2 + 18\ e^- + 18\ H^+$$

The complete equation is obtained by using the appropriate multiple of the oxidant half-cell to balance the electrons produced by the side-chain oxidation half-cell. In practice, one normally uses an excess of the oxidant.

28.2 Preparations

(A) p-Nitrobenzoic Acid

In a 100-mL round-bottomed flask, place 0.54 g (0.004 mole) of p-nitrotoluene, and a solution of 1.8 g (0.006 mole) of sodium dichromate crystals ($Na_2Cr_2O_7\cdot 2H_2O$) in 4 mL of water.

▸ *CAUTION*   Dust containing chromium salts is suspected of causing cancer. Do not grind the sodium dichromate to hasten its solution.

Add 2.6 mL (4.6 g) of concentrated sulfuric acid (**Caution**—*corrosive!*) slowly, with constant swirling and thorough mixing, and attach a water-cooled reflux condenser. Thorough mixing is essential to avoid the danger of the reaction getting out of control during the heating. Heat the reaction mixture carefully (be prepared to remove the heat source) until oxidation starts; then stop heating until the vigorous boiling subsides. When the mixture has ceased to boil (it will continue for a while from the heat of reaction), replace the heat source under the flask and reflux the material vigorously for 2 hr (**Caution**—*bumping!*). Cool the reaction mixture and pour it into 6–8 mL of cold water. Collect the precipitate of crude p-nitrobenzoic acid with suction and wash it on the filter with two 2-mL portions of water.

Transfer the precipitate to a small beaker. Add 3 mL of 5% sulfuric acid, made by adding 0.2 mL of concentrated sulfuric acid (**Caution**—*corrosive!*) to

7 mL of water. Warm on a steam bath, and stir thoroughly to extract the chromium salts as completely as possible from the *p*-nitrobenzoic acid. Cool, filter with suction, and wash the product with two 2-mL portions of water. Transfer the crude *p*-nitrobenzoic acid to a beaker and treat it with 5–6 mL of 5% aqueous sodium hydroxide. The *p*-nitrobenzoic acid dissolves, and any unchanged *p*-nitrotoluene remains undissolved; chromium salts will be converted largely to chromium hydroxide. Add a small amount of decolorizing carbon, warm to 5° with stirring for about 5 min, and filter the alkaline solution with suction.[1] Precipitate the purified acid by pouring the alkaline solution, with stirring, into 6 mL of 10% sulfuric acid (prepared by adding 0.6 mL of concentrated sulfuric acid (**Caution**—*corrosive!*) to 6 mL of water). Collect the purified acid with suction, wash it with cold water, and dry. The yield is 0.3–0.6 g. The *p*-nitrobenzoic acid obtained in this way is sufficiently pure for most purposes. To obtain a product of high purity, a small sample may be crystallized from a large volume of hot water or from glacial acetic acid.

Because the melting point of *p*-nitrobenzoic acid is fairly high (about 240°) a MEL-TEMP apparatus is preferred. The purity of the acid may be checked also by determination of its neutralization equivalent, as described in Chapter 9, but with aqueous ethanol as the solvent.

The NMR spectrum of *p*-nitrobenzoic acid is interesting because of its deceptive simplicity.[2]

(B) *o*-Nitrobenzoic Acid[3]

In a 100-mL round-bottomed flask prepare a solution of 1.8 g (0.012 mole) of potassium permanganate in 30 mL of warm water and add 0.6 mL (0.61 g, 0.005 mole) of *o*-nitrotoluene.[3] Attach an upright condenser and heat the flask until the mixture boils vigorously (**Caution**—*bumping!*). Swirl the mixture frequently to minimize risk of breaking the flask. Continue the refluxing for 2–3 hr; the time may be divided between two laboratory periods if necessary.

Filter the hot solution through a fluted filter. If the solution is colored purple by residual permanganate, add a pinch of solid sodium bisulfite and 0.6 mL of concentrated hydrochloric acid; stir the solution thoroughly. Add more bisulfite if necessary to decolorize the excess permanganate, but avoid an excess.

To complete the precipitation of the nitrobenzoic acid, pour the solution into a well-stirred mixture of 2 mL of concentrated hydrochloric acid and 5 mL of water. Dissolve the nitrobenzoic acid by heating, add about 0.1 g of decolorizing

---

[1] The addition of 0.5–0.8 g, about 2 mL in volume, of a filter aid (Celite) greatly facilitates the filtration.

[2] The two different kinds of aromatic protons of *p*-nitrobenzoic acid appear as a singlet in the NMR spectrum even though the protons are strongly coupled. The nitro group and carboxylic acid group cause similar downfield shifts of the ring protons although they are chemically nonequivalent. Such sets of protons are said to be accidently isosynchronous.

[3] *o*-Chlorobenzoic acid may be prepared by this procedure, by substituting 0.4 mL (0.64 g) of *o*-chlorotoluene for *o*-nitrotoluene.

carbon, and filter the hot solution thorough fluted filter. Collect the filtrate in a 50-mL Erlenmeyer flask. Chill the filtrate and collect the crystals with suction. Wash them with a little water, and allow them to dry thoroughly. The yield is 0.5–0.6 g.

Questions

1. What acid is formed by oxidation of **(a)** *p*-cymene (*p*-isopropyltoluene) and **(b)** *o*-xylene (1,2-dimethylbenzene)?

2. Write balanced equations for the chromic acid and permanganate oxidations of **(a)** *p*-isopropyltoluene and **(b)** 1,2-dimethylbenzene.

3. Compare the ease of oxidation of toluene, benzyl alcohol, and benzaldehyde.

4. If you were given a substance that might be *o*-, *m*-, or *p*-chlorotoluene (all three of which boil at about the same temperature), how could you identify the substance by a chemical transformation?

29 Friedel–Crafts Reactions

The Friedel–Crafts reaction, discovered in 1877, has become the most important method for introducing alkyl and acyl groups into benzene and other aromatic compounds. It depends upon the formation of a highly reactive carbocation R^+ or acylonium ion RCO^+ that adds to the aromatic species.

29.1 Alkylation of Benzene and Related Hydrocarbons

Typically, alkylation reactions generate the required cation from alkyl halides (including aralkyl types such as $Ar-CH_2Cl$, $Ar-CHCl_2$) combined with strong Lewis acids such as aluminum chloride or from alkenes with strong protic acids such as HF, H_2SO_4, or $AlCl_3-HCl$.

$$C_6H_6 + CH_3CH_2Cl \xrightarrow{AlCl_3} C_6H_5-CH_2CH_3 + HCl$$

$$C_6H_6 + CH_2{=}CH_2 \xrightarrow[HCl]{AlCl_3} C_6H_5-CH_2CH_3$$

The role of the catalyst is to generate the cation (or the equivalent carbocation salt) that attacks the pi electrons of the aromatic system by an S_E2 electrophilic substitution process. Subsequent transformations lead to formation of the alkylbenzene and regeneration of the catalyst. Any catalyst that produces the same carbocation leads to the same alkylation.

$$C_2H_5-Cl + AlCl_3 \rightleftharpoons C_2H_5^+ \, [AlCl_4]^-$$

$$C_2H_5^+ \, [AlCl_4]^- + C_6H_6 \rightleftharpoons [C_2H_5-C_6H_6]^+ \, [AlCl_4]^-$$

$$[C_2H_5-C_6H_6]^+ \, [AlCl_4]^- \rightleftharpoons C_2H_5-C_6H_5 + AlCl_3 + HCl$$

Further alkylation tends to occur because the rate of reaction (nucleophilic activity of the aromatic ring) increases with successive attachment of alkyl groups into the aromatic ring. Monoalkylation is favored by using a large excess of the aromatic hydrocarbon.

As might be expected from the carbocation mechanism, primary alkyl halides and some secondary halides give rise to rearranged products. *n*-Propyl halides give mostly isopropylbenzene; *n*-butyl halides, *sec*-butylbenzene; and isobutyl halides, *t*-butylbenzene.

The orientation of attack in the alkylation of substituted benzenes is not very selective, due in part to the extremely great reactivity of the alkyl carbocations. A further complication is that the alkylation is a reversible reaction and the products can change positions. For example, the alkylation of toluene with isopropyl chloride under very mild conditions gives essentially *ortho/para* substitution (*ca.* 63% *ortho*, 12% *meta*, 25% *para*), but under vigorous conditions all of the isomeric isopropyltoluenes rearrange to give entirely the *meta* isomer. This indicates that the initial products are the result of kinetic control, owing to the rapid attack by the alkyl carbocation, but the reversibility of the reaction leads eventually to the most stable isomer (thermodynamic control).

29.2 Friedel–Crafts Acylation

Acylation of aromatic compounds by means of the Friedel–Crafts reaction and its modifications is one of the chief synthetic methods for the preparation of aromatic ketones. Aliphatic and aromatic acid chlorides, in the presence of anhydrous aluminum chloride, react with aromatic compounds to furnish alkyl aryl ketones and diaryl ketones.

In acylations with acid chlorides a slight excess over 1 equivalent of aluminum chloride is used, since a 1:1 addition compound is formed by reaction of aluminum chloride with the ketone produced in the reaction. The slight excess guarantees that even near the end of the reaction when the amount of product is large that there will be enough catalyst to catalyze the reaction of the remaining acid chloride. Acid anhydrides react in a similar way to produce ketones, usually in better yields than are obtained from acid chlorides, but it is necessary to use 2 equivalents (plus a small excess) of aluminum chloride because the organic acid by-product formed in the reaction ties up an additional equivalent of aluminum chloride.

$$C_6H_6 + CH_3CO-Cl \longrightarrow C_6H_5-\overset{\overset{+}{}}{\underset{CH_3}{C}}=\overset{-}{O}-\bar{A}lCl_3 + HCl$$

$$C_6H_6 + (CH_3CO)_2O + 2\ AlCl_3 \longrightarrow C_6H_5-\overset{\overset{+}{}}{\underset{CH_3}{C}}=\overset{-}{O}-\bar{A}lCl_3 + CH_3CO_2AlCl_2 + HCl$$

The function of the catalyst is to generate a reactive acylcation (acylonium ion), which attacks the aromatic system in the same manner as other active electrophiles.

$$CH_3—CO—Cl + AlCl_3 \rightleftharpoons CH_3—\overset{+}{C}{=}O[AlCl_4]^-$$

Acylation does not occur in systems that have a deactivating substituent, such as NO_2, CO_2CH_3, $CO—CH_3$, or $C—N$. For this reason, Friedel–Crafts acylations do not go beyond the introduction of more than one acyl group, since the carbonyl group of the ketone deactivates the molecule for further substitution.

Unlike the alkylation reactions, where rearrangements occur readily in the alkylcarbocation intermediates, the acyl groups do not undergo rearrangement in this reaction. The intermediate acylonium cation from an acid halide or an anhydride, $[R—C—O^+ \longleftrightarrow R—C^+{=}O]$, has enhanced stability resulting from the spreading out of its pi electrons between the carbonyl carbon and the oxygen atom (or equivalently, the distribution of the positive charge between these same two atoms).

For the acylation of aromatic compounds that cannot conveniently be used in excess as the reaction medium, or for solids such as naphthalene and biphenyl, it is desirable to employ an inert solvent. Tetrachloroethane and nitrobenzene have been used, but their boiling points are relatively high, making removal from the product troublesome. Dichloromethane (bp 40°) is well suited as an inert solvent. The procedure described in Section 27.4, using acetylation of biphenyl as the example, is a general one that can be used for naphthalene, alkoxybenzenes, and other reactive hydrocarbons. With naphthalene this solvent of low polarity strongly favors acylation at the 1-position, whereas the highly polar nitrobenzene favors the 2-position.

29.3 Preparations

(A) *p*-Methoxytetraphenylmethane

The preparation of *p*-methoxytetraphenylmethane described here is an example of aromatic alkylation. The reaction conditions are quite mild because *p*-methoxybenzene (anisole) is so reactive toward electrophilic substitution.

In the hood place in a 250-mL Erlenmeyer flask 3.0 g (0.0115 mole) of triphenylmethanol (Chapter 34) and add to it 9.0 mL (9.0 g, 0.083 mole) of anisole followed by 85 mL of glacial acetic acid. Swirl the flask to bring everything into solution and then add 8.5 mL of concentrated sulfuric acid (**Caution**—*corrosive!*). Label the flask with your name and set it aside for 2–7 days.[1]

Carefully pour the brown solution into 100 mL of water and collect the white precipitate on a Büchner funnel. Wash the product with a few milliliters of diethyl ether (**Caution**—*flammable solvent!*). The melting point of the crude product is 198–200°, which can be raised to 201–202° by recrystallization from 1 : 1 (v/v) toluene–2-propanol.[2] The yield is about 1.8 g (45% of theory).

(B) 4-Acetylbiphenyl

The preparation of the ketone 4-acetylbiphenyl is a specific example of a general procedure for acylation of aromatic compounds in the presence of an inert solvent.

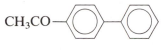

4-Acetylbiphenyl

In a hood assemble an apparatus with a 50-mL round-bottomed flask fitted with a Claisen adapter; place a separatory funnel in the central opening of the adapter and an upright condenser in the side arm. Attach a drying tube filled with calcium chloride at the top of the condenser and connect the open end of the drying tube to a gas absorption trap (see Figure 8.2) to dispose of the hydrogen chloride evolved during the reaction.

⫸ *CAUTION* Acetyl chloride is extremely corrosive and reacts vigorously with water to release toxic hydrogen chloride gas. Work only in a hood and be certain that all of the apparatus is kept scrupulously dry. If any of the acetyl chloride is spilled on your skin, wash the affected area immediately and thoroughly with water.

In the flask place 8 mL of dichloromethane and 3.3 g (0.024 mole) of anhydrous aluminum chloride.

⫸ *CAUTION* Prolonged exposure to high concentrations of methylene chloride vapors may induce cancer. As with all organic solvents, work in a well-ventilated area when using it.

Place in the separatory funnel 1.6 mL (1.74 g, 0.022 mole) of acetyl chloride, and add it dropwise with swirling while cooling the flask in an ice bath. Remove

---

[1] An alternative to this long wait is to reflux the solution in a round-bottomed flask for 1 hr. The heated reaction gives a lower yield (about 35% of theory) of a slightly less pure product.

[2] A discussion of recrystallization from solvent pairs is given in Section 5.2.

the ice bath and over a period of about 15 min, add a solution of 3.1 g (0.02 mole) of biphenyl in 10 mL of dichloromethane. Allow the reaction mixture to stand for 25 min longer at room temperature, occasionally shaking it, and then reflux it gently in a water bath for about 20 min.

To decompose the ketone–aluminum chloride complex, pour the cooled reaction mixture cautiously into a well-stirred mixture of 10 g of chipped ice and 8 mL of concentrated hydrochloric acid. Transfer the mixture to a separatory funnel, shake well to extract the aluminum chloride from the organic layer, and separate the layers. Rinse the reaction flask with about 8 mL of dichloromethane add the rinse to the organic layer. Wash the combined organic extracts with water, then with 5% aqueous sodium hydroxide, and finally with water (about 5 mL of each). Dry the dichloromethane solution with a small amount of anhydrous magnesium sulfate, remove the drying agent by passing it through a filter tube (see Figure 5.6), and distill off the solvent (bp 40°). Recrystallize the crude ketone from about 60 mL of ethanol. The yield is 2–2.2 g. The reported melting point of 4-acetylbiphenyl is 120–121°.

Questions   1. What products would be formed by the reaction of the following alkenes with benzene, in the presence of aluminum chloride/HCl?
   (a) ethylene   (b) isobutylene   (c) cyclohexene

2. Explain why *n*-propyl bromide reacts with benzene, in the presence of aluminum chloride, to give mainly isopropylbenzene. Suggest a method for obtaining *n*-propylbenzene.

3. Explain using resonance structures why the acetylation of biphenyl occurs predominantly in the *para* position.

4. How could 4-acetylbiphenyl be converted to each of the following?
   (a) 4-ethylbiphenyl
   (b) biphenyl-4-carboxylic acid
   (c) 4′-amino-4-acetylbiphenyl

5. Indicate how the following substances may be prepared by means of the Friedel–Crafts reaction.
   (a) diphenylmethane
   (b) benzyl phenyl ketone (deoxybenzoin)
   (c) *m*-bromodiphenyl ketone (*m*-bromobenzophenone)
   (d) *p*-nitrobenzophenone

6. What product is formed from the reaction of 4-acetylbiphenyl with (a) NaBH$_4$ and (b) Na$_2$Cr$_2$O$_7$? Explain.

30 Nitration of Aromatic Compounds

30.1 Mechanism of Nitration

Nitration is one of the most important examples of S_E2 electrophilic aromatic substitution. Although aromatic nitro compounds have limited usefulness as such, mainly as high explosives or booster charges (TNT, Tetryl), they are exceedingly useful as intermediates for the preparation of the corresponding amines and, indirectly, many other functional groups (—OH, —CN, —I). Nitrobenzene (bp 209°) is a good solvent for organic substances and also dissolves many inorganic compounds ($AlCl_3$, $ZnCl_2$). Most nitro compounds are *dangerously poisonous* and must be handled carefully.

Nitrations may be effected by means of pure nitric acid,[1] mixtures of concentrated nitric and sulfuric acids, and solutions of nitric acid in glacial acetic acid, acetic anhydride, or water. Selection of the appropriate nitrating agent and the conditions of reaction is based upon factors such as the reactivity of the compound to be nitrated, its solubility in the nitrating medium,[2] and the ease of isolation and purification of the product.

---

[1] Anhydrous nitric acid (sp g 1.50, sometimes called white fuming nitric acid) is a colorless liquid boiling at 86°; ordinary concentrated nitric acid (sp g 1.42) is the water–nitric acid azeotrope, bp 120°, containing 70% by weight of nitric acid; yellow fuming nitric acid contains 85–90% nitric acid with small amounts of oxides of nitrogen; red fuming nitric acid contains relatively large amounts of dissolved oxides of nitrogen.

[2] For a compound that is sparingly soluble in the nitrating mixture (aromatic hydrocarbons, aryl halides, etc.) the rate of nitration may be governed by its rate of solution in the medium; hence good agitation hastens the reactions.

The mechanism of nitration has been studied extensively, and it is known that for most aromatics the active electrophilic species is the nitronium ion, $[O\!=\!N\!=\!O]^+$. This is formed in the typical nitrating mixtures by a reversible interaction of nitric and sulfuric acids. Attack on the aromatic system by the nitronium ion, usually the slow rate-determining step, is followed by the rapid loss of a proton, leading to the nitro derivative.

$$HNO_3 + 2\ H_2SO_4 \rightleftharpoons [O\!=\!N\!=\!O]^+ + [H_3O]^+ + 2\ [HSO_4]^-$$

The ease of nitration depends on the nature of the substituents present; electron-releasing groups ($-OH$, $-NHCOCH_3$, $-CH_3$) facilitate the nitration, and electron-withdrawing groups ($-NO_2$, $-CO_2H$) retard the reaction.

Since nitration is not a reversible reaction, the distribution of *ortho*, *meta*, and *para* isomers in the product is controlled by the relative rates of substitution at each position. In general, substituents may be divided into three broad categories.

1. Activating and *ortho/para* directing

$$-OH \quad\quad -OCH_3 \quad\quad -NHCOCH_3 \quad\quad -CH_3 \quad\quad -CH_2CO_2CH_3$$

2. Deactivating and *ortho/para* directing

$$-CH_2Cl \quad\quad -Cl \quad\quad -Br \quad\quad -I$$

TABLE 30.1 *Partial Rate Factors for Nitration of Substituted Benzenes*

	Distribution of isomers			Activity	Partial rate factors		
Substituent	*ortho*	*meta*	*para*	vs. C_6H_6	f_o	f_m	f_p
$-CH_3$	56	4	40	24.5	41	2.9	59
$-C(CH_3)_3$	12	8	80	15.5	5.6	3.7	74
$-Cl$	30	1	69	0.033	0.030	0.001	0.14
$-CH_2Cl$	32	15	53	0.71	0.68	0.32	2.3
$-Br$	36	1	63	0.030	0.032	0.001	0.113
$-CO_2Et$	28	68	4	0.0037	3.1×10^{-3}	7.5×10^{-3}	8.9×10^{-4}
$-NO_2$	6	93	1	6×10^{-8}	1.1×10^{-8}	1.7×10^{-7}	3.6×10^{-9}
$-N(CH_3)_3{}^+$	0	90	10	1.2×10^{-8}	0	3.2×10^{-8}	7.2×10^{-9}

3. Deactivating and *meta* directing

—CCl$_3$ —COCH$_3$ —CO$_2$CH$_3$ —NO$_2$ —(NR$_3$)$^+$ —CN —SO$_3$H

Studies of the nitration of mixtures of benzene and substituted benzenes have led to an evaluation of substituent effects on the yields of *ortho*, *meta*, and *para* isomers. From these data, one can calculate the relative rates (partial rate factors) for a given position of the substituted compound relative to a single position in benzene. The results are shown in Table 30.1.

30.2 Preparations

(A) *o*- and *p*-Nitrophenol

This preparation uses phenol, which is both irritating and toxic. The entire procedure should be carried out in the hood to minimize exposure to phenol vapors. The preparation has a highly satisfying outcome, but it demands attention to the details of the procedure.

IIIII➤ *CAUTION*   Phenol is poisonous and caustic. Ingestion of small amounts will cause nausea and vomiting; the average fatal does is about 15 g. Phenol is absorbed through the skin so gloves should be worn when handling it.

In a large test tube (20 × 150 mm) place 3.3 mL of 30% aqueous nitric acid (prepared from 1.3 mL of concentrated nitric acid (**Caution**—*corrosive!*) and 3 mL of water). Temperature control in this experiment is critical; you should insert a thermometer into the test tube and leave it there throughout the nitration. Cool the test tube in an ice bath until the temperature of the acid is 5–10°, then add dropwise 1 mL of "liquefied" phenol.[3] Between drops swirl the test tube to ensure good mixing. Use the ice bath as required to keep the temperature in the 5–10° range. After all of the phenol has been added, allow the reaction mixture to stand at room temperature for 30 min. The reaction temperature will rise to about 35°.

Remove the thermometer, add 10 mL of water, and mix the contents well by swirling. Chill the reaction mixture, allow the semisolid organic layer to settle, and remove the upper aqueous layer with the aid of a Pasteur pipet. Repeat this washing procedure two more times with 10-mL portions of water.

Transfer the dark organic layer to a 25-mL round-bottomed flask with the aid of a total of 15 mL of water. Add a boiling chip, attach a distillation head to the

---

[3] Pure phenol melts at 41°. Addition of 8% water produces a solution that is liquid at room temperature. Use of this "liquefied" form reduces exposure to the vapors.

boiler and use a round-bottomed flask as the receiver. Do *not* pass water through the condenser since the product to be collected is a low-melting solid and will plug the condenser if it solidifies there. Instead, be sure of that the receiver flask is well-mated to the exit joint of the condenser and then immerse it up to its neck in ice water. In this way, the receiver flask will also serve as the condenser. Heat the boiler flask and collect 5–10 mL of distillate, which will consist of water and almost pure bright yellow *o*-nitrophenol. Collect the crystals on a Hirsch funnel; the yield is 250–300 mg, mp about 45°. If a purer product is desired, it can be recrystallized from mixed hexanes.

To the brown-black residue in the boiler flask, which is an unpromising mixture of *p*-nitrophenol and tarry oxidation products, add 5 mL of water and 0.3 mL of concentrated hydrochloric acid.[4] Add about 0.3 g of activated charcoal and heat the mixture to boiling. Mix well and filter the hot solution through a wetted fluted filter paper into a small beaker. Cool the dark brown filtrate in an ice bath and collect the feathery light tan needles on a Hirsch funnel. The yield is 100–150 mg, mp 112–113°. Recrystallization from 3% hydrochloric acid with added charcoal will produce an almost white product, mp 113–114°.

(B) Methyl *m*-Nitrobenzoate

Place 3 mL (5 g) of concentrated sulfuric acid (**Caution**—*corrosive!*) in a 50-mL Erlenmeyer flask, cool the acid to 0°, and add 1.3 mL (1.4 g, 0.01 mole) of methyl benzoate, with swirling. While maintaining the internal temperature at 5–15°, by cooling as needed in an ice–water bath, add drop by drop a cold mixture of 1 mL (1.8 g) of concentrated sulfuric acid and 1 mL (1.4 g, 0.017 mole) of concentrated nitric acid (**Caution**—*corrosive!*). Swirl the solution during the addition and for 10 min after all of the acid has been added.

▐▶ *CAUTION*   All nitro compounds are poisonous and must be handled carefully. Any nitro compound that comes in contact with the skin should be removed by washing with a little ethanol, followed by soap and water.

Pour the reaction mixture, with stirring, onto about 10 g of cracked ice to precipitate the crude methyl *m*-nitrobenzoate (which contains an appreciable amount of the *ortho* isomer and a trace of *para*). Collect the product with suction and wash it thoroughly on the filter with two or three 3-mL portions of water, to remove nitric and sulfuric acids. For effective washing, release the suction, mix the material thoroughly with the washing liquid, apply suction, and press the crystals firmly.

Wash the product finally with two 1-mL portions of *ice-cold* methanol, in the manner described earlier, and press the crystals thoroughly. Proper washing

---

[4] The purpose of the hydrochloric acid is to suppress the ionization of the *p*-nitrophenol and decrease its room-temperature solubility.

removes most of the more soluble *ortho* isomer. The crude product weighs about 1.0–1.4 g and melts at 74–76°. It may be purified further by recrystallization from a small volume of hot methanol.[5] Reserve a minute amount of material to be used as seed crystals. The recorded melting point of pure methyl *m*-nitro-benzoate is 78.5°

Questions

1. Why is sulfuric acid used in nitration?

2. What position will be taken by the entering nitro group when the following substances are nitrated (use Table 30.1)? Which will react most rapidly? Which is the least reactive?
   (a) bromobenzene   (b) nitrobenzene   (c) toluene
   (d) *t*-butylbenzene   (e) ethyl benzoate   (f) benzoic acid

3. Look up and write the structural formulas of
   (a) picric acid   (b) TNT   (c) Tetryl

4. In the nitration of phenols the *ortho* and *para* nitration products are separated by steam distillation. Why does the *ortho* isomer steam distill and the *para* isomer stay behind?

5. What is formed by the reduction of nitrobenzene in the presence of acids?

---

[5] To prevent contamination through ester exchange, it is good practice in recrystallization to avoid using an alcohol different from that corresponding to the alkoxyl group of the ester.

31 Nitration of Anilines: Use of a Protecting Group

31.1 Protecting Groups

In synthetic sequences it is often necesssary to temporarily convert primary and secondary arylamines into their acetyl derivatives in order to prevent them from reacting with the reagents used along the way. Unprotected aryl amines are particularly sensitive to oxidation and to attack by acid halides. At the end of the synthetic sequence, the amino group can be regenerated readily by hydrolysis with acids or bases.

Arylamines (and aliphatic amines) may be acetylated by means of acetic anhydride or acetyl chloride or by heating the amine with glacial acetic acid under conditions that permit removal of the water formed in the reaction. The last procedure is an economical one but requires a relatively long period of heating.

Acetic anhydride is the preferred acetylating agent. In some instances, the solution of the acetyl derivative in glacial acetic acid that results from the acetylation may be used for a subsequent reaction of the acetylated compound without isolating it. If the temperature and reaction time are increased in acetylations with acetic anhydride, primary amines may form a bis-acetyl derivative, $Ar—N(COCH_3)_2$, but this can be hydrolyzed under mild conditions to the monoacetyl derivative.

In spite of the high reactivity of acetic anhydride toward amines, the rate of hydrolysis is sufficiently slow to permit acetylation of amines in buffered aqueous solutions (method of Lumière and Barbier). This is a general procedure that gives a product of high purity in good yield, but it is not suitable for acetylation of the nitroanilines and other extremely weak (unreactive) bases. By a similar procedure, acetic anhydride may be used to acetylate phenols in an aqueous alkaline solution.

In the next section these three methods of acetylation are illustrated. In Sections 31.3 and 31.4 the use of an acetylated amino group to protect it from unwanted reactions is discussed and illustrated.

31.2 Acetylation of Aniline

(A) Acetylation in Water—Lumière–Barbier Method[1]

Dissolve 1.1 mL (1.12 g, 0.012 mole) of aniline in 30 mL of water to which 1 mL (0.012 mole) of concenttrated hydrochloric acid has been added. If the amine is discolored, add 0.2–0.4 g of decolorizing carbon, stir the solution for a few minutes, and filter it with suction. Meanwhile, prepare for use in the next step a solution of 1.8 g (0.013 mole) of sodium acetate crystals ($CH_3CO_2Na \cdot 3H_2O$) in 4 mL of water; if any insoluble particles are present, filter the solution.

Transfer the solution of aniline hydrochloride to a 50-mL flask. Add 1.6 mL (1.66 g, 0.03 mole) of acetic anhydride and swirl the contents to dissolve the anhydride. Add *at once* the previously prepared sodium acetate solution and mix the reactants thoroughly by swirling. Cool the reaction mixture to an ice bath and stir vigorously while the product crystallizes. Collect the crystals on a suction filter, wash with cold water, and allow them to dry. The yield is 1.0–1.4 g. The material obtained by this acetylation procedure is usually quite pure and of better quality than that prepared by the acetylation in acetic acid. If necessary, the product may be recrystallized from water, with addition of about 0.2 g of decolorizing carbon.

(B) Acetylation in Acetic Acid

In a 50-mL Erlenmeyer flask dissolve 1.4 mL (1.42 g, 0.015 mole) of aniline in 3 mL of glacial acetic acid. To the solution add 1.8 mL (1.8 g, 0.018 mole) of acetic anhydride and mix well by swirling. The solution becomes warmer from the heat of reaction. Add a boiling chip, attach a reflux condenser, and boil the solution *gently* for 15 min to complete the acetylation. To hydrolyze the excess of acetic anhydride and any bis-acetyl derivative, add cautiously through the top of the condenser tube 1 mL of water, and boil gently for 5 min longer. Allow the reaction mixture to cool slightly and pour it *slowly*, stirring thoroughly, into 7–8 mL of cold water. After allowing the mixture to stand for about 15 min with occasional stirring, collect the crystals on a suction filter, and wash them with a

---

[1] Other arylamines (toluidines, xylidines, anisidines, phenetidines) can be acetylated by this general procedure. It is advantageous to use the Lumière–Barbier method if the sample of amine is discolored because the procedure affords a preliminary treatment with decolorizing carbon. Very weak bases (nitroanilines, dihalogenated anilines) cannot be acetylated by this method. For preparation of phenacetin, see Chapter 49.

little cold water. Recrystallize the crude product from hot water (about 20 mL/g) with the addition of about 0.2 g of decolorizing carbon. The yield is 1.0–1.4 g.

(C) Direct Acetylation with Acetic Acid

This is the most economical method of effecting acetylation, but the operation requires a longer time than methods A and B, which are most suited for small-scale operations.

In a 25-mL round-bottomed flask, place 1.8 mL (1.84 g, 0.02 mole) of aniline and 2.4 mL (2.4 g, 0.04 mole) of glacial acetic acid. Provide the flask with a short fractionating column topped with a distillation head and thermometer. For the receiver use a small graduated cylinder. Add a boiling chip and heat the flask gently, so that the solution boils quietly and the vapor does not rise into the column.

After 15 min increase the heating slightly so that the water formed in the reaction, together with a little acetic acid, distills over very slowly at a *uniform* rate (vapor temperature 104–105°). After about an hour, when 1–2 mL of distillate has collected, increase the heating so that the temperature of the distilling vapor rises to about 120°. Continue the distillation slowly for about 10 min longer, to collect an additional 0.2–0.4 mL of distillate (total volume, 1.2–1.4 mL), and then discontinue the heating. The distillate, consisting of 70–75% acetic acid, may be discarded.

Since the reaction mixture will solidify upon cooling, pour it out at once into about 40 mL of ice and water in a large beaker. Stir the aqueous mixture vigorously to avoid formation of large lumps of the product. Collect the acetanilide with suction, wash with a little cold water, and press it firmly on the filter. Crystallize the moist product from hot water (about 20 mL/g) with the addition of about 0.2 g of decolorizing carbon. For filtration use a fluted filter and a funnel with a short, wide stem. Cool the filtrate rapidly while stirring vigorously to obtain small crystals. Allow the material to stand for about 10 min in an ice–water bath and then collect the crystals with suction. Wash the product with a small amount of cold water and spread it on a clean paper to dry. If the material is dark colored it should be recrystallized. The yield is 1.5–1.8 g.

31.3 Nitration of Acetanilide and Deacetylation

When aniline is nitrated with concentrated nitric and sulfuric acids, the product is largely *m*-nitroaniline, some *p*-nitroaniline, and very little of the *ortho* isomer. The yield is not high because some of the aniline is lost through oxidation.

The dominant formation of *m*-nitroaniline occurs because the amino group of the aniline is protonated by the highly acidic medium to produce the positively charged NH_3^+ group, which is strongly deactivating, especially at the *ortho* and *para* positions.

Conversion of the aromatic amine to its acetyl derivative before carrying out nitration both diminishes its susceptibility to oxidation and sharply reduces its basicity, which avoids conversion to the *meta*-directing ammonium ion. Nitration now gives predominantly the *p*-nitroacetanilide. The acyl group is removed subsequently by hydrolysis with aqueous acid or alkali to produce in good yield and good positional purity *p*-nitroaniline.

p-Nitroaniline is used to prepare azo dyes, such as para red (an ingrain color, Chapter 42). Cotton cloth may be soaked in a dilute alkaline solution of 2-naphthol (or similar coupling component), dried, and dipped into an ice-cold solution of diazotized *p*-nitroaniline; coupling takes place and the dye is formed within the pores of the cellulose fibers.

(A) *p*-Nitroacetanilide

In a 50-mL Erlenmeyer flask dissolve 1.36 g (0.01 mole) of pure acetanilide in 1.6 mL of glacial acetic acid by warming it gently. Cool the warm solution until crystals begin to form and then add slowly, while swirling the solution, 2 mL of ice-cold concentrated sulfuric acid (**Caution**—*corrosive!*). Prepare a nitrating mixture by adding 0.7 mL (1 g, 0.012 mole) of concentrated nitric acid to 1 mL of cold concentrated sulfuric acid; cool the solution to room temperature and transfer it to a small separatory funnel.

Cool the acetanilide solution to 5° in an ice bath, remove the flask from the bath, and add the nitrating mixture slowly, drop by drop. Swirl the reaction mixture to obtain good mixing in the viscous solution and do not permit the temperature to rise above 20–25°. After all of the nitrating mixture has been added, allow the solution to stand at room temperature for about 40 min (but not longer than 1 hr) to complete the reaction. Pour the solution slowly with stirring into a mixture of 20 mL of water and 4–5 g of chipped ice. Collect the product with suction, press it firmly on the filter, and wash thoroughly with more water to from a thin paste, return them to the suction filter, and wash thoroughly with more water to remove the nitric and sulfuric acids. Press the material as dry as possible. The crude, moist *p*-nitroacetanilide is sufficiently pure to be used directly for hydrolysis to *p*-nitroaniline. The moist product is equivalent to about 1.2 g of dry material.

A small portion of the material may be purified by crystallization from 80% aqueous ethanol, with the addition of a little decolorizing carbon. The melting point of *p*-nitroacetanilide is about 215–216°; use a metal block or MEL-TEMP unit.

(B) *p*-Nitroaniline

In a 50-mL Erlenmeyer flask mix the moist, crude *p*-nitroacetanilide with 3 mL of water and 4 mL of concentrated hydrochloric acid. Reflux the mixture gently for 15–20 min. The material gradually dissolves and an orange-colored solution is formed.[2] When the hydrolysis is completed, add 6 mL of cold water and cool the mixture to room temperature. Crystals of the product may separate.

Pour the *p*-nitroaniline hydrochloride slowly, stirring thoroughly, into a mixture of 4 mL of concentrated aqueous ammonia, 15 mL of water, and 5–6 g of chipped ice. The mixture must be distinctly alkaline at the end of the mixing; test with litmus, and add a little more ammonia if necessary. Collect the orange-yellow precipitate of *p*-nitroaniline with suction and wash it with cold water. Recrystallize the product from a large volume of hot water; about 30 mL of water will be required per gram of material. The yield is 0.5–0.8 g; the recorded melting point is 147°.

Questions   1. Show a detailed mechanism for the reaction of aniline with acetic anhydride, including the possibility of a cyclic intermediate for elimination of acetic acid from the initial addition product.

2. What product would you expect to obtain by reaction of aniline with the mixed anhydride acetic–formic anhydride (HCO—O—$COCH_3$)? Explain.

3. How could the presence of a small amount of nitrobenzene in a sample of aniline be detected and how could it be removed?

4. When acetic acid is used for acetylation of an amine, why is it desirable to use an excess of the acid and to distill off the water formed in the reaction?

5. In the preparation of *p*-nitroaniline, why is aniline converted to acetanilide before nitration?

6. When trimethylamine and isobutyryl chloride are mixed in dry ether, trimethylamine hydrochloride precipitates but the amine is not acylated. What is the other product?

7. What products are formed by nitration of the following?
   (a) *p*-acetotoluidide   (b) *m*-acetotoluidide
   (c) *m*-cresol   (d) *p*-toluic acid

8. Would benzoic acid undergo nitration satisfactorily under the mild conditions used in this experiment? Explain.

9. (a) Can *p*-nitroacetanilide be hydrolyzed by alkalis?
   (b) What side product may be formed by the action of hot aqueous alkalis on *p*-nitroaniline? (Consider the activating effect of the nitro group.)

10. Outline a series of reactions for each preparation.
    (a) *o*-nitroaniline from aniline   (b) *m*-nitroaniline from benzene

11. Is *p*-nitroaniline a stronger or weaker base than aniline? Is *p*-nitrophenol a stronger or weaker acid than phenol? Explain.

---

[2] If the reaction mixture has been boiled too vigorously, it may be necessary to add more hydrochloric acid to replace that lost by evaporation and to heat the solution longer to complete the hydrolysis.

32 Aldol Condensation

32.1 Introduction

In its simplest form the aldol condensation is the condensation of two molecules of the same aldehyde or ketone and results in the formation of a new carbon–carbon bond joining the carbonyl carbon of one molecule to the α-position of the second.[1]

$$CH_3CH_2-CH=O \; + \; H-\underset{\underset{CH_3}{|}}{CH}-CH=O \; \xrightarrow[\text{or } H^+]{OH^-} \; CH_3CH_2-\underset{\underset{OH}{|}}{CH}-\underset{\underset{CH_3}{|}}{CH}-CH=O$$

To react in this way it is necessary that the molecule possess a reactive carbonyl group and be able to lose a proton from the α-position to form an enolate anion. An aromatic aldehyde has no hydrogen atoms in the α-position and thus can not give this simple aldol condensation. However, aromatic aldehydes are capable of participating in mixed (crossed) aldol condensation with aliphatic aldehydes or ketones that can form enolate anions.

$$C_6H_5-\overset{\overset{O}{\|}}{CH} \; + \; CH_3-\overset{\overset{O}{\|}}{CH} \; \xrightarrow{OH^-} \; C_6H_5-\underset{\underset{OH}{|}}{CH}-CH_2-\overset{\overset{O}{\|}}{CH} \; \longrightarrow$$

$$C_6H_5-CH=CH-\overset{\overset{O}{\|}}{CH} \; + \; H_2O$$

[1] Nielsen and Houlihan, *Organic Reactions*, **16**,1 (1968).

Usually the mixed aryl aldol undergoes dehydration spontaneously to form the resonance stabilized α,β-unsaturated aldehyde or ketone. The reaction is known as the Claisen–Schmidt condensation.

Aldol-type condensations of aromatic aldehydes extend to other reactants containing an active methylene group—such as malonic ester, acid anhydrides, nitriles, nitroalkanes, and similar compounds.

Either acids or bases will catalyze the aldol condensation but basic catalysts are generally preferred. Dilute aqueous or ethanolic sodium hydroxide, sodium ethoxide, and secondary amines (diethylamine or piperidine) are effective catalysts. The first step in the process is the formation of the enolate anion of the active methylene component by proton abstraction by the base.

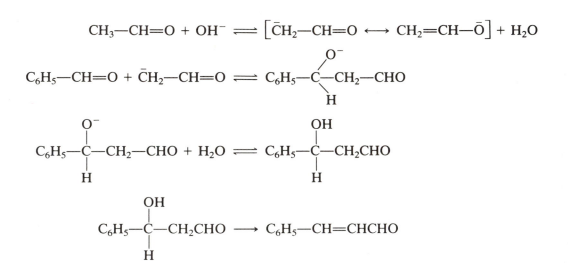

The resulting carbanion combines with the carbonyl reactant and proton interchange with the solvent leads to the mixed aldol, which then undergoes dehydration to the α,β-unsaturated compound.

The kinetics of aldol-type condensations vary with the character of the reactants and the experimental conditions. In the self-condensation of acetaldehyde under typical conditions, with aqueous sodium hydroxide as catalyst, the enolization step is slow and the second step is very fast. In the condensation of benzaldehyde and acetophenone, with sodium ethoxide as catalyst, the reversible enolization of acetophenone is fast and combination of the carbanion with benzaldehyde is the slow step. The rate of this reaction is proportional to the concentrations of the carbanion and the benzaldehyde. With less acidic ketones the rate will be slower because of the lower equilibrium carbanion concentration; with highly substituted carbanions, the rate may be slower because of steric hindrance.

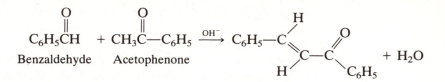

Mixed aldol condensations furnish intermediates for synthetic procedures used to obtain aromatic compounds having a variety of functional groups in the side chain and also structures having several aromatic rings attached to an aliphatic system. With ketones having two methylene groups, the aldol condensation is complicated by the possibility of mono- or disubstitution. For example, with benzaldehyde and excess acetone the monocondensation product, benzalacetone, is obtained. However, with benzaldehyde and acetone in a 2:1 mole ratio, dibenzalacetone is obtained instead.

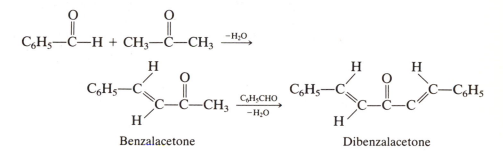

An interesting example of disubstitution is the condensation of dibenzyl ketone with benzil to produce in good yield the intensely purple tetraphenylcyclopentadieneone.

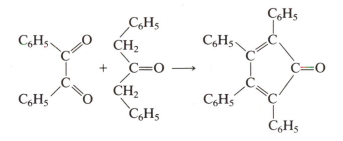

32.2 Preparations

(A) Dibenzalacetone

In a 50-mL Erlenmeyer flask prepare a solution of 0.4 g (0.001 mole) of sodium hydroxide in 2 mL and 2 mL of 95% ethanol. After the solution has cooled, add 0.3 mL (0.24 g, 0.004 mole) of acetone and then 0.8 mL (0.82 g, 0.008 mole) of benzaldehyde. A yellow turbidity will appear almost immediately, which quickly turns into a flocculent precipitate. Swirl the flask from time to time over 15-min period. Collect the mushy reaction product on a Büchner funnel and wash it first with water and then a little chilled 95% ethanol. Continue to draw air through the funnel until the product is dry and then recrystallize it from ethyl acetate (**Caution**—*flammable solvent!*) using about 2.5 g of solvent per gram of product. The yield of purified product is about 0.6 g, mp 110–111°.

(B) Tetraphenylcyclopentadienone

In a 100-mL round-bottomed flask place 1.05 g (0.005 mole) of benzil, 1.05 g (0.005 mole) of dibenzyl ketone, and 20 mL of 95% ethanol. Add about 0.15 g (about 0.004 mole) of potassium hydroxide pellets and a boiling chip to the flask. Attach a reflux condenser and boil the solution gently for 15 min. The mixture will turn deep purple as the product forms. Swirl the flask from time to time while it is being heated to insure that the solids have dissolved.

Cool the mixture to room temperature and then in an ice bath. Collect the purple crystals on a Büchner funnel, wash them well with about 10 mL of water, and then with about 10 mL of cold 95% ethanol. Allow the product to air dry. The yield of nearly pure dry product is 1.4–1.6 g (mp 218–220°). If desired, the product can be recrystallized from a 1 : 1 mixture of 95% ethanol and toluene (about 25 mL/g) to give a slightly higher melting product.

Questions

1. What products would be formed by mixed aldol condensation of benzaldehyde with propionaldehyde? with acetone (in excess)?

2. What products are formed by reaction of benzalacetophenone with the following reagents?
   (a) phenylmagnesium bromide, followed by water and dilute acid
   (b) diethylamine
   (c) phenylhydrazine, followed by cyclization with sulfuric acid

3. Write projection formulas for the stereoisomeric forms of benzalacetophenone and its dibromide.

4. Benzalacetophenone can be nitrated to give a mononitro derivative.
   (a) What structure would you expect for this compound?
   (b) If a second nitro group were introduced, where would it enter?

33 The Benzilic Acid Rearrangement

33.1 Introduction

When benzil is warmed with strong alkalis, it is converted into a salt of α-hydroxydiphenylacetic acid (benzilic acid). The 1,2-molecular rearrangement is initiated by addition of hydroxyl ion to the diketone (step 1), followed by transfer of the aryl group with its bonding electrons (carbanion rearrangement) to the adjacent carbon atom (step 2). By concurrent proton interchange (step 3) the stable benzilate anion is formed. A minor side product of this reaction is diphenylmethanol formed by decarboxylation of the benzilate anion.

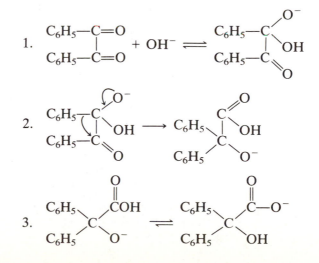

It has been established by means of oxygen exchange with ^{18}O-labeled water that step 1 is reversible and faster than step 2. At low temperatures where step 2 is slow the hydroxide adduct can be isolated. With methoxide ion in methanol, a similar rearrangement occurs, and the product is the methyl ester of benzilic acid[1]. An analogous reaction is the conversion of benzil to dilantin sodium with urea and base (see Section 48.4).

Benzilic acid can also be formed from benzoin with a basic solution of bromate ion. The benzoin is first oxidized to benzil, which then undergoes *in situ* the benzilic acid rearrangement.

The benzilic acid arrangement extends to many substituted diaryl 1,2-diketones but not to simple a aliphatic analogs (such as CH_3—CO—CO—CH_3). The latter undergo complex aldol condensations under the influence of alkaline reagents.

Cyanide ion in ethanolic solution, in catalytic amounts, causes a rapid and complete cleavage of benzil at the central carbon–carbon bond; the products are ethyl benzoate and benzaldehyde.

$$C_6H_5-C=O \atop C_6H_5-C=O \quad + C_2H_5OH \xrightarrow{CN^-} C_6H_5-C=O \atop \quad\quad\quad OC_2H_5 \quad + C_6H_5-C=O \atop \quad\quad\quad H$$

The mechanism of the cleavage is uncertain, but it is likely that the process involves addition of cyanide ion to a mono hemiketal formed by addition of ethanol to the diketone.

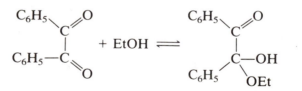

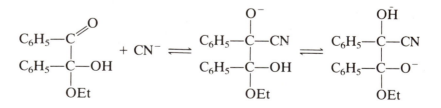

[1] For a discussion of the benzilic acid and related rearrangements see Lowry and Richardson, *Mechanism and Theory in Organic Chemistry*, 2nd ed. (New York : Harper and Row, 1981).

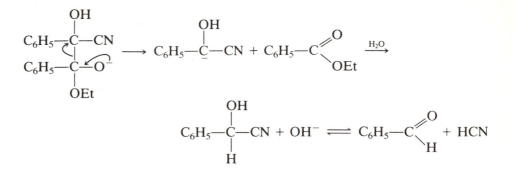

33.2 Preparation of Benzilic Acid

(A) From Benzil

In a 25-mL round-bottomed flask dissolve 1 g (0.015 mole) of solid potassium hydroxide pellets (85% pure) in 2 mL of water, add 3 mL of ethanol, and mix well by swirling. To the solution add 1 g (0.0048 mole) of benzil (a bluish coloration is developed), attach an upright condenser, and reflux the solution on a steam bath for 10–15 min. Transfer the contents of the flask to a small beaker and cover it with a watch glass. Allow the reaction mixture to stand for several hours, preferably overnight, until crystallization of the potassium salt of benzilic acid is complete. Collect the crystals on a suction filter and wash them sparingly ice-cold ethanol. The ethanolic mother liquor will furnish a small additional quantity of potassium benzilate if allowed to stand overnight.

Dissolve the potassium salt in about 30 mL of water and add to the solution, with stirring, two drops of concentrated hydrochloric acid. A reddish brown, slightly sticky precipitate of the impure diphenylmethanol is formed. Add a small amount of decolorizing carbon, stir, and filter the mixture through a fluted filter paper. If the procedure has been performed successfully, the filtrate will be colorless or only faintly yellow. Pour the clear filtrate slowly, with stirring, into a solution of 1.6 mL of concentrated hydrochloric acid in 10 mL of water. Collect the precipitated benzilic acid with suction, wash it thoroughly with water to remove the inorganic chlorides, and press it dry. The crude product is usually light pink or yellow and weighs 0.8–0.9 g. Crystallize the material from hot water, with addition of a little decolorizing carbon. The yield of purified benzilic acid is 0.7–0.8 g.

(B) From Benzoin (*Alternative Procedure*)

In a small Erlenmeyer flask prepare a solution of 1.4 g (0.035 mole) of solid sodium hydroxide and 0.3 g (0.01 mole) of sodium bromate (or 0.34 g of potassium bromate) in 3 mL of water. To the warm solution, add in several portions a total of 1.2–1.3 g of the slightly moist benzoin obtained in Experi-

ment 48 (this amount corresponds to about 0.005 mole of dry benzoin). During
and after the addition of the benzoin, heat the reaction mixture on a steam bath,
and stir it constantly. The mixture should not be heated above 90–95°, since
higher temperatures favor decomposition to form diphenylmethanol. From time
to time add small portions of water (in total about 2.4–3 mL) to keep the
mixture from becoming too thick. Continue the heating and stirring until a small
test portion is completely or almost completely soluble in water. This usually
requires 1.5–2.0 hr.

Dilute the reaction mixture with 12 mL of water, and allow it to stand,
preferably overnight. Filter the solution to remove the oily or solid side product
(diphenylmethanol). To the filtrate add slowly, with stirring, sufficient 40%
sulfuric acid (prepared by adding 1 mL of concentrated sulfuric acid (**Caution—
corrosive!**) to 3 mL of water) to reach a point just short of the liberation of
bromine.[2] Usually about 3.4 mL of the dilute acid is required. Collect the
benzilic acid with suction, wash it well with water, and press dry. The yield is
0.9–1.0 g, and the product is usually quite pure. Benzilic acid may be crystallized from hot water.

33.3 Reactions of Benzilic Acid

Benzilic acid may be characterized by acylation of its alcohol group or through
reactions of the carboxyl group (esterification, amide formation). Oxidation
with dichromate mixture converts it to benzophenone and carbon dioxide; reduction with hydroiodic acid leads to diphenylacetic acid ($(C_6H_5)_2CH—CO_2H$).

Benzilic acid is converted by warming with phosphorus pentachloride (2
moles) into diphenylchloroacetyl chloride, but a different reaction occurs upon
treatment with thionyl chloride ($SOCl_2$): benzophenone and carbon monoxide
are produced, together with sulfur dioxide and hydrogen chloride (see Question 4). Thionyl chloride (3 equivalents) in carbon tetrachloride solution
converts benzilic acid to diphenylchloroacetic acid.

(A) Benzophenone from Benzilic Acid

In a 25-mL round-bottomed flask, place 0.5 g (0.002 mole) of benzilic acid and
attach a water-cooled condenser connected to a gas trap to absorb the gaseous
by-products (SO_2 and HCl) (see Figure 8.2). Through the condenser carefully
add 1.2 mL (2.0 g, 0.016 mole) of thionyl chloride, which causes an immediate
vigorous reaction. Warm the reaction mixture gently on a steam bath until the
benzilic acid has dissolved, and then reflux the solution for 30 min longer.

[2] To minimize the danger of passing the end point, it is advisable to set aside, in a test tube,
1.5–2.0 mL of the filtrate and add sulfuric acid to the remainder until a *trace* of bromine is liberated.
This is removed by adding the small portion of the solution from the test tube.

After the heating period, add *cautiously* 3 mL of water in small portions, through the condenser, to decompose the excess thionyl chloride. Add 1 mL of methylene chloride, transfer the mixture to a test tube, and separate the organic layer. Wash the organic layer with water and then with 5% sodium bicarbonate solution (**Caution**—*frothing!*). Dry the organic layer with a small amount of anhydrous sodium sulfate and filter it into a small beaker. Remove the solvent on a steam bath, and set the beaker aside to cool. Benzophenone exists in two crystalline forms : one melting at 26° and the other at 48°. The low-melting form may require standing or a seed crystal for conversion to the higher melting form. Once they are formed, wash the crystals with a little petroleum ether (bp 30–40°) (**Caution**—*flammable!*). The yield of purified benzophenone is about 0.1 g; mp 48°.

(B) Acetylbenzilic Acid (α-Acetoxyphenylacetic Acid)

In a small test tube place 0.1 g of benzilic acid, 0.2 mL of glacial acetic acid, and 0.2 mL of acetic anhydride. Add one drop of concentrated sulfuric acid (**Caution**—*corrosive!*), mix well, and heat the tube in a bath of boiling water for 2 hr. Cool the solution to 25° and add 1 mL of water, drop by drop, while shaking it. Allow the reaction mixture to stand overnight or longer to permit crystallization of the acetyl derivative. Collect the product with suction, wash it well with water, and press it dry on the filter. The air-dried material is a monohydrate (mp 96–98°). Prolonged drying in a vacuum desiccator over sulfuric acid is necessary to obtain anhydrous acetylbenzilic acid (mp 104–105°).

(C) Methyl Benzilate

In a 25-mL round-bottomed flask place 0.1 g of benzilic acid, 2 mL of methanol, and 0.1 mL of concentrated sulfuric acid (**Caution**—*corrosive!*). Addition of the acid causes the development of a red color that disappears when the tube is shaken. Attach a condenser, add a boiling chip, and reflux the solution for 30 min. Cool the reaction mixture and pour it in small portions into 5 mL of 5% aqueous sodium carbonate solution (**Caution**—*foaming!*). Chill the mixture in an ice bath, collect the crystals with suction, and wash them well with water. The recorded melting point of methyl benzilate is 74–75°.

Questions   1. Compare the pinacol–pinacolone rearrangement (Chapter 35) with the benzilic acid rearrangement. Account for why one requires a base catalyst and the other requires acid catalyst.
2. Predict the products that would be formed in the following reactions?
   (a) benzilic acid + toluene (+ $SnCl_4$ catalyst)
   (b) benzilic amide + NaOBr + NaOH (Hofmann reaction)
   (c) methyl benzilate + excess C_6H_5 MgBr (followed by H_2O + acid)

(d) methyl benzilate + ammonia (in methanol)

(e) methyl benzilate + phenyl isocyanate (C_6H_5—N=C=O)

3. In the presence of a trace of sulfuric acid, benzilic acid reacts with acetone to form a crystalline product (mp 48°) with molecular formula $C_{17}H_{16}O_3$. Suggest a structural formula for this compound.

4. The formation of benzophenone upon treatment of benzilic acid with thionyl chloride ($SOCl_2$) presumably involves two steps: conversion of benzilic acid to diphenylhydroxyacetyl chloride, followed by decomposition of the acid chloride through a cyclic transition state. Write a detailed set of mechanisms for this transformation.

34 Triphenylmethanol

34.1 Triarylmethanols

Triphenylmethane and its derivatives may be synthesized conveniently by means of the Friedel–Crafts reaction or the Grignard reaction. Hydroxyl and amino derivatives, used in the manufacture of triphenylmethane dyes, are obtained by condensation of aromatic aldehydes and diaryl ketones with phenols and arylamines in the presence of acid catalysts.

In the presence of anhydrous aluminum chloride, chloroform reacts stepwise with benzene (in excess) to form dichlorotoluene (benzylidene chloride), diphenylchloromethane (benzohydryl chloride), and finally triphenylmethane. With carbon tetrachloride the end product is triphenylchloromethane (also called trityl chloride).

$$CHCl_3 \xrightarrow[AlCl_3]{C_6H_6} C_6H_5{-}CHCl_2 \xrightarrow{C_6H_6} (C_6H_5)_2CHCl \xrightarrow{C_6H_6} (C_6H_5)_3CH$$

$$CCl_4 \xrightarrow[AlCl_3]{C_6H_6} C_6H_5{-}CCl_3 \xrightarrow[20°]{C_6H_6} (C_6H_5)_2CCl_2 \xrightarrow[70°]{C_6H_6} (C_6H_5)_3CCl$$

The successive steps of arylation require increasingly vigorous conditions, and it is not possible to introduce a fourth aryl group by the Friedel–Crafts reaction. Tetraphenylmethane has been obtained, in low yield, by heating trityl chloride with phenylmagnesium bromide. Trityl chloride is extremely reactive and is hydrolyzed rapidly by cold water to form triphenylmethanol.

Triphenylmethanol can be synthesized readily by the Grignard reaction (see Chapter 18 for a discussion of the Grignard reaction). Phenylmagnesium bromide is prepared by direct reaction of bromobenzene with metallic magnesium in the presence of anhydrous diethyl ether (or tetrahydrofuran).

$$C_6H_5\!-\!Br + Mg + (C_2H_5)_2O \longrightarrow C_6H_5\!-\!Mg\!-\!Br \cdot 2(C_2H_5)_2O$$

Usually a crystal of iodine is added to aid in starting the reaction, which must be carried out with carefully purified reagents and under anhydrous conditions. A small amount of biphenyl ($C_6H_5\!-\!C_6H_5$) is formed through coupling of the aryl groups (Wurtz–Fittig reaction).

The magnesium bromide salt (**I**) of the tertiary alcohol triphenylmethanol may be obtained by reaction of phenylmagnesium bromide with any one of several reagents, for example, benzophenone (equation 34.1), an ester of benzoic acid (equation 34.2), and dimethyl or diethyl carbonate (equation 34.3).

$$\underset{\textbf{II}}{C_6H_5\!-\!\overset{\overset{\textstyle O}{\|}}{C}\!-\!C_6H_5} + C_6H_5\!-\!MgBr \longrightarrow \underset{\textbf{I}}{(C_6H_5)_3C\!-\!O\!-\!MgBr} \quad (34.1)$$

$$\underset{\textbf{III}}{C_6H_5\!-\!\overset{\overset{\textstyle O}{\|}}{C}\!-\!OCH_3} + C_6H_5\!-\!MgBr \longrightarrow \textbf{II} \longrightarrow \textbf{I} \quad\quad (34.2)$$

$$CH_3O\!-\!\overset{\overset{\textstyle O}{\|}}{C}\!-\!OCH_3 + C_6H_5\!-\!MgBr \longrightarrow \textbf{III} \longrightarrow \textbf{II} \longrightarrow \textbf{I} \quad (34.3)$$

Subsequent hydrolysis of the salt yields the neutral triphenylmethanol. These reactants require, respectively, 1, 2, and 3 equivalents of the Grignard reagent. Benzophenone gives the salt of triphenylmethanol directly; esters of benzoic acid add one equivalent of Grignard reagent to produce benzophenone, which then adds a second equivalent of Grignard reagent; dimethyl carbonate first produces a benzoate ester, which then reacts sequentially with two more equivalents of Grignard reagent. It is not feasible to arrest the reaction at an intermediate stage because the reactivities toward phenylmagnesium bromide decrease in the sequence: benzophenone, methyl benzoate, dimethyl carbonate, corresponding to the diminishing electrophilic activity of the carbonyl groups of these reactants. This sequence is the reverse of the reactivities observed in the Friedel–Crafts arylation of carbon tetrachloride: $CCl_4 > C_6H_5\!-\!CCl_3 > (C_6H_5)_2CCl_2$. The relative reactivities in this series correspond to the diminishing electrophilic activity of the corresponding carbocations: $[CCl_3]^+ > [C_6H_5\!-\!CCl_2]^+ > [(C_6H_5)_2CCl]^+$.

Triphenylmethanol, like other tertiary alcohols, is not acetylated by reaction with acetyl chloride, but is converted to triphenylchloromethane. The colorless alcohol dissolves in cold concentrated sulfuric acid to form a yellow solution of the halochromic salt, containing the relatively stable triphenylmethyl carbocation, $(C_6H_5)_3C^+$. On dilution with water the alcohol is regenerated.

Triphenylmethanol is the parent structure of the color bases of triphenylmethane dyes, such as malachite green and crystal violet. It is related in a similar way to the rhodamine dyes, prepared from *m*-aminophenols and phthalic anhydride, and to the phthalein and sulfonphthalein acid–base indicators (see Chapter 42).

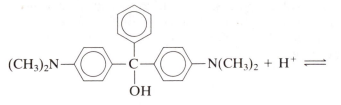

Malachite green

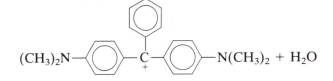

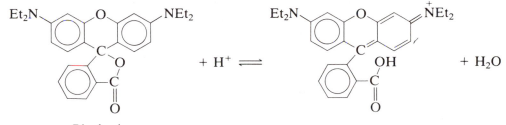

Rhodamine

The colored compounds are salts of amino or hydroxyl derivatives that have quinonoid structures of the types shown. Reduction of these compounds gives the colorless leuco bases, which are derivatives of triphenylmethane.

34.2 Preparation and Reactions of Triphenylmethanol

(A) Phenylmagnesium Bromide

In Grignard reactions it is essential that the reagents be free from ethanol and water, and the apparatus perfectly clean and dry.

▐▶ *CAUTION*   It is advisable to have a bath of ice and water at hand during this preparation as the reaction may start suddenly with vigorous boiling of the ether. Take care that no flame is nearby.

In a 50-mL round-bottomed flask provided with a Claisen adapter bearing an addition funnel and a vertical condenser, place 0.48 g (0.02 mole) of magnesium turnings. Introduce directly into the flask a mixture of 0.4 mL (0.6 g) of bromobenzene, 0.2 mL of anhydrous diethyl ether,[1] and a very small crystal of iodine. If a reaction does not start at once, warm the flask gently in a bath of warm water. *After the reaction has started*, as evidenced by disappearance of the iodine color, appearance of turbidity, and spontaneous boiling, slowly add 15 mL of anhydrous diethyl ether. For the success of the experiment it is essential that the reaction begin before the main portions of the ether and bromobenzene are added.

IIII➡ *CAUTION* Ether that has been exposed to air and light for some time while being stored may contain an unstable peroxide that can explode violently when heated, especially toward the end of a distillation. This peroxide can be detected by liberation of iodine when a test portion of the ether is shaken with a little 2% potassium iodide solution that has been acidified with hydrochloric acid.

The peroxide can be removed by washing the ether with an equal volume of dilute, weakly acidified ferrous sulfate solution.

Place in the addition funnel 1.8 mL (2.7 g) of bromobenzene (a total of 0.021 mole) and allow it to flow drop by drop into the previously activated reaction mixture at a rate such that the ether refluxes without external heating.

After all of the halide has been added, reflux the mixture gently for 30 min on a steam bath. Do not heat the material so vigorously that ether vapors traverse the condenser. The reaction is complete when the magnesium has mostly dissolved; some dark particles of impurities will remain undissolved. Remove the heating bath and proceed without delay to the next step.

(B) Triphenylmethanol[2]

Cool the reaction flask containing the Grignard reagent prepared in part A to 15–20° and place in the addition funnel a solution of 1.3 mL (1.4 g, 0.001 mole) of pure methyl benzoate in about 5 mL of anhydrous ether. Allow the methyl benzoate solution to flow slowly into the Grignard reagent, with continuous swirling, and cool the flask from time to time to control the reaction. The bromomagnesium derivative of the alcohol separates as a white precipitate. After all of the methyl benzoate has been addded, allow the mixture to stand at room temperature for 30 min or longer.

Pour the contents of the flask as completely as possible into a mixture of about 10 g of ice, 20 mL of water, and 1.0–1.26 mL of concentrated sulfuric acid (**Caution**—*corrosive!*), contained in a 100-mL flask. Add 0.8–1.0 mL of strong sodium bisulfite solution to remove any free iodine. Swirl the mixture to com-

---

[1] It is essential that the ether be anhydrous.
[2] In place of methyl benzoate, 1.6 mL (1.6 g) of ethyl benzoate may be used.

plete the decomposition of the magnesium derivative and then transfer it to a separatory funnel. Rinse the reaction flask with a few mL of diethyl ether to remove material that adheres to the wall of the flask. Add about 15 mL of ordinary diethyl ether to the separatory funnel to aid in extracting the product completely. Shake and then separate the ether layer. Wash it with two 4-mL portions of 5–10% sulfuric acid and once with saturated salt (NaCl) solution. Finally, wash the ether layer with aqueous sodium bicarbonate solution and once more with saturated salt solution.[3]

Dry the ether solution over anhydrous magnesium sulfate and transfer it to a 100-mL round-bottomed flask equipped with a distillation head. Distill off the ether as completely as possible, using a steam bath (**Caution**—*Avoid fire hazards!*). The residual crude product contains the impurities biphenyl and unreacted bromobenzene and methyl benzoate, along with triphenylmethanol.

Stir the residue with petroleum ether (about 5 mL/g of residue) and collect the solid on a suction filter; this process removes most of the impurities. Complete the purification by recrystallization from 2-propanol (about 7 mL/g), and collect the product on a suction filter. The yield is 1.4–1.6 g; mp 160–172°.

The petroleum ether extract may be evaporated to dryness and examined by chromatographic and spectroscopic methods to study its components.

(C) Triphenylmethyl Cation

In a small test tube place 1–2 mL of concentrated sulfuric acid (**Caution**—*corrosive!*). Add to this a very small pinch of triphenylmethanol and observe the color of the cation. Directions for the Friedel–Crafts reactions of triphenylmethanol (by way of the cation) with *p*-methoxybenzene to produce *p*-methoxytetraphenylmethane are given in Section 29.3.

Questions

1. Write equations for the action of phenylmagnesium bromide on the following compounds, including hydrolysis of the reaction mixture with dilute acid.
   (a) carbon dioxide (b) ethanol (c) oxygen
   (d) *p*-tolunitrile (e) ethyl formate

2. How may the following compounds be prepared from phenylmagnesium bromide?
   (a) 1,2-diphenylethanol (b) benzaldehyde
   (c) benzyl alcohol (d) benzopinacol

3. Indicate a series of reactions for the conversion of triphenylmethanol to hexaphenylethane. Cite evidence showing that hexaphenylethane undergoes dissociation (homolysis) to form triphenylmethyl radicals. Why is this radical more stable than a free methyl radical?

4. Write equations for the synthesis of a typical triphenylmethane dye, starting from simple aromatic compounds. Explain the terms leuco and color base.

[3] For a discussion of the uses of saturated salt solution as a preliminary drying agent see Ellern, *J. Org. Chem.*, **47**, 3569 (1982).

35 The Pinacol–Pinacolone Rearrangement

35.1 Introduction

Ketones are reduced by conventional reducing agents such as sodium borohydride to the corresponding secondary alcohols. However, with amalgamated magnesium, or a mixture of magnesium and iodine (magnesious iodide), ketones are reduced to form bimolecular 1,2-diol reduction products called pinacols, of the type $R_2C(OH)—C(OH)R_2$ (equation 35.1). Aryl ketones, such as diphenyl ketone (benzophenone), may be reduced photochemically by exposing a solution of the ketone in ethanol or 2-propanol to ultraviolet illumination (equation 35.2). Both the chemical and photochemical reductions are one-electron processes that involve free-radical intermediates.

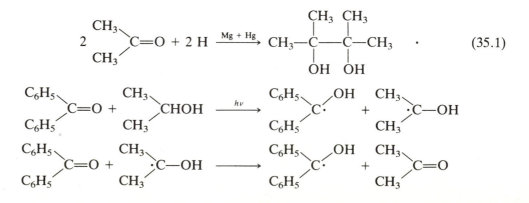

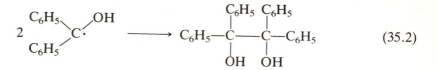

$$(35.2)$$

When heated with strong acids, or a catalyst such as iodine, substituted 1,2-diols of the pinacol type undergo dehydration and rearrangement to ketones (the pinacol–pinacolone rearrangement). The transformation involves protonation of one of the hydroxyl groups (**I**), followed by elimination of water to give a carbocation (**II**). Migration of an alkyl or aryl group from the adjacent carbon atom, with its bonding electrons, leads to the protonated form of the pinacolone (**III**).

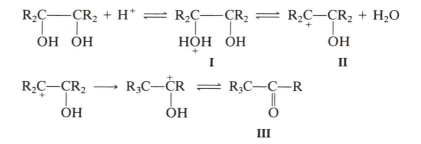

Similar carbocation rearrangements are encountered with 1,2-halohydrins, 1,2-amino alcohols, α-hydroxyaldehydes, and similar compounds. The mechanism of the rearrangement has been studied extensively.[1] The first step of the iodine-catalyzed rearrangement is believed to be formation of the covalent hypoiodite followed by loss of hypoiodite ion to give the same carbocation formed with acid.

$$R—\overset{..}{\underset{..}{O}}H + I—I \to \to R—O—I + HI$$

Hypoidoite

$$\downarrow$$

$$R^+ + OI^-$$

Studies of the rearrangement of a series of symmetrical pinacols (type **IV**) have revealed that substituents exert a marked effect on the selectivity of migration of the aryl group.

---

[1] Collins, "The Pinacol Rearrangement," *Quart. Revs.* (London), **14**, 357 (1960); Lowry and Richardson, *Mechanism and Theory in Organic Chemistry* (New York: Harper and Row, 1976); Hine, *Physical Organic Chemistry*, 2nd ed. (New York: McGraw-Hill, 1962).

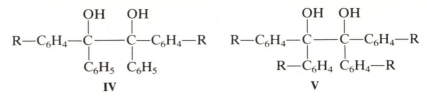

$$IV \qquad\qquad V$$

The *migration aptitudes* of a few substituted groups, relative to C_6H_5 are *p*-methoxy-, 500–1000; *p*-methyl, 15–18; *p*-chloro, 0.7; *m*-methoxy-, 1.6. The rearrangement of unsymmetrical pinacols (type **V**) does not involve selectivity of migration but is governed by selectivity in the formation of one of the two possible carbocation intermediates (**VI** and **VII**). In this situation the ease of formation of the carbocation is the dominant factor, and the effect of substituents is different from that observed with the symmetrical pinacols.

$$(R{-}C_6H_4)_2\overset{+}{C}{-}\overset{\overset{\displaystyle OH}{|}}{C}(C_6H_5)_2 \qquad (R{-}C_6H_4)_2\overset{\overset{\displaystyle OH}{|}}{C}{-}\overset{+}{C}(C_6H_5)_2$$

$$\textbf{VI} \qquad\qquad \textbf{VII}$$

Electron-releasing groups favor carbocation formation at the carbon atom to which they are directly attached, and this leads to preferential migration of a group from the adjacent carbon atom. In the unsymmetrical pinacol (type **VI**) when R is *p*-methoxyphenyl, this group migrates only to the extent of 30% and the unsubstituted phenyl group of 70%. With the *p*-chlorophenyl analog its migration is about 40% versus 60% for the phenyl group.

Benzopinacolone formed from the rearrangement of benzopinacol can be cleaved by base to yield benzoic acid and triphenylmethane by the following mechanism. With substituted benzopinacolones the cleavage reaction can be used to identify the phenyl ring that migrates.

$$\begin{array}{c}C_6H_5\\C_6H_5{-}C\\C_6H_5\end{array}\overset{\overset{\displaystyle O}{\|}}{{-}C}{-}C_6H_5 \; + \; \overline{O}H \longrightarrow \begin{array}{c}C_6H_5\\C_6H_5{-}C\\C_6H_5\end{array}\overset{\overset{\displaystyle O^-}{|}}{{-}\underset{\underset{\displaystyle OH}{|}}{C}}{-}C_6H_5 \longrightarrow$$

$$\begin{array}{c}C_6H_5\\C_6H_5{-}C^-\\C_6H_5\end{array} + \; \overset{\displaystyle O}{\underset{\underset{\displaystyle OH}{}}{\diagdown}}C{-}C_6H_5 \longrightarrow \begin{array}{c}C_6H_5\\C_6H_5{-}C{-}H\\C_6H_5\end{array} + \; \overset{\displaystyle O}{\diagdown}C{-}C_6H_5$$

35.2 Preparations

(A) Benzopinacol by Photochemical Reduction

Place 0.92 g (0.005 mole) of benzophenone in a 25-mL flask (or large test tube), add 10 mL of 2-propanol (isopropyl alcohol), and dissolve the solid by warming it on a steam bath. To the solution add exactly 1 drop of glacial acetic acid[2] and sufficient 2-propanol to fill the flask almost completely. Stopper the flask firmly with a good cork that fits tightly and projects about half its length into the neck of the flask.

Support the flask firmly by means of a condenser or buret clamp, and expose it to direct sunlight or place it in close proximity to an ultraviolet lamp. Benzophenone is activated by absorption of light in the near-ultraviolet region, which is partially transmitted through ordinary glass. As the reduction progresses, benzopinacol separates in colorless, dense crystals. By occasional tapping and swirling, the crystals may be made to settle in the neck of the flask. Five or six days' exposure to moderate sunlight will furnish an abundant crop of crystals; usually a much shorter time is required with an ultraviolet lamp. Collect the crystals with suction and, if necessary, return the filtrate for further exposure to complete the reaction. Under favorable conditions, the yield will attain at least 90%. The product is quite pure (mp 186–188°) and may be used directly for conversion to benzopinacolone.

(B) Benzopinacolone

In a 50-mL round-bottomed flask place 0.5 g (0.027 mole) of benzopinacol, 3 mL of glacial acetic acid, and a small crystal of iodine. Attach a short reflux condenser and reflux the solution for 10 min. Allow the solution to cool slightly, add 3 mL of ethanol, swirl the mixture thoroughly, and allow it to cool. Collect the crystals with suction and wash them with cold ethanol to remove iodine. Benzopinacolone forms colorless crystals (mp 179–180°). The yield is 0.40–0.45 g.

(C) Triphenylmethane by Alkaline Cleavage of Benzopinacolone

In a 50-mL round-bottomed flask place a 0.33 g (0.0010 mole) of benzopinacolone, 4 mL of *anhydrous* ethylene glycol, and 0.50 g (5 pellets, 0.0076 mole) of solid 85% potassium hydroxide. Attach a reflux condenser, add a boiling chip, and reflux the two-phase mixture fairly vigorously for 1 hr. Much of the triphenylmethane product will collect in the lower part of the condenser as the reaction proceeds. Every 15 min *temporarily* stop the flow of water in the condenser long enough to permit the collected solid to be rinsed back into the reaction flask.

Cool the mixture and transfer it to a seperatory funnel with the aid of about 15 mL of water. Rinse the flask with 15 mL of methylene chloride and pour the rinse through the condenser into the seperatory funnel (to remove any solid that may have collected in the condenser).

---

[2] The acid is added to neutralize traces of alkali from the glass vessel. Alkali is deleterious because it catalyzes cleavage of the pinacol to benzophenone and diphenylmethanol (C_6H_5—CHOH—C_6H_5).

IIII➤ *CAUTION*   Prolonged exposure to high concentrations of methylene chloride vapors may induce cancer. As with all organic solvents, work in a well-ventilated area when using it.

Shake the funnel to extract the triphenylmethane; separate the layers and save the aqueous layer for isolation of the benzoic acid.

Wash the methylene chloride extract with two 15-mL portions of water, dry it with anhydrous magnesium sulfate, and after filtration, allow the methylene chloride to evaporate in the hood. Recrystallize the triphenylmethane from 60% aqueous ethanol. The yield is about 0.10–0.15 g; mp 93–95° (literature, mp 98°).

Acidify the aqueous alkaline layer cautiously with concentrated hydrochloric acid, cool if necessary, and extract the benzoic acid with two 10-mL portions of ether (**Caution**—*flammable!*). Dry the combined ether extracts over anhydrous magnesium sulfate, and distill off the ether from a steam bath. Recrystallize the benzoic acid from a small amount of water (about 25 mL/g) and take its melting point.

Questions   1. Indicate a stepwise mechanism for the alkali-catalyzed cleavage of benzopinacol.

2. Write equations for the reaction of *p*-tolylmagnesium bromide (in excess) on the following compounds, including hydrolysis of the reaction mixture with dilute acid.
   (a) C_6H_5—CO—CO—C_6H_5 (benzil)
   (b) methyl ester of benzilic acid
   (c) dimethyl oxalate

3. Write the structures of the isomeric pinacolones that could be obtained by pinacol rearrangement of symmetrical 4,4′-dimethylbenzopinacol. This reaction actually furnishes one of the pinacolones to the extent of more than 90%. How may its structure be determined?

36 Polycyclic Quinones

36.1 Introduction

Many derivatives of anthraquinone occur in nature as pigments and active principles of plants, fungi, lichens, and insects. An example is emodin, the cathartic principle of cascara sagrada and of rhubarb, which is 1,3,8-trihydroxy-6-methylanthraquinone. Naturally occurring phenanthraquinone derivatives are less common but they do arise in some fungi.

The chemistry of polycyclic quinones has been studied extensively because of their use as dyes. Derivatives of anthraquinone are used in the manufacture of important vat dyes (Indanthrene Brown, Caledon Jade Green, etc.) that are unusually stable in light and with washing.

The hydrocarbons anthracene and phenanthrene are readily available from coal tar. They can be oxidized directly at the 9,10-positions by means of chromic acid to give the corresponding quinones. These condensed polycyclic quinones are more stable to further oxidation than the simple benzoquinones and naphthoquinones. In the oxidation reaction, some complex products are formed by coupling and disproportionation of the reactive intermediates and, with phenanthrene, some further oxidation of the quinone (**I**) leads to diphenic acid (**II**, 2,2′-biphenyldicarboxylic acid).

The crude phenanthrenequinone obtained by oxidation of technical phenanthrene can be purified by conversion to the water-soluble sodium bisulfite addition product, from which the pure quinone is regenerated easily by treatment with base. Anthraquinone, derived from anthracene present as an impurity in technical phenanthrene, has less-reactive carbonyl groups and does not form a sodium bisulfite adduct.

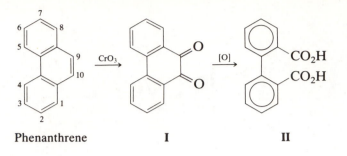

Phenanthrene   I   II

Phenanthrenequinone bears a strong resemblance to the 1,2-diketone benzil and undergoes a benzilic acid rearrangement (see Chapter 33) when warmed with concentrated aqueous alkalies; the product is 9-hydroxyfluorene-9-carboxylic acid.

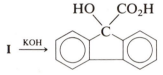

A more general method for the preparation of anthraquinone and its derivatives is to cyclize *o*-benzoylbenzoic acids, which are obtained readily by the Friedel–Crafts reaction of phthalic anhydride with benzene, toluene, chlorobenzene, and similar aromatic compounds, in the presence of anhydrous aluminum chloride. The cyclization can be achieved with concentrated sulfuric acid or polyphosphoric acid.

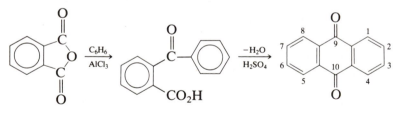

Anthraquinone

36.2 Preparations

(A) Anthraquinone

In a 50-mL round-bottomed flask place 1 g (0.0044 mole) of anhydrous *o*-benzoylbenzoic acid[1] and 5 mL of concentrated sulfuric acid (**Caution**—*corrosive!*). Heat the flask on a steam bath and swirl the mixture until the solid dissolves.

---

[1] The anhydrous acid melts at 127–128°; it forms a monohydrate (mp 94–95°). The pure anhydrous acid can be purchased from chemical supply firms.

By means of a clamp, support the flask firmly in a steam bath; place a towel or cloth around the bath and flask (to reduce heat loss), and heat the material at 100° for 30 min. With a Pasteur pipet, add 1 mL of water, drop by drop, while swirling the solution. The product will begin to separate. Remove the flask from the steam bath and allow it to cool. Dilute the reaction mixture with about 30 mL of water and transfer it to a large beaker. Add enough chipped ice to bring the total volume to 60 mL and stir the mixture thoroughly. Collect the product by suction filtration, preferably on a hardened filter paper, and wash it well with water. To remove any unchanged starting material wash the product carefully with 2 mL of concentrated aqueous ammonia diluted with 10 mL of water, followed by a washing with water. To facilitate drying, wash the quinone finally with a little ice-cold acetone, and then spread it on a clean paper. Do not determine the melting point (284–286°). The yield is 0.4–0.7 g. A small sample may be purified by sublimation, at 230–250°, using the apparatus shown in Figure 5.7 or 5.8.

(B) Phenanthrenequinone

In a 125-mL Erlenmeyer flask place 6 mL of glacial acetic acid and 20 mL of water, and carefully add 12 mL of concentrated sulfuric acid (**Caution**—*corrosive!*). Swirl the solution, add 1.2 g (0.006 mole) of technical 90% phenanthrene, and heat the mixture to 95° (internal temperature) in a bath of boiling water. Prepare a solution of 7.2 g (0.024 mole) of sodium dichromate dihydrate in 5 mL of warm water and add this dropwise, with swirling, to the hot suspension of phenanthrene.

▐▶ *CAUTION*   Dust containing chromium salts is suspected of causing cancer. Do not grind the sodium dichromate to hasten its solution.

Watch the temperature of the reaction mixture carefully to observe the onset of a strongly exothermic reaction. When this occurs, stop the addition of the oxidizing solution, remove the flask from the heating bath, and swirl the mixture vigorously. The temperature will rise to 110–120°. As soon as the temperature begins to fall, resume addition of the dichromate solution and do not allow the temperature to drop below 85–90°. Dip the flask in the boiling-water bath, when required, to maintain the desired temperature. After all of the oxidizing agent has been added, heat the mixture for 30 min in the boiling water bath, swirling it frequently.

Cool the reaction mixture and pour it into a well-stirred mixture of about 60 mL of water and 20 g of chipped ice. Break up any lumps of the product and collect the crude quinone on a large suction filter. Wash the crystals thoroughly with water, until the green chromous sulfate has been removed. Transfer the moist product to a 125-mL Erlenmeyer flask, add 13 mL of ethanol, swirl the mixture vigorously, and add a solution of 6 g (0.032 mole) of sodium bisulfite in 12 mL of water. The yellow color of the quinone is discharged as the bisulfite adduct is formed. Allow the mixture to stand for about 20 min, occasionally swirling it, to complete the reaction. Add water until the flask is nearly filled, cork it firmly, and shake the mixture thoroughly to dissolve the sodium bisulfite addition compound. Filter the mixture through a fluted filter to remove anthraquinone and other impurities. After the insoluble residue is washed with 4 mL of 50% aqueous ethanol, it may be discarded.

Place the filtrate and washings in a large beaker and add saturated aqueous sodium carbonate in small portions (**Caution**—*foaming!*), stirring thoroughly, until the solution is distinctly alkaline. Collect the phenanthrenequinone on a suction filter, wash it thoroughly with cold water, and press it as dry as possible on the filter. Place the damp product in a 50-mL round-bottomed flask, add 11–12 mL of glacial acetic acid, and reflux the mixture to dissolve the quinone. Cool the solution to room temperature and allow the material to stand for 20 min or longer, occasionally swirling it before collecting the orange crystals of the purified product. Wash the crystals thoroughly with cold water and allow them to dry in the air. The yield is about 0.6 g; the recorded mp is 206°.

Questions   **1.** What is the source of technical anthracene and phenanthrene? What methods are used to obtain these hydrocarbons in a state of high purity?

2. Write equations for the following syntheses of 2-methylanthraquinone.
  (a) starting from naphthalene and toluene
  (b) starting from naphthalene and aliphatic compounds, making use of the Diels–Alder reaction (Chapter 39)

3. Write a balanced equation for the oxidation of phenanthrene to phenanthrenequinone.

4. Write equations for the reaction of phenanthrenequinone with the following reagents.
  (a) hot aqueous potassium hydroxide solution
  (b) *o*-phenylenediamine
  (c) hydrogen peroxide, in glacial acetic acid

Enamine Synthesis of a Diketone: 2-Acetylcyclohexanone

37.1 The Enamine Reaction

The acylation and alkylation of aldehydes and ketones form an important class of carbon–carbon bond-forming reactions. The key step is the formation of an intermediate enolate anion, which reacts as a nucleophile toward acyl and alkyl halides.

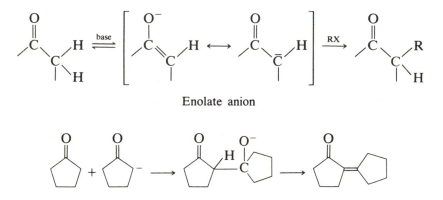

Enolate anion

The scope of these reactions is limited by three factors. To form the enolate anion it is necessary to use a strong base such as sodium amide, triphenylmethide ion, or *t*-alkoxides, which may react competitively with the acylating or alkylating reagent. A second complication is the unwanted base-catalyzed aldol condensation of the carbonyl compound. Finally, owing to the rapid proton equilibration of the intermediate anion with the product as it is formed, further acylation or alkylation may occur. An extreme case is 6-methoxy-2-tetralone, which on attempted monomethylation results largely in a one-to-one mixture of dimethylated ketone and unreacted starting material.

In 1954 Stork[1] introduced a new method for the acylation and alkylation of aldehydes and ketones via enamines (**I**), derived by condensation of the carbonyl compound with a secondary amine. In many ways the enamine reacts as though it were an enolate anion. It undergoes preferential attack at carbon as would the enolate anion, but being neutral it does not condense with itself nor exchange a proton with the product to form an intermediate that can react further.

The enamines are formed readily by refluxing a mixture of the amine and the carbonyl compound in a solvent such as dioxane, acetonitrile, toluene, or ethyl alcohol. If the solvent is toluene, as it is in this experiment, the water formed in the reaction is removed by azeotropic distillation (Chapter 2). After acylation or alkylation of the enamine, the substituted carbonyl product is liberated by heating with water.

$$R_2NH + R{-}CO{-}CH_2{-}R' \rightleftharpoons H_2O + \left[\begin{array}{c} R_2N{-}\underset{R}{C}{=}\underset{R'}{CH} \longleftrightarrow R_2\overset{+}{N}{=}\underset{R}{C}{-}\underset{R'}{\overset{-}{CH}} \end{array} \right] \xrightarrow{CH_3{-}X}$$

$$R_2\overset{+}{N}{=}\underset{R}{C}{-}\underset{R'}{CH}{-}CH_3 \xrightarrow{H_2O} O{=}\underset{R}{C}{-}\underset{R'}{CH}{-}CH_3$$

$$\mathbf{I}$$

With the enamine synthesis it is possible to prepare compounds that would otherwise require circuitous routes.

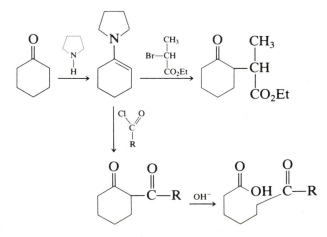

[1] Stork, Terrell, and Szmuszkovicz, *J. Am. Chem. Soc.*, **76**, 2029 (1954); Stork, Brizzolara, Landesman, Szmuszkovicz, and Terrell, *ibid.*, **85**, 207 (1963).

Enamines also can participate in Michael-type additions to α,β-unsaturated carbonyl or nitrile derivatives. This reaction differs from the earlier acylation and alkylation reactions in that a proton shift occurs in the adduct to regenerate a new enamine, which in ethanol (but not in toluene) can react with an excess of the α,β-unsaturated ketone.

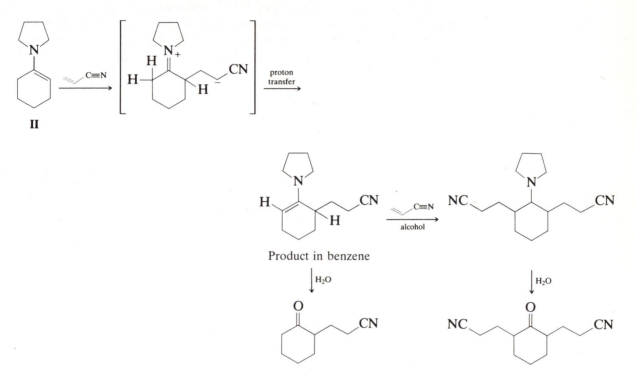

Product in benzene

37.2 Preparation of 2-Acetylcyclohexanone

(A) *N*-1-Cyclohexenylpyrrolidine (II)

Arrange an apparatus for fractional distillation using a 50-mL round-bottomed flask as a boiler and an unpacked fractionating column (or a Claisen head with the central arm stoppered). In the flask place 2.0 mL (2.0 g, 0.02 mole) of cyclohexanone,[2] 1.9 mL (1.6 g, 0.025 mole) of pyrrolidine, and about 0.020 g of *p*-toluenesulfonic acid. Add 15 mL of toluene, and a boiling chip. Heat the flask until the reagents begin to boil and then adjust the heat control so that the ring of condensate doesn't make it up to the thermometer in the head. Maintain this condition for about 30 min and then increase the heat so that the mixture begins to distill slowly (it may be necessary to wrap the column with glass wool). Collect in a graduated cylinder about 5 mL of distillate, which will consist of a mixture

[2] The material prepared in Chapter 21 is suitable.

of water and toluene. Then reduce the heat and reflux the reaction mixture for an additional 30 min. Finally, increase the heat again and collect another 5 mL of distillate. The point of the interrupted distillation is to allow time for the reaction mixture to come to equilibrium. An alternate procedure would be to carry out a slow distillation of 10 mL over a 60-min period.

Allow the reaction mixture to cool, disassemble the apparatus, and replace the adapter, attached water trap, and condenser by a distillation head. Remove the excess pyrrolidine by distilling the mixture until the thermometer reaches 105–108°. The residual enamine can be used in the next step without further purification.[3]

(B) 2-Acetylcyclohexanone

Add a solution of 2.1 mL (2.24 g, 0.022 mole) of acetic anhydride in 4 mL of toluene to the enamine at room temperature. Mix the reagents well, stopper the flask, and set it aside for 24 hr or longer.

Add 2 mL of water and a fresh boiling chip to the reaction flask, and heat the mixture under reflux for 0.5 hr. Wash the cooled toluene solution with water, then with 5% hydrochloric acid, and finally with water (10 mL of each). Dry the solution with about 0.5 g of anhydrous magnesium sulfate, filter the solution directly into a 25-mL flask using a filter tube (see Figure 5.6), and remove the toluene (bp 111°) and any unreacted cyclohexanone (bp 155°) by distillation. Distill the residual liquid under reduced pressure using a fresh boiling chip; bp 97–104° at 12–14 mm, 135–147° at 74–76 mm. The yield of 2-acetylcyclohexanone is 1.0–1.6 g.

It is instructive to determine the purity of your product by gas chromatography (5 ft of 3% Carbowax on Chromosorb works well; see Appendix for column preparation).

Questions 1. In the enamine preparation of 2-acetylcyclohexanone:
   (a) What is the purpose of the azeotropic distillation?
   (b) Why do you carry out a protracted distillation?
   (c) How is unreacted pyrrolidine removed?
2. 2-Ethylbutanal can be prepared from butanal in 41% yield by using the enamine reaction with ethyl iodide. Describe the sequence of mechanistic steps involved in this preparation.
3. Devise syntheses for the following molecules that make use of the enamine reaction.
   (a) 2-benzoylcyclopentanone   (b) 2-carbomethoxycyclopentanone

---

[3] Somewhat better yields are obtained in the acylation if the cyclohexenylpyrrolidine (bp 106° at 13 mm) is purified by distillation.

38 Wagner–Meerwein Rearrangements: Camphor from Camphene

38.1 Introduction

An implicit guiding principle used in predicting products of organic reactions is that minimum structural change occurs in the reaction. Important exceptions to the principle are those carbocation reactions in which an alkyl or aryl group adjacent to the developing positive charge migrates to the positive carbon atom and gives rise to products with rearranged carbon skeletons (Wagner–Meerwein rearrangements).

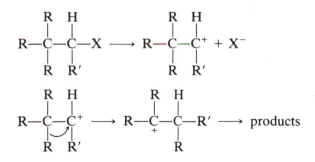

The confusion that can arise from an unsuspected Wagner–Meerwein rearrangement is amplified in many of the reactions of bicyclic molecules because the rearranged products may retain the same bicyclic structure and differ from the nonrearranged products only in the position of substituents. The commercial conversion of camphene (3,3-dimethyl-2-methylenebicyclo[2.2.1]heptane) to isobornyl acetate (1,7,7-trimethylbicyclo[2.2.1]heptan-*exo*-2-ol acetate) is a classic example of a reaction of a bicyclic molecule proceeding with a Wagner–Meerwein rearrangement.[1]

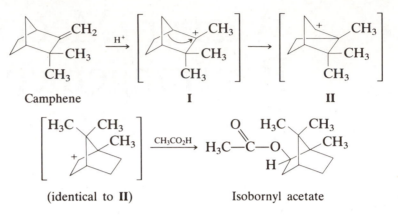

Camphene   I   II

(identical to **II**)   Isobornyl acetate

The product formed by addition of acetic acid to camphene is a secondary acetate rather than the expected tertiary acetate. The explanation is that the first-formed tertiary cation rearranges to a secondary cation that then captures the acetate group. This seemingly uphill rearrangement (tertiary cations are normally more stable than secondary ones) occurs because these acid-catalyzed reactions are reversible and the isobornyl acetate is more stable as a result of lower steric congestion. Some chemists prefer to look at cations **I** and **II** as resonance structures describing a single cation in which there is a partial charge simultaneously at both the secondary and tertiary carbons. If such a "nonclassical" ion description is used, no uphill rearrangement is involved. Instead, the question becomes why does the acetate anion attack the secondary center even though most of the charge is presumably on the tertiary center.

The remaining steps of the synthesis of camphor do not involve Wagner–Meerwein rearrangements since neither the saponification of isobornyl acetate to isoborneol nor the chromic acid oxidation of isoborneol to camphor produces a carbocation.

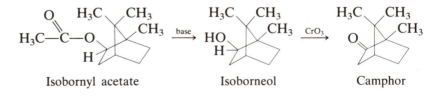

Isobornyl acetate   Isoborneol   Camphor

---

[1] In studying this reaction, you are urged to prepare ball-and-stick models of each stage. It is difficult to see from two-dimensional drawings how the transfer of one bond causes the movement of so many groups.

38.2 Preparation of Camphor

(A) Isobornyl Acetate

In a 50-mL round-bottomed flask dissolve 1.5 g (0.011 mole) of camphene in 3.5 mL (3.8 g, 0.065 mole) of glacial acetic acid and add 0.2 mL of 30% (v/v) sulfuric acid. Warm the flask on a steam bath or water bath at 90–95° for 15 min with frequent swirling. Add 2 mL of water, mix well, and allow to cool.

Transfer the cooled reaction mixture to a large test tube; rinse the flask with a little water and add the rinse to the test tube. Withdraw, using a Pasteur pipet, most of the lower aqueous layer and wash the residual ester first with water, then with saturated aqueous sodium bicarbonate, and finally again with water. In these washes use the "squirting" technique (see Section 6.5) with about a dozen squirts for each wash reagent. The crude product is suitable for conversion to isoborneol without further purification. The yield is about 1.2–1.6 g.

It is instructive to determine the IR of the product and note the appearance of the ester carbonyl absorption accompanying the loss of the *exo*-methylene group. If the IR is to be determined, the sample must be dry ($CaCl_2$ works well).

(B) Isoborneol

In a 50-mL round-bottomed flask place 1.25 g (0.0065 mole) of isobornyl acetate and add 4 mL of 2 *M* solution of potassium hydroxide in 80% (v/v) ethanol/water. Add a boiling chip and heat the mixture under reflux for 1 hr. Pour the solution onto about 5 g of ice contained in a small beaker. Rinse the flask with water and add the rinse to the beaker. Stir the mixture for several minutes until the isoborneol solidifies. Collect the solid on a Büchner funnel, wash it well with cold water, and press it dry. The yield of crude isoborneol is about 0.8 g. The crude product is sufficiently pure for oxidation to camphor.

The IR can be examined for loss of the carbonyl group on a mulled sample of the *dry* isoborneol.

(C) Camphor by Oxidation of Isoborneol with Jones' Reagent

In a 50-mL round-bottomed flask prepare a solution of 0.5 g (0.0033 mole) of isoborneol in 1 mL of acetone and cool the solution in a beaker of ice water to about 15°. Add to the solution, drop by drop, 0.8 mL (0.0023 mole) of Jones' reagent.[2] Allow the reaction mixture to stand for 30 min at room temperature after the addition has been completed.

Add 10 mL of water to the reaction flask, attach a distillation head with a water-cooled condenser, and collect about 3 mL of distillate. This first portion of

---

[2] Jones' reagent is available on the side shelf. It is prepared by dissolving 27 g of chromium trioxide. (**Caution**—*solid form is extremely hazardous!*) in 23 mL of concentrated sulfuric acid (**Caution**—*corrosive!*) followed by cautious dilution with water to 100 mL.

distillate contains most of the acetone and should be discarded. Stop the distillation at this point. Turn off the water flow to the condenser and allow the condenser to drain. Attach a small standard taper flask to the lower end of the condenser and immerse the receiver flask in an ice bath. Continue the distillation until no more product is collected. If the camphor begins to collect in the condenser to the point where the condenser might become plugged, you should stop the distillation and clear the condenser before continuing the distillation.[3] The solid product is collected on a Büchner funnel and pressed dry. The yield is about 0.5 g. Camphor may be purified by sublimation (see Figure 5.7 or 5.8). Pure camphor has a melting point (sealed tube) of 175°.[4]

A TLC with silica gel plates and methylene chloride as a developing solvent gives the following R_f values: camphor = 0.7; isoborneol = 0.55. With I_2 staining, the camphor spot turns bright yellow and the isoborneol is dark brown.

(D) Camphor by Oxidation of Isoborneol with Sodium Hypochlorite

This oxidation of isoborneol to camphor uses "swimming pool chlorine," which is a 12.5% by weight aqueous sodium hypochlorite solution. Because of the equilibrium between aqueous hypochlorite and chlorine gas, this reaction must be carried out in the hood or other well-ventilated area.

In a 125-mL Erlenmeyer flask prepare a solution of 0.5 g (0.0033 mole) of isoborneol in 5 mL of glacial acetic acid. Cool the solution in a beaker of ice water to about 15°. Add to the solution, drop by drop using a Pasteur pipet, 2.8 ml (about 0.35 g, 0.0056 mole) of aqueous sodium hypochlorite.[5] Swirl the mixture during the addition to insure thorough mixing and observe its temperature. The temperature of the mixture should be maintained between 25–30° by controlling the addition rate and the degree of cooling. The yellow color of the hypochlorite solution will be discharged as it reacts, but a pale yellow color will persist near the end of the addition and a white precipitate will form. Swirl the mixture occasionally for 20 min and then add saturated sodium bisulfite solution until the yellow color disappears.

Transfer the mixture of solid and liquid into a 50-mL round-bottomed flask with the aid of an additional 10–20 mL of water as a rinse. Add a

---

[3] An alternative isolation procedure is to stop the distillation after the acetone has been removed, add water, and collect the solidified camphor by filtration. The product should be washed well with water. This procedure does not work as well because of the volatility of the camphor.

[4] Because camphor has an unusually large change of melting temperature with pressure, the observed melting point in an open capillary is several degrees below that taken in a sealed tube. For the same reason, the melting point is particularly sensitive to traces of impurities; the melting points of mixtures of camphor with known amounts of foreign materials are useful in determining molecular weights (Rast method).

[5] If the "swimming pool chlorine" is old, there may not be sufficient hypochlorite (as indicated by the yellow color of the chlorine in equilibrium with it) to complete the oxidation. Add more hypochlorite solution (no more than 0.5 mL) until a persistent pale-yellow color develops.

boiling chip; attach a distillation head with a small standard taper flask attached to the lower end of the condenser. Immerse this receiver in an ice bath, but do *not* pass cooling water through the condenser because this will cause the solid camphor to condense and plug the condenser. In this setup the ice bath around the receiver serves as the condenser. Steam distill the mixture until no more solid product is collected (about 5 mL of distillate). If camphor collects in the condenser to the point where the condenser might become plugged, you should stop the distillation and clear the condenser by passing hot water through the jacket before continuing the distillation.

Collect the solid product on a Hirsch funnel and press it dry. The yield is about 0.5 g. Camphor may be purified by sublimation (see Figure 5.7 or 5.8). Pure camphor has a melting point (sealed tube) of 175°.[4]

A TLC with silica gel plates and methylene chloride as a developing solvent gives the following R_f values: camphor = 0.7; isoborneol = 0.55. With I_2 staining, the camphor spot turns bright yellow and the isoborneol is dark brown.

Questions

1. How many asymmetric carbon atoms (chiral centers) are present in camphor? How many optically active forms of camphor exist?

2. Explain how dehydration of borneol (1,7,7-trimethylbicyclo[2.2.1]heptan-*endo*-2-ol) with aqueous acid gives camphene.

3. Explain why optically active *exo*- and *endo*-bicyclo[2.2.1]heptan-2-ol (nor-borneol) undergo racemization when treated with acids.

4. Draw the structure for 1,5,5-trimethylbicyclo[3.1.1]heptan-2-ol acetate. Explain how it could arise from addition of acetic acid to camphene and why it does not.

39 The Diels–Alder Reaction

39.1 Introduction

One of the most interesting synthetic reactions of unsaturated compounds is the 1,4-addition of a conjugated diene to a molecule containing an active ethylenic or acetylenic bond (the dienophile), to form an adduct having a six-membered unsaturated ring, a 4 + 2 cycloaddition. This cyclization process, known as the Diels–Alder reaction[1] is of exceedingly broad scope and has been applied to syntheses of important medicinal products, insecticides, terpene derivatives, and intermediates for the manufacture of industrial chemicals.

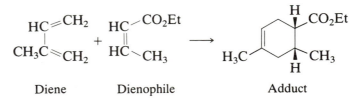

Diene   Dienophile   Adduct

Almost any diene will undergo the Diels–Alder reaction. Diene reactants include alkyl, halogen, alkoxy derivatives of 1,3-butadiene, and also cyclic 1,3-dienes such as cyclopentadiene and 1,3-cyclohexadiene. Even a few aromatic species that have formal diene substructures will serve. For example, furan and a number of furan derivatives, and the inner ring of anthracene, participate in

---

[1] For reviews of the Diels–Alder reaction, see Kloetzel, *Organic Reactions*, **4**, 1 (1984); Holmes, *ibid.*, **4**, 60 (1948); Butz and Rytina, *ibid.*, **5**, 136 (1949).

the reaction as dienes. With the commonly used dieneophiles, the most reactive dienes are those that have electron-releasing substituents (CH_3—, CH_3O—, etc.).

The most typical dienophiles are α, β-unsaturated carbonyl compounds and nitriles. Examples are acrolein, p-benzoquinone,[2] and 1,4-naphthoquinones, maleic anhydride, esters of acetylene–dicarboxylic acid, acrylic esters, and acrylonitrile. Under vigorous conditions vinyl ethers and halides, and even ethylene and acetylene, can be made to act as dienophiles toward the more reactive dienes. A particularly useful example is acetylene–dicarboxylic acid which reacts with dienes to give a cyclic diene that can be readily converted to a benzene ring, either by oxidation or bromination followed by dehydrohalogenation.

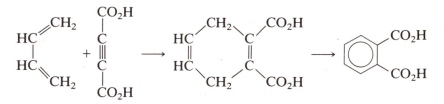

The Diels-Alder reaction is stereospecific: the diene is obliged to assume a *cis* conformation to permit ring closure and it is observed that the configuration of substituents (*cis* or *trans*) in the dienophile is retained in the adduct. Thus, 1,3-butadiene reacts with maleic anhydride to give 4-cyclohexene-*cis*-1,2-dicarboxylic anhydride (**I**).

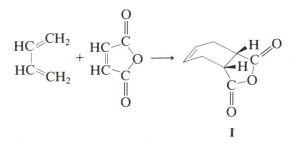

I

Since butadiene (bp $-4.4°$) is a gas at room temperature, it is awkward to work with. However, it can be prepared conveniently for small-scale reactions right in the reaction vessel by thermal decomposition of 3-sulfolene (butadiene sulfone, **II**), which gives butadiene and sulfur dioxide. This source of butadiene is used in the present experiment.

II

[2] For benzoquinone reactions see Chapter 40.

Active dienophiles are useful reagents for detecting the presence of a conjugated diene system and for analytical purposes. Diels–Alder adducts have been of value in establishing the structure of 1,3-dienes and in characterizing known dienes. *N*-Phenylmaleimide (**III**) is a convenient reagent for identification purposes since it forms crystalline adducts (**IV**) that can be easily isolated and purified.

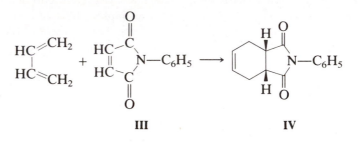

III   **IV**

N-Phenylmaleimide is readily synthesized from maleic anhydride and aniline. In the present experiment it is prepared and used to form adducts with 1,3-butadiene and with cyclopentadiene.

With cyclic dienes such as cyclopentadiene and also furan derivatives, the 4 + 2 cycloaddition product might have either of two possible configurations. The ring system of the dienophile could have a *trans* (*endo*) disposition (**V**) or a *cis* (*exo*) relationship (**VI**) to the newly formed methylene (or oxygen) bridge of the adduct. The case of cyclopentadiene and maleic anhydride is illustrated here. In general, it is observed that the *endo* configuration is favored in Diels–Alder reactions.

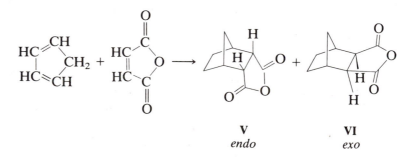

V   **VI**
endo   *exo*

39.2 Preparations

(A) *N*-Phenylmaleimide

N-Phenylmaleamic Acid. To a solution of 1 g (0.01 mole) of maleic anhydride in 8 mL of toluene warmed to 80–100° on a water or steam bath (**Caution—** *flammable solvent!*) add slowly with swirling, a solution of 0.82 mL (0.85 g,

0.01 mole) of aniline in 2 mL of toluene. Reaction occurs rapidly and a white crystalline precipitate of the phenylmaleamic acid separates. After cooling the solution to 20°, collect the product on a suction filter and allow it to dry. The yield is 1.9–2.4 g; mp 200–201°(dec). It is unnecessary to purify it.

N-Phenylmaleimide (III). In a small Erlenmeyer flask prepare a slurry of 1.6 g (0.008 mole) of N-phenylmaleamic acid in 3 mL of acetic anhydride and add 0.3 g of anhydrous sodium acetate. Close the flask with a stopper bearing a drying tube and swirl it, occasionally, for 30 min. During this time the maleamic acid dissolves, and a brown solution is formed. Pour the solution, while stirring it, into an ice–water mixture, and allow it to stand while the N-phenylmaleimide separates as a yellow flocculent mass. Wash the product with two 3-mL portions of cold water, then with one 2-mL portion of cyclohexane, and allow it to dry. The yield is 1.0–1.2 g; mp 86.5–88°.

(B) *N*-Phenyl-4-cyclohexene-1,2-dicarboximide (IV)

Place 0.7 g (0.006 mole) of 3-sulfolene, 0.7 g (0.004 mole) of N-phenylmaleimide, and 1.4 mL of technical mixed xylenes (**Caution**—*flammable solvent!*) in a 10-mL round-bottomed flask. Attach a reflux condenser, and arrange a gas trap to take care of the sulfur dioxide evolved. Heat the mixture gently until the solids dissolve, and then heat under reflux for 45 min. Cool the solution to a few degrees below its boiling point, add about 2 mL of xylene, and transfer the solution to a small beaker. Cool the solution to room temperature, and add 6 mL of petroleum ether (bp 60–90°). Collect the solid on a suction filter, and wash it with a small amount of cold methanol. The yield of off-white product is 0.5–0.7 g; mp 115–116°. Recrystallization from 70% aqueous methanol gives fine white crystals of the same melting point.

If an IR instrument is available, determine the absorption frequency of the imide carbonyls and compare the result with the frequency of a ketone and an amide carbonyl.

(C) *N*-Phenyl-*endo*-norbornene-5,6-dicarboximide

Prepare about 1 g of cyclopentadiene from 5 g of cyclopentadiene dimer by the procedure described in Chapter 41. This should be sufficient to make the adduct from phenylmaleimide and from maleic anhydride (see procedure E). The monomer must be kept cool and used promptly as it dimerizes on standing.

Dissolve 0.7 g (0.005 mole) of N-phenylmaleimide in 1.2 ml of mixed xylenes (**Caution**—*flammable solvent!*) in a 10-mL round-bottomed flask and add 0.5 g (0.006 mole) of cyclopentadiene. Attach a reflux condenser and warm the mixture gently, or until the phenylmaleimide has dissolved. Cool the solution, add 3 mL of petroleum ether (bp 60–90°), and allow it to stand until the white crystals of the adduct have separated. The yield is 0.3–0.4 g; mp 139–142° (literature, mp 144–145°).

This is the N-phenylimide related to compound **V**.

(D) 4-Cyclohexene-*cis*-1,2-dicarboxylic Anhydride (I)

Place in a 10-mL round-bottomed flask 0.75 g (0.0063 mole) of 3-sulfolene, 0.4 g (0.004 mole) of finely powdered maleic anhydride and 1 mL of mixed technical xylenes (**Caution**—*flammable solvents!*). Attach a condenser, arrange a gas trap to remove the evolved sulfur dioxide, and warm the flask gently while swirling it until the solids dissolve. Heat the solution under reflux for 25–30 min and then cool it to room temperature. Add about 2 mL of toluene and a little decolorizing carbon, and heat the material, swirling it, on a steam bath. Filter the hot solution through a folded filter (**Caution**—*flammable solvents!*) and add petroleum ether (bp 60–90°) in small portions until a turbidity develops. Allow the solution to cool and stand until crystallization is complete. The yield is 0.2–0.35 g; mp 101–102° (literature, mp 103°).

(E) *Endo*-Norbornene-*cis*-5,6-dicarboxylic Anhydride (V)

In a 10-mL round-bottomed flask dissolve 0.4 g (0.004 mole) of maleic anhydride in 1.4 mL of ethyl acetate by warming on a steam bath and add 1.4 mL of petroleum ether (bp 60–90°) (**Caution**—*flammable solvents!*). Cool the solution thoroughly in an ice bath, and add carefully 0.4 mL (0.32 g, 0.005 mole) of cyclopentadiene. Swirl the solution in the cooling bath until the initial exothermic reaction is over and the adduct has separated. Heat the flask on a steam bath until the product has dissolved, and then let the solution cool slowly and stand undisturbed. The yield of white crystals is 0.4–0.5 g; mp 163–164°.

The anhydride can be hydrolyzed to the *endo-cis*-dicarboxylic acid by boiling it with 5–6 mL of distilled water until all of the solid and oil has dissolved and then letting the solution cool undisturbed. Usually it is necessary to induce crystallization by scratching with a glass rod or by adding a seed crystal. The yield is about 0.3 g, mp 178–180° (dec).

Questions   1. Write the structures of the Diels–Alder adducts formed by 4 + 2 cycloaddition of the following.

   (a) ethyl propiolate (H—C≡C—CO$_2$Et) and 2-ethoxy-1,3-butadiene (regioisomers are possible)

   (b) 2,5-dimethylfuran and acrylonitrile (CH$_2$=CH—C—N)

   (c) 1,4-dimethoxy-1,3-butadiene and 1,4-naphthoquinone

2. The structure of the adducts formed in parts (a) and (b) of the preceding question might be established by conversion to known benzene derivatives. How could this be done ?

3. Write the structure of the dimers formed through self-addition of the following, in a reaction of the Diels–Alder type (4 + 2 cycloaddition).

   (a) isoprene   (b) cyclopentadiene   (c) acrolein (CH$_2$=CH—CH=O)

40 Benzoquinone and Dihydroxytriptycene

40.1 Diels–Alder Reactions of Benzoquinone

Controlled oxidation of 1,2- and 1,4-dihydroxybenzene (catechol and hydroquinone) leads to *o*- and *p*-benzoquinone. The quinones possess a strong chromophore arising from extended conjugation of the carbonyl groups, for which the generic term quinoid structure is used.

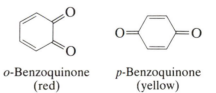

o-Benzoquinone
(red)

p-Benzoquinone
(yellow)

Because of the strength of the carbon–oxygen double bond, the carbonyl groups of quinones are largely localized and the ring system has lost most of its aromatic character; the quinones behave as highly reactive α,β-unsaturated ketones and show typical 1,4-addition reactions. Another consequence of their lost aromatic character is that quinones are oxidizing (dehydrogenating) agents and can readily be reduced to their colorless, benzenoid precursors.

On a small scale, *p*-benzoquinone may be prepared conveniently by oxidation of hydroquinone with dichromate or potassium bromate in acid solution. Commercially, *p*-benzoquinone is made by oxidation of aniline with manganese dioxide and aqueous sulfuric acid. 2-Methylbenzoquinone is prepared in a similar way from *o*-toluidine. The yields in the amine oxidations are not high but the starting materials are readily available and cheap.

Benzoquinone acts as a dienophile in the Diels–Alder reaction, 4 + 2 cy-cloaddition (see Chapter 39). With one molecule of 1,3-butadiene it gives 5,8,9,10-tetrahydro-1,4-naphthoquinone; a second molecule of the diene can be added to obtain octahydro-9,10-anthraquinone. Dehydrogenation of the latter is a route to anthraquinone.

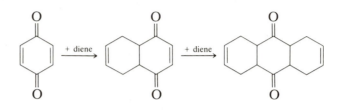

The central ring of anthracene (**I**) functions as a diene in Diels–Alder reactions even though its two double bonds are part of a formal aromatic structure. The Diels–Alder adduct has two independent benzene rings with a combined resonance stabilization that matches the stabilization of the starting anthracene. A striking example is the condensation of *p*-benzoquinone with anthracene to yield the bridged polycyclic system (**II**) containing two benzene rings and a cyclohexene-1,4-dione ring. The dione ring corresponds structurally to

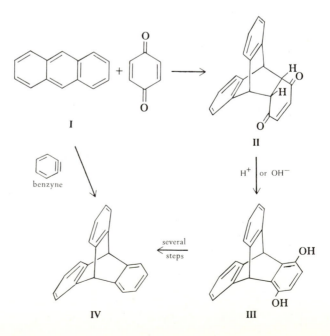

the unstable diketone tautomer of hydroquinone. Treatment of the adduct with either acids or bases transforms it rapidly to the more stable, aromatized hydroquinone form (**III**), which is dihydroxytriptycene. The hydroxyl functions of **III** can be removed by several means to yield the highly symmetric triptycene (**IV**). This unusual name is based on the ancient Roman three-leaved book called a triptych.

In the Diels–Alder reaction of this procedure, *p*-benzoquinone is added to anthracene and the transformation of the adduct to triptycene requires several steps. In an alternate Diels–Alder reaction, a synthetically efficient dienophile, can be used to obtain triptycene (**IV**) in one step by direct addition of anthracene. Unfortunately, benzyne is extremely unstable and substances that yield it often tend to be potentially explosive. One source of benzyne is the decomposition of benzenediazonium-2-carboxylate (from anthranilic acid) under controlled conditions (equation 40.1), but even this has occasionally caused explosions. Another method involves the thermal decomposition of diphenyliodonium-2-carboxylate (equation 40.2).

$$\text{(structure with } CO_2^- \text{ and } N_2^+ \text{)} \longrightarrow \text{(benzyne)} + CO_2 + N_2 \qquad (40.1)$$

$$\text{(structure with } \overset{+}{I} \text{ and } CO_2^- \text{)} \longrightarrow \text{(benzyne)} + I\text{—(phenyl)} + CO_2 \qquad (40.2)$$

40.2 Preparations and Reactions

(A) *p*-Benzoquinone

In a 50-mL Erlenmeyer flask place 1.1 g (0.01 mole) of hydroquinone and a warm solution of 0.6 mL of concentrated sulfuric acid (**Caution**—*corrosive!*) in 10 mL of water. Dissolve the hydroquinone completely by gentle warming and then cool the solution at once in an ice bath, vigorously swirling it, to obtain a suspension of fine crystals.

➤ *CAUTION*   Benzoquinone is extremely volatile and toxic. Its vapor is irritating to the eyes and the solid is a skin irritant. Manipulate quinone in the hood to minimize contact with the vapor and the crystals.

Remove the flask from the cooling bath, and add dropwise, while swirling the flask, a solution of 1.5 g (0.005 mole) of sodium dichromate dihydrate in 2 mL of water.

▐▶ *CAUTION*   Dust containing chromium salts is suspected of causing cancer. Do not grind the sodium dichromate to hasten its solution.

Cool the reaction mixture as needed to maintain its temperature at 20–25°. During the early stages of oxidation, a greenish black precipitate of quinhydrone (a 1:1 complex of quinone and hydroquinone) separates. As the oxidation to quinone is completed, the color of the precipitate becomes yellowish green, About 20 min is required to complete the oxidation.

  Cool the reaction mixture in an ice bath and collect the crystals of quinone with suction. Wash them with two or three 1-mL portions of ice-cold water, press firmly on the filter, and suck them as dry as possible. Purify the crude product by sublimation, using one of the methods shown in Figures 5.7 and 5.8. Transfer the sublimed crystals quickly to a dry, weighed sample tube. The yield is about 0.5 g. The bright yellow crystals of quinone become discolored upon standing.

Reactions. *p*-Benzoquinone undergoes 1,4-addition reactions with halogen acids, alcohols, and amines. The initial adducts undergo isomerization to form substituted hydroquinones; hydrogen chloride, for example, gives chlorohydroquinone. Acetic anhydride, in the presence of acid catalysts, adds in the same way and the intermediate diacetoxy compound is acetylated further to yield the triacetate of hydroxyhydroquinone (1,2,4-trihydroxybenzene).

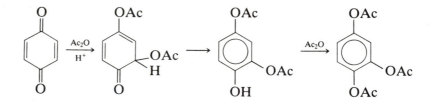

  1,2,4-Hydroxyhydroquinone triacetate. In a small beaker, place 0.2 g of benzoquinone and 6 mL of acetic anhydride. Mix the reactants thoroughly and add one drop of concentrated sulfuric acid (**Caution**—*corrosive!*). The temperature of the reaction mixture should not be allowed to rise above 50°. After swirling for 5 min, add 7 mL of cold water to the solution. Collect the crystals with suction and crystallize them from about 1 mL of ethanol in a small test tube or centrifuge cone. The yield is about 0.2 g. Hydroxyhydroquinone triacetate forms colorless crystals, mp 96–97°. The product becomes discolored when it stands exposed to the air.

(B) 1,4-Dihydroxytriptycene

p-**Benzoquinone-Anthracene Adduct.** In a 25-mL round-bottomed flask dissolve, in 2 mL of xylene, 0.36 g (0.002 mole) of anthracene and 0.22 g (0.002 mole) of *p*-benzoquinone. Attach a water-cooled reflux condenser and boil the solution for 45 min. Cool the solution to 15–20° and allow the adduct to crystallize. Collect the pale yellow crystals with suction and press them firmly on the filter. The yield of adduct is about 0.5 g. The crude product is sufficiently pure for the conversion to dihydroxytriptycene. Do not attempt to determine its very high melting point using an oil bath.

1,4-Dihydroxytriptycene. Place the *p*-benzoquinone-anthracene adduct (about 0.5 g) in a 50-mL round-bottomed flask and add a solution of 0.1 g of potassium hydroxide pellets in 10 mL of ethanol. Warm the flask on a steam bath for about 5 min, or until the adduct has dissolved. Dilute the ethanolic solution with 20 mL of water, cool the flask in an ice–water bath, and carefully neutralize the base by adding 20% hydrochloric acid, drop by drop with swirling, until the liquid is acidic to pH paper (about 0.4 mL of acid will be required). Collect the precipitated dihydroxytriptycene by suction filtration, wash it well with water, and spread it to dry in the air. The yield is about 0.4 g.

The crude product is almost colorless; it shows no carbonyl absorption in the 1700 cm^{-1} region of the infrared. To obtain pure dihydroxytriptycene, you may crystallize the crude material from 95% ethanol or from ethanol–water mixtures. Since it melts at a very high temperature, do not attempt to determine its melting point using an oil bath.

If an IR instrument is available, record the IR spectrum of a Nujol mull of the product.

Questions

1. Write equations for the reaction of benzoquinone with the following.
   (**a**) hydrogen chloride   (**b**) cyclopentadiene
2. What is quinhydrone?
3. Suggest a method, starting from *o*-toluidine, for the preparation of 2-methyl-1,4-naphthoquinone. Use your lecture text to discover the structural relationship of this quinone to the antihemorrhagic (blood-clotting) factors, vitamins K_1 and K_2.
4. Devise a synthesis of 3,6-dimethylphenanthrenequinone from *o*-benzoquinone by a 4 + 2 cycloaddition.
5. What product would form from a 4 + 2 cycloaddition of benzyne to furan? Predict its behavior on treatment with strongly acidic reagents.

41 Ferrocene and Acetylferrocene

41.1 Metallocenes

A fortunate accident in 1951 led to the discovery of an extraordinary organoiron compound,[1] dicyclopentadienyliron (**I**), which later became known as ferrocene. Studies of this remarkably stable and atypical organometallic compound led to its formulation as a combination of two cyclopentadienide anions with a ferrous cation.[2] The bonding involves the six π-electrons of each anion in a way that binds every carbon atom equally to the metal. In this rather flat "sandwich-type" structure (**II**), 12 electrons from two anions and 6 electrons from the ferrous cation lead to the stable 18-electron configuration of the inert gas krypton. Ferrocene is stable to 450° and is soluble in organic solvents and insoluble in water.

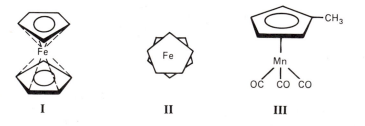

I   **II**   **III**

[1] Kealy and Pauson, *Nature*, **168**, 1039 (1951).
[2] Wilkinson et al., J. Am. Chem. Soc., **74**, 2125 (1952).

Metallocenes have been prepared with a great variety of metals as the central atom—ruthenium, osmium, platinum, chromium, and many others. Some of these have made significant contributions to theoretical chemistry and some have proved to have highly useful catalytic properties for industrial chemical transformations. It is interesting to note that ferrocene is electronically analogous to dibenzenechromium, in which the central atom has zero charge; other metal derivatives of this type are known.

In addition to the metallocenes, there are many cyclopentadienyl derivatives of metal carbonyls, such as methylcyclopentadienylmanganese tricarbonyl (**III**). This compound is an effective antiknoock additive for motor fuel and has been approved for use in nonleaded gasoline. Ferrocene itself also has antiknock properties.

The chemistry of ferrocene resembles that of benzene: it undergoes various electrophilic substitution reactions, similar to the Friedel–Crafts reaction, even *more easily* than benzene. It is somewhat limited in its synthetic applications by its susceptibility to oxidation (even by air), which leads to the blue ferricinium cation. In general, it is desirable to carry out reactions of ferrocene with minimal exposure to air (and to moisture).

Among simple hydrocarbons cyclopentadiene is relatively acidic (pK_a = 15.5) but still weak compared to inorganic acids, and quite strong bases are required to abstract a proton to form the cyclopentadienide anion. With potassium hydroxide, the choice of solvent is important. Two suitable solvents of moderate polarity that produce particularly strong base solutions are used in the present experiment: 1,2-dimethoxyethane (Glyme) and dimethyl sulfoxide (DMSO). Chromatographic methods are presented as a possible way to purify the reaction products.

41.2 Preparations

(A) Cyclopentadiene

Place 5 mL (5.4 g, 0.04 mole) of dicyclopentadiene in a 50-mL round-bottomed flask, attach a Vigreux column, and then a distillation head leading to an ice-cooled receiver. Heat the dimer gently until brisk refluxing occurs (the dimer "cracks" slowly on heating) and the monomer begins to distill steadily in the range of 40–42°.

IIII➧ *CAUTION*   Cyclopentadiene is extremely volatile and flammable.

Continue the distillation at a rapid rate but do not allow the temperature to rise above 43–45°. This should yield about 2.2 mL (1.8 g) of cyclopentadiene in 10 min. Keep the distillate cooled to 0–5°; if it has become cloudy from moisture, add 1–2 g of anhydrous calcium chloride. On standing, cyclopentadiene reverts to the dimer. It is important that the apparatus be disassembled and cleaned as soon as the distillation is completed; on standing the cyclopentadiene forms a polymer that is difficult to remove and could irretrievably cement the joints together.

(B) Ferrocene (Dicyclopentadienyliron)

Prepare an assembly consisting of a 100-mL round-bottomed flask fitted with a Claisen adapter bearing a 125-mL separatory funnel on the central arm and a one-hole stopper on the side arm. *Use grease on all joints!*

In the flask place 8 g (0.14 mole) of potassium hydroxide pellets, followed by 20 mL of 1,2-dimethoxyethane (Glyme) and 2 mL (1.6 g, 0.024 mole) of freshly cracked cyclopentadiene. Swirl the mixture occasionally for 10 min as the black solution of cyclopentadienide anion is formed. Not all of the hydroxide pellets will dissolve.

Meanwhile, in a 125-mL Erlenmeyer flask, prepare a solution of 2.4 g (0.074 mole) of ferrous chloride hydrate in 12 mL (13 g) of dimethylsulfoxide (DMSO, CH_3—SO—CH_3) by swirling the flask over a gentle steam bath for a few minutes (*not over 5 min*). The idea here is to warm the mixture without getting moisture in it.

➤ *CAUTION*   Handle DMSO carefully. It can be absorbed directly through the skin; also it is flammable when hot.

Cool the solution to 15–20° in an ice bath and transfer it to the separatory funnel of the assembled apparatus. Allow the ferrous chloride solution to flow drop by drop into the black solution of potassium cyclopentadienide, over a period of 8–10 min, while frequently agitating the mixture. Disconnect the reaction flask, swirl the liquid for 5 min longer, and then pour it into a beaker (250-mL or larger) containing 40 g of ice and 35 mL of 6 N hydrochloric acid.[3] Rinse the flask three times with 20-mL portions of water, pour the washings into the beaker, and stir with a glass rod until the ice has melted. *At this point, but not earlier, the experiment may be interrupted.*

Filter the reaction mixture with suction using a Büchner funnel. Wash the crude ferrocene on the filter with about 20 mL of water to remove the blue color from the crystals. This color is attributed to oxidation of ferrocene to the blue cation $[Fe(C_5H_5)_2]^+$.

Dry the product by drawing air through the filter; the yield of crude, dry ferrocene is 1.0–1.3 g. The product can be purified by either recrystallization or sublimation.

Recrystallization. Place all of the crude product in a 125-mL Erlenmeyer flask and add 30 mL of hexane (**Caution**—*flammable solvent!*). Heat the hexane–ferrocene mixture until it boils gently and refluxes on the lower wall of the flask.

---

[3] This is approximately 20% hydrochloric acid by weight (22 g/L, *d* 1.10 g/mL), which can be made from 90 mL of concentrated hydrochloric acid and 85 mL of water.

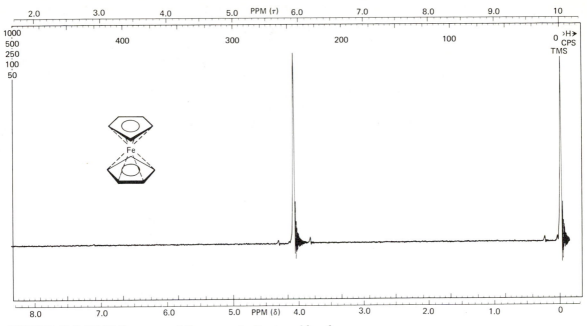

FIGURE 41.1 *NMR Spectrum of Ferrocene in Deuterochloroform*

Not all of the solid will dissolve. Filter the mixture through a filter tube containing a wad of glass wool at the constriction into another 125-mL flask, add a boiling chip, and reduce the volume *in the hood* to 10 mL by heating on a steam bath. Chill the concentrated solution in an ice bath for 10–15 min, and collect the crystalline product on a Hirsch funnel (typical yield is 0.7–0.9 g) and the filtrate in a clean, dry filter flask. Additional product can be obtained by further concentration of the filtrate. Record the weight of your purified product and determine its melting point (after the sample has been introduced into the capillary melting point tube, seal the upper end to prevent sublimation) (literature, mp 174–176°).

Analyze the sample by TLC (hexane is appropriate).

The NMR of ferrocene is shown in Figure 41.1.

Sublimation. Place the crude, dry ferrocene in a large side-arm test tube,[4] and insert in the mouth of the tube a rubber stopper bearing a 16 × 150-mm test tube (see Figure 5.8). Attach the tube to a water aspirator through a trap, and tap the apparatus gently to distribute the solid evenly. Apply suction, and heat the tube gently to drive off the last traces of water. Fill the test tube with chipped ice and increase the heat on the bottom of the tube until yellow crystals begin to form on the walls of the tube and the inner, cold test tube. Continue heating at this rate

---

[4] Alternatively, ferrocene can be sublimed in a covered Petri dish on a gently heated hot plate. No special cooling of the cover dish is required.

until only a black carbonaceous residue remains (10–20 min). Allow the side-arm
test tube to cool and disconnect the suction line. Scrape the sublimed material from
the tubes onto a clean, smooth paper and transfer it to a tared vial.[5]

Record the weight of your purified product and determine its melting point
(after the sample has been introduced into the capillary melting-point tube, seal
the upper end to prevent sublimation) (literature, mp 174–176°).

Analyze the sample by TLC (hexane is appropriate).

The NMR of ferrocene is shown in Figure 41.1.

(C) Acetylferrocene

The conditions of this acetylation are designed to enhance conversion to *mono*
acetylferrocene and diminish further reaction to give 1,1'-diacetylferrocene.
Although an excess of acetylating reagent is used, the extent of reaction is
controlled by limiting the reaction time. Unchanged ferrocene and any diacetyl
derivative are removed by column chromatography.

Fit a 25-mL round-bottomed flask with a stopper bearing a drying tube filled
with anhydrous calcium chloride. In the flask place 0.20 g (0.0011 mole) of
purified ferrocene and add 2.0 mL (0.021 mole) of acetic anhydride. To the
reaction mixture slowly add 0.4 mL of 85% phosphoric acid and then swirl the
reaction mixture. Heat the flask, with drying tube attached, on a steam bath for
10 min. Allow the reaction mixture to cool for 10 min and then pour it onto about
10 g of ice in a 250-mL beaker. Neutralize the mixture by portionwise addition of
6 N sodium hydroxide solution (about 7 mL) until pH paper indicates the solution is neutral. When the mixture has cooled to room temperature, collect the
product with suction and dry it by drawing air through the filter for about 15 min.
Purify the crude acetylferrocene by column chromatography on alumina.

Column Chromatography of Acetylferrocene. Before starting this purification read the discussion of column chromatography in Section 7.5 and look at
Figure 7.3.

Clamp the chromatography column vertically on a ring stand using clamps at
both the top and bottom and make certain that the drainage tube at the bottom is
clamped shut. Fill the column about two-thirds full of cyclohexane (**Caution—**
highly flammable!). Drop a plug of glass wool onto the surface of the solvent. Use
a glass rod or wooden dowel to push the plug to the bottom of the column and
tamp it firmly to make it compact so that adsorbent will not fall out the bottom of
the column. Add a small amount of clean sand to form a 5-mm layer on top of the

---

[5] The outer test tube can be cleaned by first washing it with acetone, then *thoroughly* with soap and
water, and finally with *aqua regia* (in a hood), prepared by putting 3 mL of concentrated hydrochloric
acid in the tube and carefully adding 1 mL of concentrated nitric acid (**Caution—***noxious fumes!*).
Undiluted *aqua regia* is extremely corrosive and a powerful oxidizing agent. It can be deactivated by
addition to a large volume of water; it must *never* be added to an organic waste jar (*explosion hazard*)
or used to remove large amounts or organic tars.

glass wool plug. Tap the column gently to produce an even surface on the sand layer. Next, introduce the dry alumina adsorbent (Fisher 80-200 mesh, Brockman activity 1) in a *slow, steady* stream at the top of the column so that it sinks through the petroleum ether and makes an evenly packed column about 7–8 cm high (about 8–10 g of alumina). Beginning students tend to add the adsorbent in bursts, which produces uneven columns that streak badly. Draw off some of the solvent from the bottom of the column, if necessary, to avoid overflow as the alumina is added. One secret of success is to avoid trapping air bubbles in the column (by tapping the column as the solid is added) as they will produce poor resolution. When the desired length of column has been attained, add enough sand to produce about a 5-mm layer at the top. Drain off the solvent until the level falls just to the upper surface of the packing. The column is now ready for operation.

Dissolve the crude acetylferrocene in a minimum volume of toluene (about 1 mL), ignoring any resinous insoluble by-products, and pipet the solution slowly onto the top of the column without disturbing the top sand layer. Open the drainage clamp briefly to draw the liquid down to the top of the alumina layer. To complete the transfer of the crude acetylferrocene onto the column, pipet in carefully about 1 mL of toluene. It is a good idea at this point to place a small circle of filter paper on top of the sand layer to prevent the surface from being disturbed as solvent is added.

Do not allow the level of solvent to be drained below the upper surface of the alumina, as this will draw air into the packing and disrupt further uniform flow.

Add a few milliliters of cyclohexane carefully by pipet to the top of the column, and drain the solvent down to the top of the alumina. Add 50 mL of petroleum ether to the column and allow it to flow into a 125-mL Erlenmeyer flask. The unreacted ferrocene will move down the column and (in part) into the flask. Now add 50 mL of a 1 : 1 mixture of petroleum ether and ethyl acetate. This much more polar solvent will cause the rest of the ferrocene to flow out the bottom and will move the orange-red band of acetylferrocene down the column behind the ferrocene. Collect the acetylferrocene fraction in another 125-mL Erlenmeyer flask and evaporate the solvent on a steam bath in a hood (**Caution**—*flammable solvents!*). Obtain the weight of the crude product and then recrystallize it from petroleum ether. Determine the weight, melting point, and TLC analysis of the purified material. Typical yields of pure material are 30–40%.

The recorded melting points are ferrocene, 173°; acetylferrocene, 85°; 1,1'-diacetylferrocene, 130–131°.

Questions   1. Cyclopentadiene dimer cracks slowly to give monomeric cyclopentadiene, but on standing, even at elevated temperatures, the monomer reverts to the dimer. Explain.

2. It is observed that addition of hydrogen-bonding solvents like water or alcohol to DMSO solutions of potassium hydroxide sharply reduces the effective base strength. Explain.

3. Explain how the NMR spectrum of ferrocene supports the assigned π-complex structure rather than a σ-bonded structure in which the iron atom is bound to only one carbon atom of each ring.

4. Ferrocene undergoes electrophilic substitution more readily than benzene. Why cannot ferrocene be nitrated successfully by the usual mixture of nitric and sulfuric acid?

5. Uranocene is another stable π-complex between uranium and two neutral cyclooctatetraene ligands. How many electrons does the uranium atom contribute to the bonding that holds this complex together?

42 Dyes and Indicators

42.1 Diazonium-Coupling Reactions

Methyl orange belongs to a class of dyes known as "azo colors," which contain the —N=N— group linked to two aromatic nuclei. In addition to the azo group the dyes must contain salt-forming groups such as hydroxyl, amino, sulfonic acid, or carboxyl groups (auxochromes), which usually intensify the color and at the same time enable the molecule to bind to the fabric. If the dye is used alone, it is called a *direct dye*; if it is precipitated in the fabric chelated with an inorganic ion (typically copper, cobalt, or chromium), it is called a *mordant dye* (L. *mordere* to bite). Two typical commercial azo dyes are Ponceau 2R and Chicago Blue.

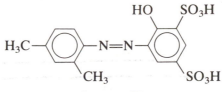

Ponceau 2R
(a brilliant scarlet)

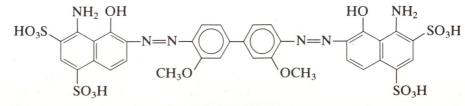

Chicago Blue

Azo dyes are formed by coupling a diazonium ion with a phenol or an aromatic amine. Since many diazonium ions decompose rapidly in solution, it is desirable that the coupling reaction be completed quickly.

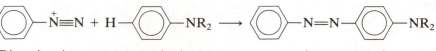

| Diazonium ion | An arylamine (or a phenol) | | Azo compound |

The rate at which a diazonium couples with an aromatic amine is proportional to the product of the concentrations of the diazonium ion and the free (unprotonated) amine. At high pHs the diazonium ion is converted into the unreactive diazoate anion and at low pHs the free amine is converted into the unreactive ammonium salt. Only at intermediate pHs will there be a sufficient concentration of both required species to give a significant coupling rate.

$$ArN_2^+ + H_2O \overset{K_1}{\rightleftharpoons} ArN_2O^- + 2\ H^+$$
$$Ar\overset{+}{N}HR_2 \overset{K_2}{\rightleftharpoons} ArNR_2 + H^+$$

The pH dependence of the relative rate of coupling can be expressed quantitatively by solving the equilibrium equations for the concentrations of free amine and diazonium ion in terms of $[H^+]$ and K_1 and K_2. It is found that the rate is nearly constant for pHs between $\frac{1}{2}pK_1$ and pK_2 (pK is defined as $-\log K$ by analogy to pH) and diminishes sharply for pHs outside this range. There exists a broad plateau of maximum coupling rate that is of great practical significance, for it is within the range of pHs defining this plateau that the coupling reaction should be carried out. The plateau limits for the coupling of p-diazobenzenesulfonate and dimethylaniline to give methyl orange are about pH 4.4 (pK_2) and 10.7 ($\frac{1}{2}pK_1$). To ensure that the reaction solution is not outside this range the dimethylaniline may be converted into its acetate salt, which buffers the solution to about pH 4.6.

The influence of pH on the coupling rate of diazonium ions with phenols is qualitatively similar to the situation for coupling rate with amines. With phenols the active coupling species is the phenoxide anion, which, because of the greater availability of electrons, is more rapidly attacked than the free phenol (or than an aromatic amine). Because of the greater basicity of phenoxide anions compared to aromatic amines (the pK_2 values for phenols are 7–10), the acceptable range of pHs for coupling with phenols is much narrower than with amines and more careful control of the pH is required.

Benzenediazonium salts in a medium buffered with sodium acetate will undergo coupling reactions with certain primary and secondary amines to form diazoamino compounds. With aniline, diazoaminobenzene (C_6H_5—N=N—NH—C_6H_5) is formed. To avoid this type of reaction, diazotization is carried out in a strongly acid medium, since it is the free amine that reacts to form the diazoamino compound.

Reduction of azo compounds leads to the corresponding primary aromatic amines. For example, reduction of methyl orange with sodium hydrosulfite ($Na_2S_2O_4$) in neutral or alkaline solution gives p-aminodimethylaniline (N,N-dimethylphenylenediamine) and sulfanilic acid.

$$HO_3S-\langle\bigcirc\rangle-N{=}N-\langle\bigcirc\rangle-N(CH_3)_2 \xrightarrow{4\,H^+} HO_3S-\langle\bigcirc\rangle-NH_2 + H_2N-\langle\bigcirc\rangle-N(CH_3)_2$$

Methyl orange

42.2 Preparation of Azo Dyes

(A) Methyl Orange

Diazotization. In a 250-mL beaker place 5 mL of a 5% solution of sodium carbonate, dilute it with water to about 12 mL, and add 1.3 g (0.0075 mole) of anhydrous sulfanilic acid (or 1.5 g of the hydrate). Warm slightly on a water bath and if the sulfanilic acid does not dissolve completely add 1–2 mL more of 5% aqueous sodium carbonate (do not add more than 2 mL). If necessary, filter the solution through a fluted filter to remove any undissolved residue. Weigh out 0.45 g (0.0063 mole) of sodium nitrite, dissolve it in about 3 mL of water, and add the nitrite solution to the sodium sulfanilate. Cool the solution in a slush of water and ice until the temperature is between 3 and 5°, then stir vigorously, and add, drop by drop, a solution of 1 mL of concentrated hydrochloric acid diluted with 1–2 mL of water. Do not allow the resulting diazonium solution to stand any longer than necessary; proceed at once to the next step.[1]

Coupling. To 0.8 mL (0.75 g, 0.0063 mole) of dimethylaniline in a test tube, add 0.4 mL (0.4 g, 0.007 mole) of glacial acetic acid and mix thoroughly. To the diazonium salt solution add quickly, with vigorous mixing, the dimethylaniline acetate and allow the mixture to stand with occasional stirring for 5–10 min. Finally make the solution alkaline by adding a solution of 0.9 g of solid sodium hydroxide in about 3 mL of water. This causes the deep red color to change to yellowish orange. The methyl orange separates at once; it can be made to precipitate more completely by adding 2–3 g of table salt. Collect the precipitate with suction on a Büchner funnel, and crystallize the impure product from hot water. Usually 5–6 mL of hot water will be required for each gram of material to be crystallized. Do not use too much solvent; filter the hot solution if necessary.

---

[1] It is advisable to test for the presence of free nitrous acid (to ensure complete diazotization and to avoid a large excess of nitrous acid) by placing a drop of the solution on potassium iodide–starch test paper. In the presence of nitrous acid, iodine is liberated and the starch is colored *immediately*. The test paper may be prepared by dipping strips of filter paper into 1% aqueous potassium iodide, then into colloidal starch solution (prepared readily from "instant starch"), and allowing them to dry.

Collect the crystals with suction, wash them with ethanol, and allow them to dry. The yield is 1–1.5 g purified methyl orange. Do not attempt to determine the melting point of this substance.

Strips of fabric (wool, cotton, polyester, nylon) can be tested for their ability to be dyed by methyl orange. In a 250-mL beaker place 0.1 g of your methyl orange and add 50 mL of water, 2 mL of 15% aqueous sodium sulfate, and 3 drops of concentrated sulfuric acid (**Caution**—*corrosive!*). Heat the solution until it begins to boil and then remove the heat. With the aid of a pair of tongs insert the fabric samples in the dye bath for 5–10 min. Remove the samples (use tongs), rinse well with water, and note which samples are dyed.

Dissolve a little methyl orange in water, add a few drops of dilute hydrochloric acid, then make alkaline again with dilute sodium hydroxide solution. Observe the color changes. The effect of acids and alkalies is probably represented by the following structural changes.

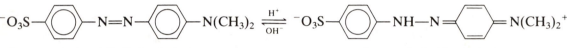

Anion (alkaline solution)   Inner salt (acid solution)
(yellow)   (red)

Many types of organic molecules can be used as indicators. All these have the property of undergoing practically instantaneous change (or changes) in structure in going from acid to alkaline solution, or the reverse, generally within a narrow pH range.

(B) Para Red

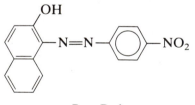

Para Red

In a 250-mL beaker dissolve 0.5 g of (0.0125 mole) of sodium hydroxide, 4.8 g (0.0125 mole) of trisodium phosphate ($Na_3PO_4 \cdot 12H_2O$), as a buffer, and finally 0.7 g (0.005 mole) of 2-naphthol in 100 mL of water, in the order given. Chill the solution by allowing it to stand in an ice bath for 5 min, stirring intermittently. If you wish to test the dyeing properties of Para Red soak test strips of fabric (wool, cotton, polyester, nylon) in the basic 2-naphthol solution while you prepare the next solution.

Prepare a solution of *p*-nitrophenyldiazonium chloride by warming 0.7 g (0.005 mole) of *p*-nitroaniline with 1.5 mL of concentrated hydrochloric acid diluted with 15 mL of water. Pour the solution into a 125-mL Erlenmeyer flask containing about 5 g of chopped ice.

Swirl the mixture to obtain a fine suspension of the crystals of p-nitroaniline hydrochloride. While maintaining the mixture at 5–10° and agitating it vigorously, add as quickly as possible a cold solution of 0.4 g (0.005 mole) of sodium nitrite in 1–2 mL of water. Swirl the mixture until most of the hydrochloride has dissolved (2–3 min), allow it to stand a few minutes to complete the diazotization, and then proceed at once to the next step.

If you are testing the dyeing properties of Para Red, divide the cold diazotized p-nitroaniline solution into two portions in two beakers. Dip the test strips of fabric into one of the portions. Allow them to stand for a few minutes, and then with the aid of tongs remove the strips, wash them well with water, and note which are dyed. Meanwhile pour the other portion of diazonium solution all at once into the chilled alkaline solution of 2-naphthol. Stir the material vigorously for a few minutes to ensure complete reaction and then, after adding 5 mL of concentrated hydrochloric acid, raise the temperature to about 30° on a steam bath and stir for 30–40 min.

Collect the bright red dye with suction. Allow the solid to dry completely and then extract the inorganic salts by stirring with 100 mL of water. Collect the solid with suction and wash it with water until the filtrate is essentially free of chloride ion.[2] Wash the crystals finally with small portion of ethanol. After it is dried, the product weighs about 1.0 g (if no dyeing experiments are performed and all of the reagents are used to form solid dye). Do not attempt to determine its melting point.

42.3 Phthalein and Sulfonphthalein Indicators

Phthalic anhydride undergoes a stepwise condensation with two molecules of phenol, in the presence of acid catalysts such as sulfuric acid or zinc chloride, to form phenolphthalein (**I**).

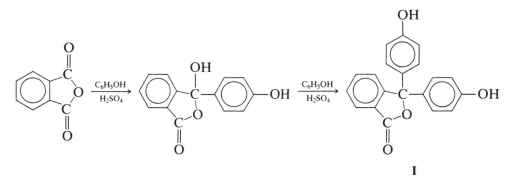

[2] For testing, acidify a 1-mL portion of the filtrate with 0.5 mL of concentrated nitric acid and add 1 mL of dilute (5–10%) aqueous silver nitrate solution.

The phthaleins are derivatives of triphenylmethanol and are related structurally to the triphenylmethane dyes. In neutral or acidic solutions phenolphthalein exists in the colorless, lactone form (**I**). In basic solutions, in the 8.3–10 pH range, it is converted to the red dianion (**II**); in very strongly alkaline solutions, hydroxyl ion is taken up at the central carbon atom (destroying the quinoid structure) and the resulting trianion (**III**) is colorless. Similar phthaleins can be obtained by using *o*-cresol or thymol instead of phenol. If phthalic anhydride is replaced by the anhydride of *o*-sulfobenzoic acid, sulfonphthaleins are produced. Both types of phthaleins are useful acid–base indicators.

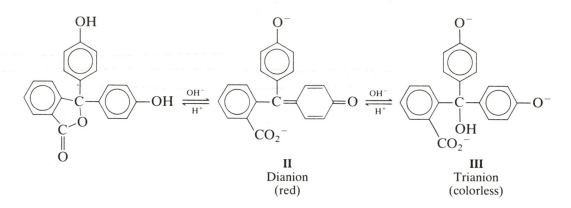

	II	III
	Dianion	Trianion
	(red)	(colorless)

The sulfonphthaleins are excellent pH indicators because they are moderately soluble in water and give brilliant color changes over narrow pH ranges. The parent compound, phenolsulfonphthalein, is formed by condensation of *o*-sulfobenzoic anhydride with phenol.

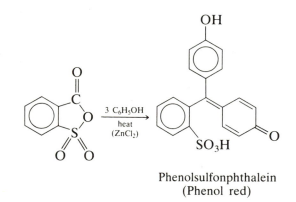

Phenolsulfonphthalein
(Phenol red)

The neutral form of phenol red is yellow, but above pH 8 the phenolic and sulfonic acid hydrogens dissociate to give the red dianion. In extremely strong acidic solutions the red cation is formed.

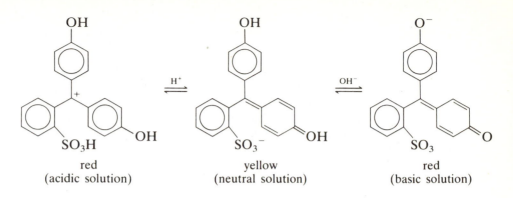

If electron-withdrawing groups are added to the phenol rings both color changes occur at lower pHs; electron-releasing groups shift the color changes to higher pHs. The pH changes with different groups can be understood in terms of the relative stability of the cationic and anionic forms of the indicator. Table 42.1 lists the properties of a number of sufonphthalein indicators. The visible spectra of *o*-cresol red at three different pHs are shown in Figure 42.1. The general method for preparing sulfonphthaleins is to heat *o*-sulfobenzoic anhydride with the appropriate phenol in the presence of a Lewis acid. The preparation of *o*-cresolsulfonphthalein described below is a typical procedure. Bromine substituents can be introduced by direct bromination of the sulfonphthaleins in acetic acid solution. An alternative procedure for preparing sulfonphthaleins is to use the imide of *o*-carboxybenzenesulfonic acid (insoluble saccharin) in place of *o*-sulfobenzoic anhydride. Substitution of the imide requires condensation catalysts such as sulfuric acid and more vigorous reaction conditions.

TABLE 42.1
Properties of Sulfonphthalein Indicators

Sulfonphthalein indicators		pH at color change	Acid color	Base color
Cresol red	*o*-Cresolsulfonphthalein	1.0–2.0	Red	Yellow
Thymol blue	Thymolsulfonphthalein	1.2–2.8	Red	Yellow
Meta cresol purple	*m*-Cresolsulfonphthalein	1.2–2.8	Red	Yellow
Bromophenol blue	3′,3″,5′,5″-Tetrabromo-phenolsulfonphthalein	3.0–4.7	Yellow	Blue
Bromocresol green	3′,3″,5′,5″-Tetrabromo-*m*-cresolsulfonphthalein	3.8–5.4	Yellow	Blue
Bromocresol purple	5′,5″-Dibromo-*o*-cresol-sulfonphthalein	5.2–6.8	Yellow	Purple
Bromothymol blue	3′,3″-Dibromothymol-sulfonphthalein	6.0–7.6	Yellow	Blue
Phenol red	Phenolsulfonphthalein	6.6–8.0	Yellow	Red
Cresol red	*o*-Cresolsulfonphthalein	7.0–8.8	Yellow	Red
Meta cresol purple	*m*-Cresolsulfonphthalein	7.4–9.0	Yellow	Purple
Thymol blue	Thymolsulfonphthalein	8.0–9.6	Yellow	Blue

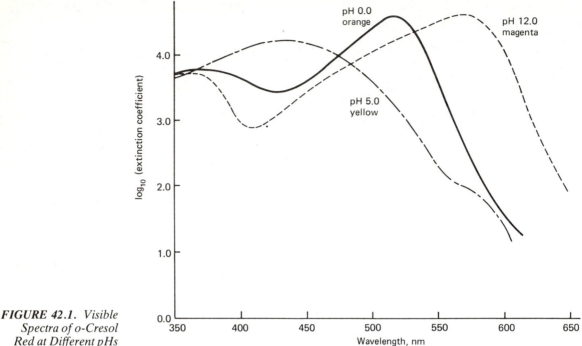

FIGURE 42.1. *Visible Spectra of o-Cresol Red at Different pHs*

42.4 Preparation of *o*-Cresol Red

o-Cresol red

In a 125-mL Erlenmeyer flask, place 1 g of anhydrous zinc chloride, 1 g (0.0055 mole) of *o*-sulfobenzoic anhydride, and 1.6 g (0.015 mole) of *o*-cresol. Mix the reactants thoroughly with a stirring rod and then heat the mixture in the hood in a heating bath at 145° (bath temperature) for 1 hr. At the end of the heating period add 50 mL of 10% aqueous sodium hydroxide and boil the solution in the hood for about 10 min to dissolve the solid mass. Acidify the resulting wine-colored solution with concentrated hydrochloric acid until the product precipitates as dark crystals (about 12 mL of acid). During the addition of acid, zinc salts will precipitate and then redissolve. If the material in the flask remains liquid from unreacted o-cresol, boil the solution in the hood until the volume has been reduced by half. The unreacted o-cresol will steam distill and the product should solidify. If any oily material remains add more water and repeat the steam distillation. Collect the iridescent, green crystals with suction and allow them to air dry. The yield is about 1–1.5 g. Do not attempt to determine the melting point.

Questions

1. What effect will the following substituents have on the rate of coupling of the aryldiazonium ion, relative to the benzenediazonium ion?
   (a) *p*-nitro   **(b)** *m*-nitro   **(c)** *p*-methoxy   **(d)** 2,4-dimethyl

2. What is a diazoamino compound? How may it be converted to an aminoaryl azo compound?

3. At pH 5–6, 7-amino-2-naphthol undergoes coupling at the 8-position, but at pH 10 the coupling occurs at the 1-position. Explain this behavior.

4. What compounds are formed by the reduction of methyl orange with strong reducing agents, such as sodium hydrosulfite ($Na_2S_2O_4$) or stannous chloride? Suggest a method for separating and characterizing the reduction products.

5. What is meant by (a) a chromophore group and (b) an auxochrome group ? Give example of each.

6. Define or explain the following terms. (a) direct dye (b) mordant dye (c) vat dye (see Chapter 44 or your lecture text). Cite an example of each.

7. Explain the color change that occurs when a phthalein, such as phenolphthalein or fluorescein, is treated with sodium hydroxide solution.

8. Compare the structure of phenolphthalein with the structure of phenol red and comment on the colors of these two indicators at high pH (dianion forms).

9. Phenolphthalein is colorless at pH 6, but phenol red is yellow. Suggest a reason for this difference in color.

10. Why does bromophenol blue change color at lower pH then phenol red?

11. What color would be expected for the tetrabromophenolphthalein dianion (high pH form)?

43 Solvatochromic Dyes

43.1 Merocyanin Dyes

Dyes are highly colored compounds that usually contain polar functional groups, which serve both to help bind the dye to the fiber and to determine its color. The class of dyes known as the merocyanin dyes have the general structure represented by resonance structures **Ia** and **Ib**. Three other resonance structures of type **Ia**, involving different arrangements of the double bonds in the pyridine and benzene rings, can be drawn.

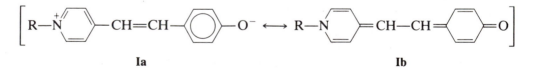

The merocyanin dyes are neutral. Structures of type **Ia** contain two charges, but they are of opposite sign and cancel; structure **Ib** has no formal charges. As structures of type **Ia** imply, however, the merocyanin dyes do possess a large dipole moment. The merocyanin dyes are particularly interesting because their color is strongly dependent on the polarity and hydrogen bonding ability of the solvent. Depending on the solvent the color can be purple, blue, green, orange, red, or yellow. Such substances are called solvatochromic dyes. A proposed commercial use of these dyes is to detect adulteration of solvents. In this preparation you will synthesize the merocyanin dye with R equal to methyl and examine the solvent dependence of its color.

43.2 Theoretical Basis for Solvatochromism

In the ground state of the merocyanin dyes, the electron density is higher at the oxygen end of the molecule than at the nitrogen end. The resulting dipole points in the direction indicated by structure **Ia** and is quite large because of the great distance between the charges. In a polar or hydrogen-bonding solvent, molecules interact strongly with the merocyanin dipole, as shown at the left of Figure 43.1, and stabilize the ground state. Less polar solvents produce less stabilization. On electronic excitation of a merocyanin molecule by absorption of visible light, electrons move from the negatively charged end toward the positively charged end. This reduces the merocyanin dipole moment and reduces the stabilization provided by the solvent. As a consequence (see Figure 43.1, right) the energy gap between the ground and excited state is greater in a polar solvent such as ethanol than in a nonpolar solvent such as acetone. From the Einstein relationship described in Chapter 10, $E = h\nu = hc/\lambda$, a larger excitation energy corresponds to a shorter wavelength for the absorbed light and longer wavelengths for the light that is not absorbed. Table 43.1 illustrates this relationship between energy of

FIGURE 43.1
Interaction of Solvent with Merocyanin Dye

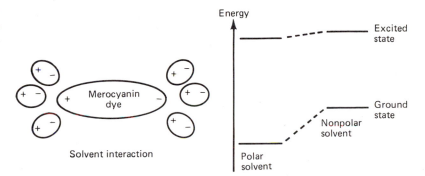

TABLE 43.1
Relationship Between Absorption Energy and Color

Energy kcal/mole	Wavelength (nm)	Color of absorbed light	Color of transmitted light
>75	<380	Ultraviolet	Colorless
75–66	380–435	Violet	Greenish yellow
66–60	435–480	Blue	Yellow
60–58	480–490	Greenish blue	Orange
58–57	490–500	Bluish green	Red
57–51	500–560	Green	Purple
51–49	560–589	Greenish yellow	Violet
49–48	589–595	Yellow	Blue
48–44	595–650	Orange	Greenish blue
44–37	650–780	Red	Bluish green
<37	>780	Infrared	Colorless

absorption and solution color. Note particularly that the color of a dye solution is the color of the light that was *not* absorbed. In a nonpolar solvent the merocyanin is purple (absorption of low-energy long-wavelength yellow light), and in polar solvents it is yellow (absorption of high-energy short-wavelength purple light).

43.3 Synthesis of Merocyanin Dyes

The synthesis of the merocyanin dye with R = CH$_3$ proceeds in three steps.

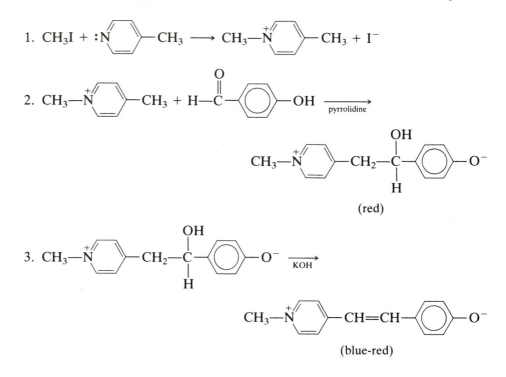

In step 1 the tertiary amine, 4-methylpyridine, is quaternarized with methyl iodide in an S$_N$2 reaction. Other merocyanin dyes can be prepared by using different alkylating agents. With methyl iodide the reaction is rapid but proceeds somewhat more slowly with the larger alkyl iodides because of greater steric hindrance in the transition state. One general word of caution is that some alkylating agents, including methyl iodide, are suspected of causing cancer on prolonged exposure. All alkylating agents should be treated as hazardous substances and used only in the hood or other well-ventilated areas.

The second and third steps are the two parts of an aldol condensation. In step 2 the pyrrolidine first removes a proton from the ring methyl of the methyl 4-methylpyridinium salt to produce a low equilibrium concentration of the methyl anion.

Normally pyrrolidine is far too weak a base to remove protons from alkyl groups, but in this instance the acidity of the ring methyl protons is enhanced by

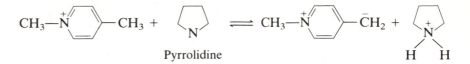

Pyrrolidine

the presence of the positive charge in the pyridine ring. The methyl anion then adds to the carbonyl group of the 4-hydroxybenzaldehyde to produce, after an intramolecular proton transfer, the red alcohol shown in step 2.

In step 3 the red product is boiled briefly with dilute aqueous potassium hydroxide to complete the aldol reaction by the E2 elimination of water to yield the blue-red merocyanin dye.

43.4 Preparation and Measurements[1]

(A) 1,4-Dimethylpyridinium Iodide

To a 25-mL round-bottomed flask containing a solution of 1.0 mL (0.96 g, 0.01 mole) of 4-methylpyridine (4-picoline) (**Caution**—*stench!* Use only in hood) in 2 mL of 2-propanol, slowly add 0.65 mL (1.48 g, 0.010 mole) of methyl iodide (**Caution**—*cancer suspect agent!* Use only in hood). Attach a reflux condenser, add a boiling chip, and boil the solution gently for 30 min. Cool the solution in an ice bath and collect on a Büchner funnel the yellow crystals that separate. Recrystallize the product from 1 : 1 95% ethanol: acetone (it is necessary to chill the solution to get a good recovery). The yield is 0.7–1.5 g (30–60% of theory), mp 144°. The product decomposes slowly on standing, and it is therefore best to proceed to the next stop without delay.

(B) Merocyanin Dye[2]

In a 25-mL round-bottomed flask dissolve 0.7 g (0.003 mole) of 1,4-dimethylpyridinium iodide prepared above, 0.741 g (0.006 mole) of 4-hydroxybenzaldehyde,[3] and 0.5 mL of pyrrolidine (1.11 g, 0.0016 mole) in 8 mL of 1-propanol (*Note*: Do not use isopropanol). Attach a reflux condenser and boil the solution gently for 1 hr. Remove the heat, allow the reaction to cool, add 5 mL of water, and boil the solution for an additional 20 min. Allow the solution to cool overnight,[4] and collect the red precipitate on a Büchner funnel.

---

[1] Adapted from the experiment described by Minch and Sadiq Shah, *J. Chem. Educ.*, **54**, 709 (1977).

[2] The proper name is 1-methyl-4-[(oxycyclohexadienylidine) ethylidene]-1,4-dihydropyridine.

[3] 4-Hydroxybenzaldehyde is usually discolored from air oxidation. It can be recrystallized from 1 : 3 water : 95% ethanol; the solution must be chilled to increase the recovery. Satisfactory yields of dye can be obtained without recrystallization.

[4] The boiling with added water and overnight standing is to induce the product to crystallize. Without this treatment the product is sometimes difficult to filter. Once crystalline, it seems to behave well subsequently.

Suspend the solid in 20 mL of 0.1 M KOH and boil the solution gently for 30 min. Cool the solution to room temperature and collect the blue-red crystals and recrystallize them from hot water. The yield of dye is about 1.0 g (85% of theory). The product decomposes near 150°.

(C) Color Experiments

Determine and record the color of solutions of small samples of the merocyanin dye in acetone, water, and ethanol. Pyridine and 4-methylpyridine give interesting colors although both solvents have unpleasant odors and should be used in the hood. It is instructive to prepare an acetone solution of the dye and to determine the amount of ethanol or water required to produce a visible change in color. If a UV–visible spectrophotometer is available, determine the absorption spectrum of a 2×10^{-5} M solution of the dye in water over the wavelength range of 300–700 nm.

Questions

1. If ethyl iodide were used in place of methyl iodide in the alkylation step, how would the rate of reaction change?

2. Normally the protons of ethyl groups attached to benzene rings are not sufficiently acidic to be abstracted by a mild base such as pyrrolidine. What is the enhanced acidity in the merocyanin dye preparation?

3. Predict the color of solutions of the merocyanin dye in the following solvents.
   (a) toluene   **(b)** acetic acid
   (c) triethylamine   **(d)** 2-propanol

4. The color of the merocyanin dyes does not change much as the alkyl side chain is varied from methyl to hexadecyl. Why?

5. Would you expect the solvent sensitivity of the merocyanin dyes to change if the phenoxide anion were converted to a phenyl ether by alkylation? Explain briefly.

6. If a solution of a dye absorbed at 495 nm, what would be the color of the solution?

44 Synthesis of Indigo

44.1 Introduction

Indigo is a plant dyestuff with a history that stretches back to the ancient Egyptians, who used it as a blue dye for mummy cloths. Today, synthetic Indigo is used to give "blue jeans" their characteristic color. The ancient method of dyeing was indirect and reveals a surprisingly high level of chemical technology. It starts with a colorless natural product, *indican*, that occurs in the leaves of tropical plants of the *Indigofera* species. Indican is a glucoside of another colorless natural product, *indoxyl*. The leaves are cut and allowed to ferment in water. During the fermentation, the extracted indican is cleaved to glucose and indoxyl. The fermented brew is then transferred to a large open vat, the cloth to be dyed is added, and the mixture is beaten with bamboo sticks. During this process the water-soluble indoxyl permeates the fibers and is air oxidized to form the intense blue water-insoluble dimer Indigo that precipitates inside the fibers of the cloth.

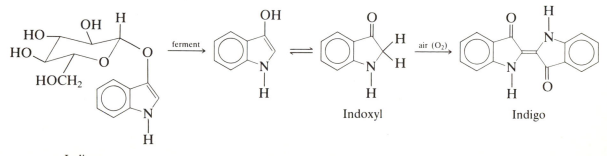

Indican

Indoxyl

Indigo

Today the dyeing process is chemically more sophisticated, but the same underlying principal persists. Synthetic Indigo is chemically reduced to a base-soluble yellow dihydro derivative (called "indigo white"). The fabric is dipped into the alkaline solution of the reduced dye (known generically as a *vat*), the vat is absorbed, and the vat in the impregnated cloth on exposure to air is oxidized back to the original blue Indigo. Indigo, today, goes by the appropriate technical name of Vat Blue 1.

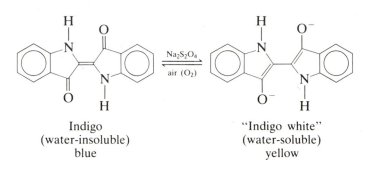

Indigo
(water-insoluble)
blue

"Indigo white"
(water-soluble)
yellow

The first synthesis of Indigo was reported by J. F. W. Adolf von Baeyer (of aspirin fame) in 1880 after 15 years of investigation. The first commercially viable synthesis was developed in 1897. In this experiment you will prepare Indigo by an almost forgotten method developed by Baeyer in 1882. Although the method was a commercial failure, it is direct and rapid and for the small quantities used here is a good way to prepare Indigo and try dyeing samples of cloth.

44.2 Indigo Synthesis

The 1882 Baeyer synthesis of Indigo starts with *o*-nitrobenzaldehyde, which is heated with acetone and base. The acetone enolate anion adds to the aldehyde to produce a hydroxy ketone, **I**, that has been isolated and identified. The next steps

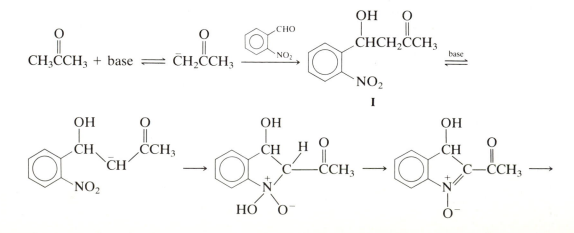

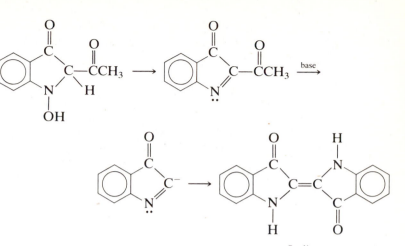

Indigo

are unknown and what is outlined is pure speculation. The sequence shown is chemically plausible but there may be better explanations.

44.3 Preparation and Reactions

(A) Indigo

In a 25-mL round-bottomed flask place 6 ml of acetone and add to it 0.75 g (0.0017 mole) of *o*-nitrobenzaldehye. Attach a reflux condenser, and heat the flask until the solution refluxes gently. Add dropwise through the condenser over a period of 5 min a solution of 0.25 g (3 pellets) of sodium hydroxide in 5 mL of water (~12 N NaOH). Continue to reflux the mixture for another 20 min. Remove the heat and allow the flask to cool for 5 min. Add 10 mL of water and swirl the contents to mix them. Remove the condenser, chill the flask in an ice bath, and collect the solid on a Büchner funnel. Wash the dark blue precipitate with a little water and then several *small* portions of acetone. Transfer the filter paper with the indigo on it to a paper towel and allow it to air dry while you prepare for the next step.

(B) Indigo White and Vat Dyeing

In a 250-mL beaker place 50 mL of water and add 2.5 mL of 10% sodium hydroxide and 1.0 of sodium dithionite ($Na_2S_2O_4$) (do not confuse with sodium thiosulfate, $Na_2S_2O_3$). Add all but a small sample of the Indigo prepared above and stir the mixture until a yellow to green solution of the "indigo white" forms. Sodium dithionite does not have a long shelf life and if the mixture remains blue it will be necessary to add more reducing agent. Place strips of cotton and polyester cloth in the solution and heat the solution gently for 5 min. Remove the cloth strips with a pair of tweezers, rinse them with water and allow them to dry. The blue color of Indigo will be fully formed in about 5 min.

Questions   1. What product is first formed when acetone is heated with dilute aqueous base?

2. Fill in the missing steps in the postulated mechanism for formation of Indigo from *o*-nitrobenzaldehyde and acetone.

3. What would happen if you spilled sodium dithionite solution on your blue jeans?

45 Preparation and Analysis of Deuterated Toluene

45.1 Introduction

In studying organic reaction mechanisms there is frequently a need to label some portion of a molecule so that its position can be followed from starting material to products. One widely used method is to substitute selected protons by deuteriums. As long as the reaction site is not too close to the deuteriums, the deuterated species will have chemical properties very similar to its all-proton parent.

As an example of the possible use of a deuterium label consider the reaction of chlorobenzene with potassium amide in liquid ammonia to give aniline.

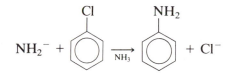

Conceptually, the simplest mechanism for this reaction would be the direct nucleophilic substitution of the chorine by amide. However, this simple mechanism is now known to be wrong and that the reaction proceeds instead by first eliminating HCl to produce the unstable acetylenic intermediate, benzyne, that then adds ammonia to give aniline. That the first mechanism is wrong could be demonstrated by starting with *p*-deuterochlorobenzene. If the first mechanism were correct only a single deuterated aniline would be produced. The second mechanism requires, as is observed, a nearly equal mixture of *meta*- and *para*-deuterated anilines.

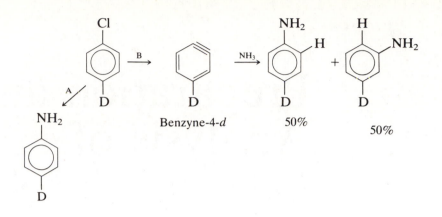

In the experiment of this chapter you will prepare toluene-α-d by formation of tolylmagnesium chloride from benzyl chloride and then quench it with deuterium oxide.

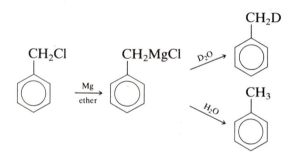

You will isolate the deuterated toluene, demonstrate qualitatively the presence of deuterium by IR, and determine quantitatively by NMR the percentage of deuterium incorporation. To the extent water is present in the deuterium oxide or adsorbed on the walls of your apparatus, it will give rise to ordinary undeuterated toluene. The deuterated toluene will not be used to determine the mechanism of any reaction, but the preparative and measurement techniques employed are the same as those used in a mechanistic study.

45.2 Quantitative Determination of Deuterium Content by NMR

Even though the chemical behavior of deuterium is similar to that of a proton, when it comes to NMR the two are quite different. Chemistry depends almost exclusively on the behavior of valence electrons; NMR is basically a nuclear property that only indirectly involves the surrounding electrons. One important difference in the NMR is that deuteriums absorb at much lower frequencies than protons.[1] Another difference is that the coupling constant between protons and deuteriums is just about $\frac{1}{7}$ of the coupling constant between two protons in the same positions. A further difference is that while protons have two spin states ($+\frac{1}{2}$ and $-\frac{1}{2}$), deuteriums have three spin states ($+1, 0,$ and -1). A proton coupled to a single proton is split into two peaks of equal height at $+J_{HH}/2$ and $-J_{HH}/2$ relative to the uncoupled proton; a proton coupled to a single deuterium is split into three peaks of equal height at $+J_{DH}$, 0, $-J_{DH}$ relative to the uncoupled proton. A consequence of these differences is that a proton NMR spectrum will show no deuterium absorptions, and the proton peaks will show closely spaced splittings that are as likely as not to appear as featureless, broadened peaks on all but the most carefully tuned NMR machines.

In principle, the number of deuteriums in a molecule could be determined by careful analysis of the peak shape, but in practice this is too tedious and uncertain. Instead, one takes advantage of the fact that deuteriums do *not* absorb in the proton region. Integration of a proton spectrum of a deuterated molecule will show no contribution to the peak areas from the deuterium. If the molecule has one or more protons that can be used as an internal area reference, the relative areas of the other peaks can be meausred and the "missing" area ascribed to the deuterium. For example, the NMR spectrum of ordinary toluene will show two peaks for the methyl and phenyl protons of relative areas $3:5$. With toluene containing one deuterium in the methyl group, the relative areas will be $2:5$. If a labeling experiment designed to introduce one deuterium into the methyl group were only 50% successful, the area ratio would be $2.5:5$. This idea can be generalized and turned around to give a simple expression for the fraction of deuterium incorporation into the methyl group:

$$X = 3 - 5R \tag{45.1}$$

where X is the mole fraction of deuterium incoporation and R is the ratio of observed methyl area to phenyl area. For pure toluene R is $\frac{3}{5}$, which gives a calculated mole fraction of 0.0; for pure mono deuterated toluene R is $\frac{2}{5}$, which gives a calculated mole fraction of 1.0.

In this experiment you will measure the NMR of your deuterated toluene and then use equation (45.1) to calculate the percent incorporation of deuterium in your sample.

45.3 Preparation and Analysis of Toluene-α-d

(A) Tolylmagnesium Bromide

Assemble an apparatus like that shown in Figure 8.1b with reflux condenser, separatory funnel for addition, and a 50-mL reaction flask. Prepare a bath of ice and water to permit rapid cooling if the reaction should become too vigorous.

---

[1] In a magnetic field of 14,000 gauss, protons resonate at about 60 MHz. In the same field deuteriums resonate at 9.2 Mhz.

During all of the operations that follow make certain that there are no flames *anywhere nearby* that could ignite ether vapor. Because of the special hazard of benzyl chloride all of the operations except the final purification and analysis must be conducted in a hood.

CAUTION Benzyl halides are powerful lachrymators (tear gas agents). These halides and apparatus that might still have traces of them on it must be handled only in a hood. People wearing contact lenses could suffer eye damage from prolonged exposure to benzyl halide vapors and, if at all possible, prescription glasses should be worn instead of contact lenses. If you are exposed and your eyes begin to water, *do not rub them*; instead, leave the area immediately and wash your eyes thoroughly with water. Removal and hydrolysis of the halide requires considerable washing.

It is essential that the all of the apparatus be clean and *dry* since traces of impurities or water can markedly reduce the yield (for example, do not handle the magnesium turnings with your fingers) and lower the percent deuterium incorporation. If an oven is available, it is helpful to dry the apparatus and the magnesium turnings in it for about 20 min before starting.

CAUTION Ether is extremely volatile and highly flammable. Extreme care must be taken to avoid flames and electrical sparks.

In the reaction flask place 2.3 g (0.10 mole, a large excess) of magnesium turnings. Place 2–3 mL of *anhydrous* diethyl ether[2] in a small dry test tube and stopper the tube with a cork. In a dry 25-mL Erlenmeyer flask place 2.3 mL (2.77 g, 0.022 mole) of benzyl chloride and add 15 mL *anhydrous* diethyl ether. When you are ready to start, temporarily lift the separatory funnel and add a few drops of methyl iodide[3] and then reassemble the apparatus quickly. Pour the ether–chloride solution into the separatory funnel and allow 1–1.5 mL to flow onto the magnesium in the flask. Under favorable conditions, the reaction will start within 5 min without heat, accompanied by vigorous boiling of the ether.

CAUTION Methyl iodide is a cancer suspect agent and should be handled in the hood.

If the reaction does not start promptly, warm the flask gently in a bath of tepid water, but be prepared to cool the reaction quickly with the ice water bath if the reaction starts suddenly. If the warming does not initiate reaction within 5 min, the yield will be low and it is best to start over with more carefully dried equipment and reagents. As soon as this occurs, introduce the 2.5 mL portion of dry ether previously set aside directly through the top of the condenser to moderate the vigor of the reaction. For the success of the experiment it is *absolutely essential*

---

[2] The ether used at this stage *must* be anhydrous.
[3] Methyl iodide is used here because it forms a Grignard reagent with great ease. The resulting methylmagnesium iodide aids the formation of benzylmagnesium iodide ("entrainment method").

that the reaction begin before the extra ether and remainder of the benzyl chloride solution is added to the magnesium.

When the initial vigorous reaction has moderated, allow the remainder of the benzyl chloride solution to flow dropwise into the flask at a rate such that the ether refluxes gently without external heating (about 10 min). Swirl the flask frequently to avoid a buildup of local concentrations of Grignard reagent. After all of the chloride has been introduced, reflux the mixture gently on a steam bath or hot water bath for 20–30 min. Do not heat so vigorously that the ether vapor escapes through the condenser. The volume of solution should not be less than 15 mL; if it is, add more dry ether to bring the volume to about 15–20 mL.

(B) Toluene-*α*-*d*

Before adding deuterium oxide to the Grignard solution, cool the reaction flask in an ice–salt mixture to as low a temperature as possible (at least to −5°).

In the separatory funnel place 3 mL (3.33 g, 0.17 mole) of deuterium oxide and allow it to drop *very slowly* into the cooled Grignard solution, while swirling the solution to insure good mixing and effective cooling. Each drop of the solution reacts vigorously, producing a hissing sound and forming a white precipitate that usually redissolves when the solution is swirled. After all the deuterium oxide has been added, remove the cooling bath and allow the reaction mixture to stand at room temperature with occasional shaking for 20 min or longer. Disassemble the apparatus and carry *just the reaction flask* to your bench. Before doing anything else, rinse the remaining apparatus *in the hood* with acetone and then water to remove traces of unreacted benzyl chloride.

Decant the reaction mixture from the residual magnesium chips slowly onto a mixture of chipped ice and dilute sulfuric acid (prepared by adding 1.1 mL of concentrated acid (**Caution**—*corrosive!*) to 10 mL of water, and adding about 10 mL of chipped ice). Rinse the reaction flask with a little of the dilute sulfuric acid and a little ether[4], and add these washings to the main product. Transfer the mixture to a separatory funnel and separate the two layers; *save both layers*. Extract the aqueous layer with two 5-mL portions of ether and combine the ether extracts with the ether layer from the first separation. The *aqueous* layer may now be discarded.

Wash the ether layer with 2–3 mL of cold water, then with 2–3 mL of saturated sodium bicarbonate, and once again with water. Separate the layers carefully and dry the ethereal layer over anhydrous magnesium sulfate. Pass the dried solution through a filter tube containing a small wad of glass wool[5] to remove the drying agent and let it drain into a small distilling flask. Remove most of the ether by distillation through a short Vigreux column (about 10 cm in length) using a steam or hot water bath.

---

[4] From here on the ether used may be ordinary wet diethyl ether.

[5] You can use a small fluted filter to remove the drying agent, but there will be greater loss from evaporation.

Replace the bath with an electrical heating device (*no flames or sparks allowed when ether vapor is present in the lab*) and continue the distillation. The deuterated toluenes boil close to the boiling point of undeuterated toluene, 111°, and can not be separated from each by the distillation equipment available. However, the observed boiling point may be depressed by the presence of some residual diethyl ether. The purpose of the fractional distillation is to obtain a deuterotoluene sample as free of diethyl ether as possible in order to avoid complications in the NMR analysis. Collect three fractions boiling at A: 105–109°; B: 109–110°; and C: 110–112°. If fraction C is not large enough to analyze (about 0.3 mL is desirable), fractions B and C may be combined. If this is still not enough, it is best to combine A, B, and C and redistill them. Do not forget that toluene-α-*d* is volatile and will evaporate overnight if not stored in a tightly sealed container.

(C) NMR Analysis

The fraction of deuterated toluene selected for analysis should be given a final drying by passing it through a micro column of florisil or silica gel. The simplest technique, which is widely used in NMR spectrometry, is to place a very small wad of glass wool in a Pasteur pipet, push it down to the tapered end, and then add about 1.5 cm of drying agent. The toluene is added to the top of the column (no solvent) and forced through the column into an appropriately small receiver using gentle pressure from a rubber bulb.

In the hood prepare a solution of the deuterated toluene in carbon tetrachloride by dissolving about 0.15 mL of the toluene in 0.6 mL of the solvent and add a few drops of a 1% solution of tetramethylsilane in carbon tetrachloride as an internal marker. The solution may be prepared directly in an NMR tube (invert the capped tube several times to mix the contents). Record the spectrum and determine the relative areas under the phenyl peak (δ 7.2 ppm) and the methyl peak (δ 2.3 ppm). It is best to use the average of several area integrals obtained with the recorder traveling in both directions. Using equation 45.1, calculate the percentage of deuterium incorporation. Note the shape of methyl peak. If time permits prepare a carbon tetrachloride solution of undeuterated toluene and repeat the analysis on its spectrum.

(D) IR Analysis

Determine the infrared spectrum of your deuterated toluene on a thin-film sample held between two salt plates. Remember that toluene is quite volatile and will evaporate quickly. Examine the 2300 cm^{-1} region for evidence of C—D absorption. If time permits, it is a good idea to first record the IR spectrum of undeuterated toluene, both to have a reference spectrum and to gain practice with determining thin-film spectra of volatile samples.

Questions   1. In the preparation of tolylmagnesium chloride, what reaction is minimized by the dilution with additional ether? Why is the benzyl chloride added dropwise?

2. If one didn't have diethyl ether available for the Grignard reaction, explain why each of the following solvents would or would not be a suitable substitute.
   (a) petroleum ether   (b) tetrahydrofuran
   (c) methanol   (d) acetonitrile

3. The flammability of a solvent is determined in part by its volatility and in part by its rate of reaction with oxygen. What intermediate is formed in the reaction of ether with oxygen that causes it to react so rapidly?

4. Derive equation 45.1.

5. Indicate how you could prepare the following deuterated compounds.
   (a) *p*-deuterotoluene
   (b) *p*-deuterobenzoic acid
   (c) α,α,α-trideuterotoluene

46 Binaphthol

46.1 Oxidative Coupling of Phenols

A characteristic property of phenol is its easy oxidation to *p*-benzoquinone by many oxidizing agents. With chromic acid the mechanism appears to follow an ionic course much like the chromic acid oxidation of alcohols.

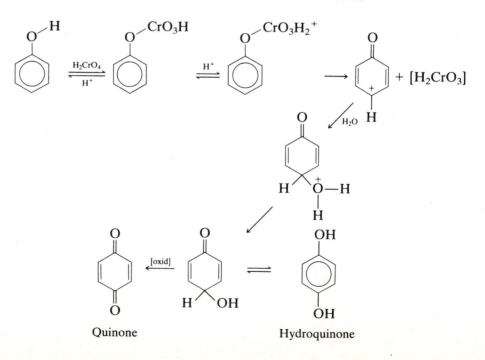

Quinone   Hydroquinone

As the mechanism suggests, hydroquinone, which is formed as an equilibrating side product, is also oxidized to *p*-quinone.

Some other oxidizing agents such as oxygen appear to go by a path involving free-radical intermediates. These agents lead not only by *p*-quinone, but if the concentrations are high enough, to complex mixtures of dimers and their possible oxidation products. The first step is the formation of a radical cation that leads to a resonance-stabilized phenoxy radical. Two phenoxy radicals can couple to produce a dimeric cyclohexadienone that enolizes to give a dihydroxy biphenyl. If the coupling occurs ortho in both of the phenoxy radicals, the 2,2' dihydroxy biphenyl is produced; coupling at two para positions and at one para and one ortho position leads to the 4,4' and 2,4' isomers.

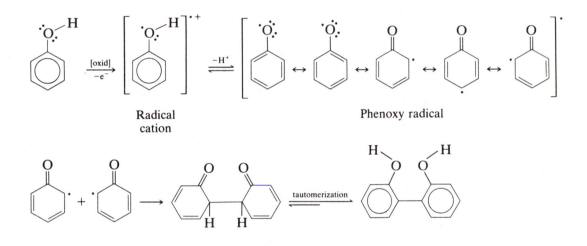

Radical
cation

Phenoxy radical

The oxidative-coupling reaction is not restricted to phenol. It also occurs with substituted phenols at any unsubstituted ortho or para position; coupling also occurs with naphthols and other arenols. The experiment to be described here is with 2-naphthol to form (*R*)-(+)-binaphthol.

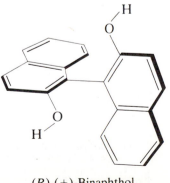

(*R*)-(+)-Binaphthol

46.2 Optical Properties of Binaphthol

When 2-naphthol is oxidized under free-radical conditions, the 1,1'-dimer is formed in moderate yields. This material is optically active by virtue of restricted rotation about the single bond joining the two rings. Such isomerism arising from restricted rotation is known as *atropisomerism*.

Binaphthol has proved to be a remarkably valuable chiral substance for organic synthesis. When equimolar amounts of chiral binaphthol and ethanol react with lithium aluminum hydride, a chiral complexed aluminum hydride is produced that has the property of reducing prochiral ketones[1] in good yield to chiral alcohols of high optical purity (i.e., high *enantioselectivity*). For example, this complex reduces, in 61% yield, acetophenone to 1-phenylethanol that is 97.5% optically pure.

46.3 Preparation of 2,2'-Binaphthol

In a 100-mL round-bottomed flask place 0.72 g (0.003 mole) of cupric nitrate ($Cu(NO_3)_2$) and add 30 mL of water followed by 0.71 g of pyridine and a solution of 0.43 g (0.003 mole) of 2-naphthol in 10 mL of methanol. Add a boiling chip, attach a reflux condenser, and heat the mixture in a boiling-water bath for 5 min. Cool the dark-colored mixture and then acidify it to about pH 5 (use pH paper) with 10% hydrochloric acid. Collect the white precipitate on a Hirsch funnel and wash it well with water. The yield of nearly pure product is about 0.3 g, mp 214–216° (literature, mp 216–218°).

Questions 1. What coupling product would be expected from the free-radical oxidation of 2,6-dimethylphenol?

2. Draw the structure for the complex that results from the reaction of one equivalent of lithium aluminum hydride with one equivalent of 2,2'-binaphthol and one equivalent of ethanol. Why does reduction of prochiral ketones with this reagent produce chiral alcohols?

---

[1] Prochiral ketones are ketones that are themselves optically inactive but on reduction produce alcohols that are chiral (enantiomeric). If the reducing agent produces an alcohol with optical activity close to the theoretical maximum, it is said to show high enantioselectivity. Prochiral ketones and *achiral* reducing agents produce racemic mixtures.

47

Carbene Additions with Phase-Transfer Catalysis

47.1 Carbene Generation

Carbenes are remarkable species. With only six valence electrons and two atoms bonded to the carbene carbon one might have supposed that carbenes would be much too unstable to ever be significant in organic chemistry. Indeed, the simplest carbene, methylene (CH_2), was first discovered in the hot corona of the sun. Surprisingly, dichlorocarbene (CCl_2), while still much too unstable to isolate, is nonetheless easy to generate and use as a transitory reagent in solution.

Carbenes can exist in two electronic configurations. One, a singlet state, is planar and employs two approximately sp^2-hybridized orbitals for its two bonds. The third sp^2 orbital holds two paired electrons, and the remaining p orbital is empty. In a sense the carbene is simultaneously a carbanion (the electron pair in the sp^2 orbital) and a carbocation (the empty p orbital). Of course, a critical distinguishing feature is that while each of these latter two species bears a charge, the carbene is uncharged overall and thus has no long-range electrostatic interactions with ions.

The other electronic configuration for carbenes is a triplet state in which the angle between the two bonded groups is nearly 180°. The two nonbonded electrons have the same spin and lie in two separate orbitals, which are approximately p orbitals (they would be pure p orbitals if the bonded groups were at exactly 160° relative to the central carbene carbon). Depending on the nature of the bonded groups either the singlet or the triplet state can be more stable.

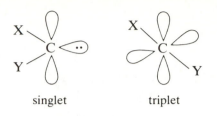

singlet   triplet

As J. Hine discovered, an easy way to prepare dichlorocarbene is to treat chloroform ($CHCl_3$) with a strong aqueous base. The base abstracts a proton from the chloroform to give the trichloromethyl carbanion. Normally, aqueous base would not produce a detectable concentration of carbanion, but here the three chlorines help stabilize it by withdrawal of the charge.

$$CHCl_3 + OH^- \rightleftharpoons \underset{\text{stabilized}}{\bar{C}Cl_3} + H_2O$$

The carbanion is unstable and ejects a chloride ion (α-elimination) to produce dichlorocarbene that reacts with whatever nucleophiles are nearby. If only water and base are present, the carbene will react to produce the hydroxy derivative that reacts further to eventually yield carbon monoxide. If carbon–carbon double bonds are present in high enough concentration, they will add to the carbene to produce a three-membered ring compound bearing two chlorines.

$$\bar{C}Cl_3 \longrightarrow :CCl_2 + Cl^-$$

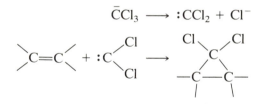

Since the chlorines frequently can be removed by subsequent chemistry, this carbene generation and addition provides an easy entry into cyclopropane derivatives.

47.2 Phase-Transfer Catalysis

A serious complication in dichlorocarbene additions is the usual insolubility of alkenes in aqueous media and the inherent slowness of two-phase bimolecular reactions in which the reagents are segregated in different phases. A fascinating solution is to introduce a quaternary ammonium salt phase-transfer catalyst such as tricaprylmethylammonium chloride (known commercially as Aliquat 336). The lipophilic (hydrophobic) C_8H_{17} capryl groups of the catalyst will mix with hydrocarbons such as alkenes, whereas the charged ammonium ion and the chloride counterion are more stable in an aqueous environment. As a consequence of

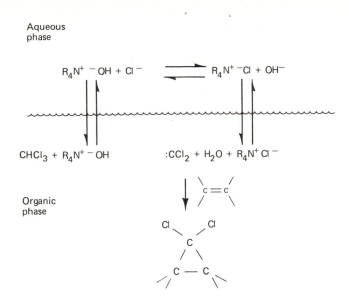

FIGURE 47.1 *Phase-Transfer Catalysis*

these two opposing factors the catalyst can exist in both phases and easily transfer from one to the other.

When the catalyst is in the strongly basic aqueous phase, it can exchange its chloride ion for a hydroxide ion. On transfer back to the organic phase the hydroxide is carried along (as a tight ion pair). In the organic phase the hydroxide ion reacts with the chloroform to produce dichlorocarbene, which preferentially reacts with the high local concentration of alkene present. The water and chloride ion by-products are transferred back to the aqueous phase along with the catalyst where they are exchanged for another hydroxide ion. These reactions are summarized in Figure 47.1. Other phase-transfer catalysts that are commonly used include benzyltriethylammonium chloride, tetrabutylammonium bisulfate, and tricaprylmethylammonium chloride (Aliquat 336).

47.3 Preparation of 7,7-Dichlorobicyclo[4.1.0]heptane

In this reaction the cyclohexene prepared in Section 17.5 (A) will be converted to a bicyclic dichlorocyclopropane.

This experiment presents a number of hazards but these are easily controlled by proper technique.

Because of the small scale of the experiment, you will need to distill about 1 mL of product. This is too small for a regular distillation apparatus but can be done conveniently in a Hickman still (Chapter 2) or the substitute apparatus shown in Figure 2.12.

7,7-Dichlorobicyclo[4.1.0]heptane. Place 2.0 g of sodium hydroxide pellets (0.05 mole) in a 125-mL Erlenmeyer flask and add 2.0 mL of cold water. Swirl the flask to dissolve the base. Considerable heat will develop as the sodium hydroxide dissolves and the solution should be cooled in an ice bath to bring it down to room temperature.

▸ *CAUTION*   Sodium hydroxide is extremely corrosive. Wear goggles and gloves. If any is spilled on the skin, wash the affected area thoroughly with water. If any is splashed in the eyes flush them with water as rapidly as possible and see a physician as soon as possible.

Add 10 drops (0.4 g, 0.001 mole) of tricaprylmethylammonium chloride (Aliquat 336) to the aqueous base followed by a solution of 2.0 mL (1.62 g, 0.02 mole) of cyclohexene in 2.0 mL (2.98 g, 0.025 mole) of chloroform. Swirl (*do not shake*) the solution vigorously for about 20 min; the idea is to form a thick emulsion. During this time the solution will become warm as the reaction proceeds.

▸ *CAUTION*   Chloroform is a cancer suspect agent and should be used only in the hood.

Transfer the reaction mixture to a separatory funnel with the aid of about 15 mL of water, rinse the flask with 4 mL of methylene chloride, and add the rinse to the funnel. Shake the funnel and draw off the lower organic layer into a 50-mL Erlenmeyer flask. Extract the aqueous layer remaining in the funnel with a second 4-mL portion of methylene chloride. Combine the two organic extracts and extract them with a fresh 5-mL portion of water. Dry the extract over a little anhydrous magnesium sulfate, transfer the dried solution through a filter tube into 50-mL Erlenmeyer flask, and evaporate the bulk of the methylene chloride on a steam bath in the hood. Transfer the residue with a Pasteur pipet into a 50-mL round-bottomed flask (use about 1 mL of pentane to aid the transfer), attach a Claisen adapter to the flask and set the equipment on an angle as shown in Figure 2.12. Loosely wrap glass wool around the joint between the flask and the Claisen adapter; distill slowly and collect the distillate in the bend of the adapter. Withdraw from the well and discard any material that boils below 100° (pentane, cyclohexene and chloroform) and save the fraction that collects between 180 and 200°. Typical yields are in the range of 0.5–1.0 g.

▸ *CAUTION*   Prolonged exposure to high concentrations of methylene chloride vapors may induce cancer. As with all organic solvents, work in a well-ventilated area when using it.

If IR or NMR instruments are available, the product can be checked for contaminating cyclohexene.

Questions

1. Rationalize the observation that diphenylcarbene is essentially linear.

2. What product would be formed by reaction of dichlorocarbene with phenanthrene?

3. Thermolysis of 1,1-dichlorocyclopropane produces 2,3-dichloropropene. Write a mechanism for this transformation. Write the structure of the product that would be formed by heating 7,7-dichloro[4.1.0]heptane.

IV

Compounds of Medicinal and Biological Interest

48 Dilantin, an Antiepileptic Drug by a Biomimetic Synthesis

48.1 Introduction

This chapter describes the synthesis of the antiepilepsy drug dilantin. First, however, a few words about the causes of epilepsy and its control.

In the body a signal from a sensory organ is transmitted along a nerve fiber until it reaches the end of the cell, called the synapse. At the synapse the arriving electrical signal induces the release of a *chemical neurotransmitter* that diffuses across the *synaptic gulf*, a short distance of about 100Å that separates the end of one nerve cell and the beginning of another, where it triggers another electrical signal in the next nerve fiber. Thus nerve transmission is a combination of chemical and electrical signals.

In the brain many simultaneous signals are involved and the signal flow is controlled, in part, by the plasma level of γ-aminobutyric acid (GABA). The GABA level, to offer a simplistic analogy, is like the gain control on a public address system in an auditorium. If the GABA level is high, the gain of the nervous system is low; low GABA levels product high gains and increased sensitivity to small signals. In the public address system, if the gain is too high, even the smallest output feeds back to the input microphone and the system breaks into uncontrolled squeals. In animals the equivalent to the squeal is a random, uncontrolled firing of neurons that induce violent spastic movements. This apparently is what happens in at least some forms of epileptic seizures and is associated with lower than normal levels of GABA in the brain. One might hope to control the seizures by administering GABA but this fails because GABA does not pass the blood–brain barrier. Instead, one can take advantage of the fact that much of the natural GABA is absorbed on cell walls; if it is

427

displaced by some agent, the level of GABA in the plasma can rise to an adequate control level.

In 1938 a compound that controls epileptic convulsions in this fashion was discovered. This material, now called dilantin, proved particularly valuable because in ordinary does it it not a sedative and does not impair consciousness, unlike phenobarbital that had been used earlier.

You will be preparing dilantin by the three-step synthesis shown below. As it happens, each of these steps contains some particularly pretty chemistry that will be described separately in the next few sections.

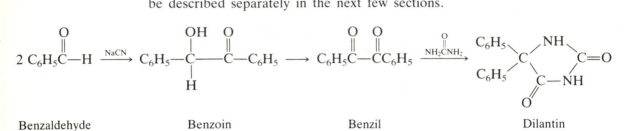

Benzaldehyde   Benzoin   Benzil   Dilantin

48.2 The Benzoin Condensation

Two molecules of an aromatic aldehyde, when heated with a *catalytic* amount of sodium or potassium cyanide in aqueous ethanol, react to form a new carbon–carbon bond between the carbonyl carbons. The product is an α-hydroxy ketone (a class of compounds with the generic name benzoin).

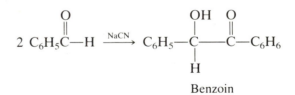

Benzoin

This remarkably facile condensation was discovered accidently by Wohler and Liebig in 1832 when they attempted to extract the cyanohydrin of benzaldehyde with base to remove acid impurities.

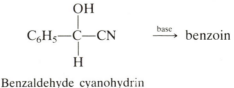

Benzaldehyde cyanohydrin
(Mandelonitrile)

The mechanism for cyanide-catalyzed benzoin formation involves a rather long sequence of steps. It starts with reversible cyanide ion addition to the carbonyl group of one benzaldehyde to form the anion of the cyanohydrin (step 1), which in aqueous ethanol rapidly equilibrates with the neutral cyanohydrin (step 2). The acidity of the C—H bond adjacent to the cyano group is enhanced by resonance stabilization of the anion and under the basic conditions of the reaction (NaCN is basic) the isomeric carbanion is formed (step 3). This adds to a second molecule of benzaldehyde (step 4); proton interchange and loss of cyanide ion (steps 5 and 6) lead to benzoin. The rate-determining step appears to be step 4.

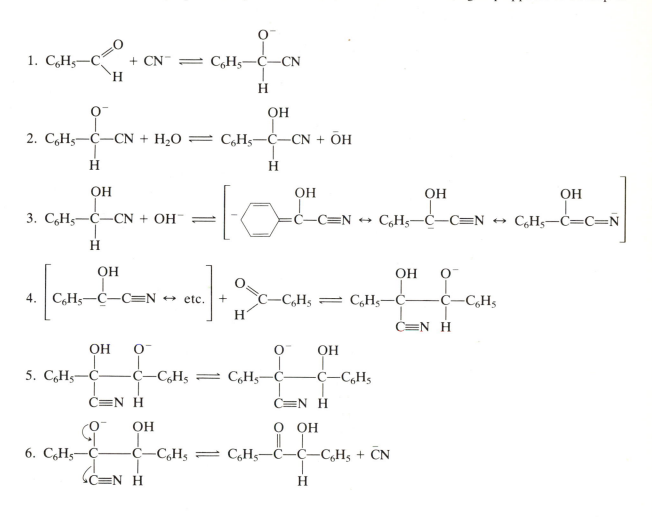

There are two requirements for an effective catalyst of the benzoin condensation. First, the catalyst must give significant amounts of carbonyl adduct (steps 1 and 2), but not form such a strong bond that the catalyst is not easily lost in the last step. Second, the catalyst must stabilize the anion sufficiently to allow the C—H bond to be broken readily, but not so much that the anion becomes unreactive. For more than a 100 years, the only species that had been found that satisfied these requirements was the cyanide ion. However, in 1958 Breslow discovered that the conjugate base of a thiazolium salt also was an effective catalyst; it added reversibly to aldehydes and stabilized the α-anion by resonance.

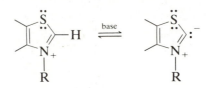

Thiazolium salt   Conjugate base

Resonance-stabilized α-anion

What gave Breslow's study broader significance was his recognition that thiamine (vitamin B_1) contains a thiazole unit and that a number of important biochemical reactions requiring it as coenzyme could be understood as analogs of the benzoin condensation.

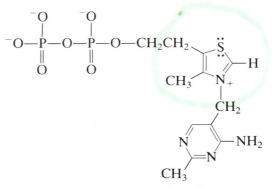

Thiamine pyrophosphate (Cocarboxylase)

The preparation of benzoin using thiamine hydrochloride as the catalyst is described below. The conversion of benzoin to benzil and the use of these compounds are intermediates for syntheses are discussed later.

Benzoin. In a 50-mL Erlenmeyer flask prepare a solution of 1.04 g (0.003 mole) of thiamine hydrochloride in 3 mL of water. When all of the thiamine hydrochloride has dissolved, add 8 mL of 95% ethanol, 3 mL of 10% sodium hydroxide (0.006 mole), and 3 mL (3.2 g, 0.03 mole) of benzaldehyde, with thorough mixing between each addition. Stopper the flask and allow it to stand at room temperature at least overnight (longer periods do no harm).

At the end of the reaction period, the benzoin should have separated as fine crystals. Cool the flask in an ice–water bath to complete the crystallization, collect the product on a Hirsch funnel and wash the crystals thoroughly with two 7-mL portions of cold 50% ethanol and several portions of water. Press the crystals as dry as possible and spread them on a fresh filter paper to dry in the air. The yield is 1.8–2.2 g (dry weight).

The product may be used, without careful drying or recrystallization, for the preparation of derivatives or for conversion to benzil or benzilic acid. Benzoin may be purified, with a loss of 10–15%, by recrystallization from methanol (12 mL/g of benzoin) or from ethanol (8 mL/g). Determine the infrared spectrum of the purified product and demonstrate its purity by TLC.

48.3 Oxidation of Benzoin to Benzil

Benzoin can be oxidized to the diketone benzil in a number of ways, of which the most interesting is by a "coupled oxidation" that uses Cu^{2+} as the catalytic-transfer oxidant. In a coupled oxidation the overall oxidation proceeds in two distinct stages. In the present procedure, cupric acetate is used in catalytic amount (less than 1% of the stoichiometric requirement) and is continuously reoxidized from the reduced (cuprous) state by ammonium nitrate, which is present in excess. The latter is reduced to ammonium nitrite, which decomposes in the reaction mixture into nitrogen and water. It is convenient to represent this two-stage oxidation in the manner used by biochemists, who commonly deal with multiple coupled reactions.

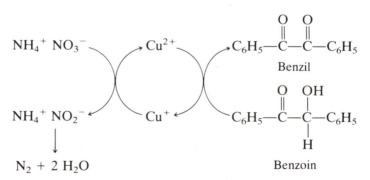

Cupric salts are mild oxidizing agents that do not attack the diketone product. In the absence of Cu^{2+}, ammonium nitrate will not oxidize benzoin (or benzil) at a significant rate. The reaction is general for α-hydroxyketones (acyloins) and is the basis for the Fehling's test for reducing sugars (Chapter 54).

Oxidation of Benzoin by Cupric Salts. In a 25-mL round-bottomed flask place 1.75 g (0.008 mole) of unrecrystallized benzoin, 5 mL of glacial acetic acid,

0.8 g (0.01 mole) of pulverized ammonium nitrate, and 1 mL of a 2% solution of cupric acetate.[1] Add one or two boiling chips, attach a reflux condenser equipped with a gas trap (see Figure 8.2), and bring the solution to a gentle boil. As the reactants dissolve, evolution of nitrogen begins. Boil the blue solution for 1.5 hr to complete the reaction. Cool the solution to 50–60° and pour it into 10 mL of ice-water, while stirring it. After crystallization of the benzil is complete, collect the crystals on a suction filter and wash them thoroughly with water. Press the product as dry as possible on the filter. The yield is 1.4–1.6 g (dry weight). Benzil obtained in this way is sufficiently pure for conversion to dilantin. If desired, it may be purified by recrystallization from methanol or 75% aqueous ethanol.

Benzil α-monoxime. The benzil can be characterized by preparing the oxime derivative. In a test tube place 200 mg of benzil, 1 mL of ethanol, 0.2 mL of 35% aqueous solution of hydroxylamine hydrochloride, and 0.4 mL of 30% aqueous sodium hydroxide. Swirl the mixture and allow it to stand at room temperature for 1 hr. Add 4 mL of water to the reaction mixture, mix, and filter the mixture through a filter tube. Add 20% sulfuric acid dropwise to the clarified filtrate until it is acidic. Chill the mixture and collect the product on a Hirsch funnel. Wash the product with water and recrystallize from 30% ethanol or from methanol. The recorded melting point is 137–138°; it has a configuration with the α-oxime hydroxyl group *anti* to the —CO—C_6H_5 group. The stereoisomeric *syn* form melts at 105–108°.

48.4 Condensation of Benzil with Urea to Form Dilantin

Benzil and urea when heated together with base as catalyst condense to form dilantin. The process involves a skeletal rearrangement with both phenyls ending up on the same carbon atom. One possible sequence is shown below.

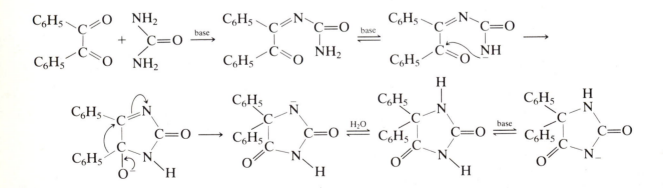

---

[1] The catalyst solution may be prepared by dissolving 2.5 g of cupric acetate monohydrate in 100 mL aqueous acetic acid, stirring well, and filtering to remove any copper salts that have precipitated.

The first step involves base abstraction of an amide proton followed by addition to one of the carbonyl groups of benzil. Subsequent proton transfer and loss of water give the indicated condensation intermediate. This step is analogous to the aldol condensation discussed in Chapter 32. The second step starts off similar to the first, but after the addition to the carbonyl, the attached phenyl group migrates to produce the dilantin skeleton. Protonation then gives dilantin. The rearrangement is analogous to the benzilic acid rearrangement discussed in Chaper 33. It is not clear in this case why the rearrangement occurs. Perhaps it is the stability of the imide (—CO—NH—CO—) group that drives it.

Dilantin. In a small round-bottomed flask place 400 mg of unrecrystallized benzil, 200 mg of urea, 6.0 mL of ethanol, and 1.2 mL of 30% aqueous sodium hydroxide. Attach an upright condenser, add a boiling chip, and boil the mixture gently for 1 hr. Cool the reaction mixture, add 10 mL of water, and filter the solution to remove a sparingly soluble side product that sometimes forms. Acidify the filtrate with hydrochloric acid, collect the product on a suction filter, and wash it thoroughly with water. The product may be recrystallized from ethanol. The yield is 0.28–0.40 g. The recorded melting point is 286–295°; do not attempt to determine the melting point with an oil bath. (**Caution**–*Dilantin is a powerful therapeutic agent and must be taken only on the advice and supervision of a physician!*)

Questions
1. Draw the structure of the product formed by addition of vitamin B_1 and benzaldehyde.
2. Ammonia adds to carbonyl groups but it is ineffective as a catalyst for the benzoin condensation. Explain.
3. Write a balanced oxidation–reduction equation for the oxidation of benzoin by ammonium nitrate (the cupric ion *catalysis* need not be considered).
4. In the benzil oxidation the initial blue color changes to green as the reaction proceeds. Why?
5. In dilantin sodium it is the imide proton that is abstracted rather than the amide proton. Why?

49 Analgesics and Anesthetics

49.1 Introduction

Analgesics are drugs that relieve pain while the patient retains full consciousness. The nonnarcotic group of analgesics includes acetylsalicyclic acid and *p*-ethoxyacetanilide (phenacetin). Aspirin, unlike phenacetin, also has antiinflammatory properties, which makes it useful in the treatment of the symptoms of arthritis. The mechanism of pain-relieving action of these drugs is not clear although it appears to be related to the production or delivery of hormones used to signal pain. Our understanding of the relationship between structure and activity of these drugs is even more obscure. Consequently, such drugs are not designed but are developed by systematic structural modification of compounds discovered by chance. By now, the kinds of modifications likely to produce improved drugs are well known but humans are so complex that extensive trial and error is still required.

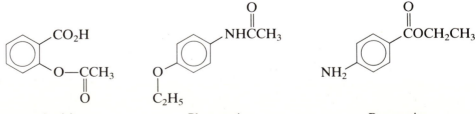

Aspirin   Phenacetin   Benzocaine

Local anesthetics, such as *p*-aminobenzoic acid (PABA) and its many esters, are substances that reversibly block peripheral sensory receptors (or the signals emanating from them) and prevent "pain" information from reaching the central nervous system. Here too, such drugs are developed by trial and error.

The experiments of this chapter include the synthesis of the two analgesics, aspirin and phenacetin, and the local anesthetic benzocaine.

49.2 Synthesis and Reactions of Acetylsalicylic Acid (Aspirin)

Derivatives of salicylic acid have been used in medicine for many years. Salicylic acid occurs in nature in the form of esters in a variety of glycosides and essential oils. The methyl ester is present in oil of wintergreen and in many other fragrant oils from flowers, leaves, and bark. The phenyl ester of salicylic acid, known as Salol, is used as an intestinal antiseptic.

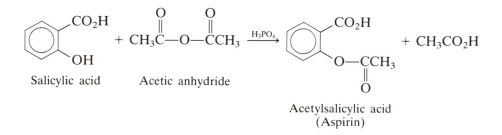

Aspirin forms a beautifully colored blue complex with copper (II) acetate. Although this salt has no medicinal value (in fact, it is probably poisonous because of the copper), it is illustrative of the many complexes that *ortho*-hydroxybenzoic acids and esters form with divalent cations. It has been suggested that it is the formation of such a complex with Ca(II) ions that lies behind the analgesic properties of aspirin.

(A) Acetylsalicylic Acid

In a 50-mL Erlenmeyer flask place 1.4 g (0.01 mole) of salicylic acid. Add 3 mL (3.1 g, 0.03 mole) of acetic anhydride in such a way as to wash down any material adhering to the walls of the flask, and then 5 drops of syrupy (85%) phosphoric acid. Heat the flask for 5 min on a steam bath or in a beaker of water heated to 85–90°. Remove the flask from the bath and, without allowing it cool, add 2 mL of water in one portion. The excess acetic anhydride decomposes vigorously and the contents of the flask come to a boil (**Caution**—*hot acid vapors!*).

When the decomposition is complete, add 20 mL of water and allow the flask to stand at room temperature until crystallization begins. The crystallization can be hastened by occasionally scratching the walls of the flask at the surface of the solution with a glass stirring rod. When crystals begin to appear, place the flask in an ice bath, add 10–15 mL of cold water, and chill thoroughly until crystallization is complete. Collect the product on a Büchner funnel and press it firmly to remove the mother liquor. Wash the material with cold water and allow it to dry thoroughly. The yield is 1.5–1.6 g, mp 140–142°.

The crude product can be recrystallized from toluene (**Caution**—*flammable solvent!*), approximately 8 mL/g.

(B) Copper Aspirinate (Copper Acetylsalicylate)

In a 100-mL beaker prepare a solution of 0.20 g (1.0 mmole, an excess) of copper(II) acetate monohydrate in 40 mL of water that has been heated to about 55° on a steam bath. In a large test tube dissolve 0.28 g (1.8 mmole) of acetylsalicylic acid in 5 mL of 95% ethanol and add the solution to the warm solution of copper acetate. Stir the mixture well and set the beaker aside to cool slowly. After the mixture has cooled to room temperature, collect the dark blue crystals of copper(II) acetylsalicylate on a Hirsh funnel and wash them with a little water to remove any adhering solution of the excess copper acetate.

If an IR machine is available, determine and compare the IR spectra of acetylsalicylic acid and the copper salt.

49.3 *p*-Ethoxyacetanilide (Phenacetin)

Acetanilide is one of the oldest synthetic medicines (1886) and was used for many years as an antipyretic (fever-reducing) and analgesic (pain-relieving) drug under the name Antifebrin. It was abandoned because of the toxicity of its metabolite aniline. *p*-Ethoxyacetanilide, called phenacetin, has similar activity and is less toxic than acetanilide. Another drug of similar type, *p*-acetaminophenol (Tylenol, *p*-hydroxyacetanilide; generic drug name, acetaminophen), has a free phenolic group instead of an ethoxy group. In general these drugs have the effect of reducing the oxygen-carrying capacity of the bloodstream. The widely used antipyretic and analgesic drug acetylsalicylic acid (aspirin) remains one of the least toxic medicinals of this type.

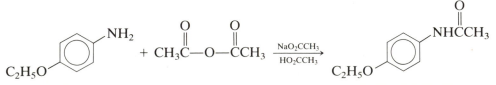

p-Phenetidine   Acetic anhydride   p-Ethoxyacetanilide
                                                   (Phenacetin)

Preparation of Phenacetin[1]

In a small beaker dissolve 3.3 mL (3.5 g, 0.025 mole) of *p*-ethoxyaniline (*p*-phenetidine) in 60 mL of water to which 2.5 mL of concentrated hydrochloric acid has been added. When the amine has dissolved, stir the solution with 1 g decolorizing carbon for a few minutes, and filter the solution through a fluted filter paper. Meanwhile prepare a solution of 5 g of sodium acetate crystals ($CH_3CO_2Na \cdot 3H_2O$) in 15 mL of water.

Transfer the filtered solution of the amine hydrochloride to a 250-mL flask and warm it to 50°. Add 3 mL (3.25 g) of acetic anhydride and swirl the liquid to dissolve the anhydride. Add at once the previously prepared sodium acetate solution, and mix the reactants thoroughly by swirling them. After a few minutes cool the reaction mixture in an ice bath and stir vigorously during crystallization of the product. Collect the crystals with suction, wash them with cold water, and allow them to dry. The yield is 2.5–3 g; recorded mp is 134–135°. The phenacetin prepared in this way is usually quite pure. If desired, it may be recrystallized from hot water with the addition of a little decolorizing carbon.

49.4 *p*-Aminobenzoic Acid (PABA) and Esters

p-Aminobenzoic acid is a member of the group of substances associated with the vitamin B complex. It is present as the central unit of folic acid, vitamin B_{10}, which is made up of a pteridine unit, a *p*-aminobenzoic unit, and a glutamic acid unit. *p*-Aminobenzoic acid is required for folic acid synthesis by some bacteria, and the sulfa drugs are thought to interfere with this synthesis.

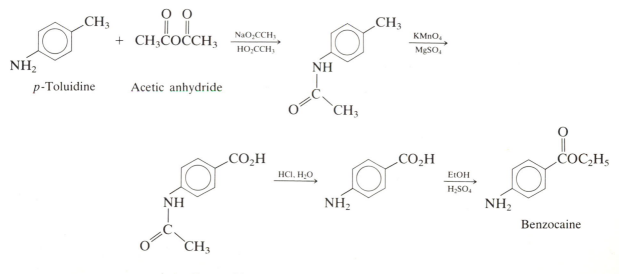

[1] See Chapter 31.

The ethyl ester of *p*-aminobenzoic acid is a local anesthetic (Benzocaine) and a sunburn preventive. There is evidence that sodium *p*-hexadecylaminobenzoate ($C_{16}H_{33}NH—C_6H_4CO_2Na$) raises the level of a desirable cholesterol-controlling mechanism in the bloodstream of experimental animals.

p-Aminobenzoic acid usually made from *p*-nitrotoluene by oxidation to *p*-nitrobenzoic acid (see Chapter 26) and subsequent reduction of the nitro group catalytically or by iron or zinc and hydrochloric acid. The present method is simpler for small-scale laboratory preparations and starts from *p*-toluidine ($CH_3C_6H_4NH_4$). This is acetylated, and the resulting *p*-acetotoluidide is oxidized by potassium permanganate, buffered with magnesium sulfate to avoid hydrolysis of the protective acetyl group. This furnishes *p*-acetamidobenzoic acid, which is hydrolyzed by heating with hydrochloric acid to give the amino acid.

(A) *p*-Acetotoluidide

In a 125-mL Erlenmeyer flask place 2.1 g (0.02 mole) of *p*-toluidine, 50 mL of water, and 1.7 mL (0.02 mole) of concentrated hydrochloric acid in that order. Prepare for use in the next step a solution of 2.8 g (0.022 mole) of sodium acetate crystals ($CH_3CO_2Na \cdot 3H_2O$) in 7 mL of water.

When both solutions are prepared, add 2.7 mL (2.8 g, 0.05 mole) of acetic anhydride and swirl the contents to dissolve the anhydride. Add *at once* the previously prepared sodium acetate solution and mix the reactants thoroughly by swirling. Cool the reaction mixture in an ice bath and stir vigorously while the product crystallizes. Collect the crystals on a suction filter, wash with cold water, and allow them to dry. The yield is 2–3 g and should be somewhat more than is required for the next step.

(B) *p*-Acetamidobenzoic Acid[2]

In a 250-mL flask place 1.9 g (0.013 mole) of *p*-acetotoluidide, 5 g (0.02 mole) of magnesium sulfate crystals ($MgSO_4 \cdot 7H_2O$) and 125 mL of water. Heat the material to about 85° on a steam bath. Meanwhile prepare a solution of 5.1 g (0.032 mole) of potassium permanganate in 20 mL of boiling water in a beaker or Erlenmeyer flask.

While swirling the acetotoluidide solution vigorously, add the hot permanganate in solution in small portions over a period of about 30 min. Avoid a local excess of the oxidizing agent since this tends to destroy the product. After all of the permanganate has been added, swirl the mixture vigorously. Filter off the precipitated manganese dioxide from the hot solution, using a fluted filter, and wash the manganese dioxide with a little water. If the filtrate is colored by excess permanganate, add 1–3 mL of ethanol and boil the solution until the color has been discharged, and filter it again through a fresh paper.

Cool the colorless filtrate and acidify it to pH 3–4 with 20% aqueous sulfuric

[2] This method is a modification of the procedure described by Kremer, *J. Chem. Educ.*, **33**, 71 (1956).

acid. Collect the *p*-acetamidobenzoic acid in a suction filter and press it as dry as possible. Dry a small portion for melting-point determination (literature, mp 250–252°); the yield is 60–75%. It is not necessary to dry the main portion of the product thoroughly for the next step. Because of the high melting point, do not use an oil bath to determine the melting point.

(C) *p*-Aminobenzoic Acid

Weigh the acetamidobenzoic acid from the preceding step and for each gram use 5 mL of 18% hydrochloric acid (1 volume of concentrated acid to 1 volume of water) for the hydrolysis. Place the materials in a 50-mL round-bottomed flask, add a reflux condenser, and boil the mixture gently for 25–30 min. Cool the reaction mixture, add half its volume of cold water, and make the solution just barely alkaline (pH 7–8 measured with wide-range pH paper) with 10% aqueous ammonia; do not go beyond the end point. For each 30 mL of solution add 1 mL of glacial acetic acid, stir vigorously, and cool the solution in an ice bath. If necessary, induce crystallization by scratching with a glass rod or adding a small seed crystal. Collect the product with suction, allow it to dry, and take the melting point (literature, mp 186–187°). The yield is about 40–50% of the weight of the *p*-acetamidobenzoic acid used.

(D) Benzocaine

In a 25-mL round-bottomed flask place 0.66 g (0.005 mole) of *p*-aminobenzoic acid, 5 mL of ethanol, and 0.5 mL (0.95 g, 0.005 mole) of concentrated sulfuric acid (*add cautiously*). Attach a reflux condenser and heat under reflux for 1 hr. Cool the solution to room temperature, neutralize with sodium carbonate (*foaming*), and extract with two 5-mL portions of methylene chloride. Extract the combined organic layers twice with 15-mL portions of water and dry them over anhydrous magnesium sulfate. Remove the methylene chloride by distillation using a steam bath as a heat source and recrystallize the residue from methanol–water. The yield of white crystals is about 0.4 g, mp 91–92°.

⬛➡ *CAUTION*   Benzocaine may cause dermatitis in sensitive individuals.

The methyl ester (mp 114–115°), the *n*-propyl ester (propaesin, mp 73–74°), and the *n*-butyl ester (butesin, mp 57–58°) can be prepared in a similar fashion with about the same yield by substituting 5 mL of the appropriate alcohol. The lower members of the series are easier to work with because of their higher melting points and the greater solubility of the unreacted alcohol in water.

Questions   1. Write an equation and mechanism for the reaction of salicylic acid with acetic anhydride.

2. What major change in the IR spectrum would be expected upon conversion of salicylic acid to acetylsalicylate? Give frequencies as precisely as you can.

3. In the phenacetin preparation *p*-ethoxyaniline is solubilized with hydrochloric acid and the aqueous solution clarified by treatment with charcoal. What is the structure of the water-soluble species?

4. Why is it necessary to add sodium acetate to the solution of *p*-ethoxyaniline hydrochloride and acetic anhydride?

5. PABA is used as a sunscreen agent. How does it block the UV light and where does the energy go?

6. Write a balanced equation for the oxidation of *p*-acetotoluidide with potassium permanganate.

7. In the hydrolysis of *p*-acetamidobenzoic acid, why is the reaction faster with aqueous hydrochloric acid than with water alone?

50 Pheromones and Insect Repellents

50.1 Chemical Communication

Pheromones are compounds excreted by animals for communication among members of the same species. Such chemical messages include attraction of the opposite sex, marking trails to food, and warning of danger. In the case of sex pheromones, the amount of material required can be phenomenally small. A particularly striking example is the female gypsy moth's ability to attract a mate by releasing as little as 1×10^{-9} g of pheromone, an amount so small that it would go undetected by ordinary analytical instruments. Structurally, sex pheromones differ widely from species to species and can be complex or simple, as can be seen from the following examples.

Sex Pheromones

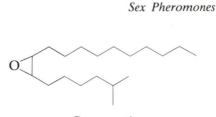

Gypsy moth

American cockroach

Sugar beet wireworm

Beagle

There is evidence that some animals may require more than one chemical for transmission of a chemical message and that the relative amount of each component is important as well. The use of a multicomponent message would reduce the "cross-talk" between different species. It is known that although the chemical signal of one species can be detected by members of a different species, the response is usually different and much less intense than an intraspecies response.

50.2 Insect Repellents

Strictly speaking, current commercial insect repellents should not be classified as pheromones because they are artificial substances that are not used for chemical communication in nature. It is not even clear that they interact with the pheromone receptors. However, because their effects are transmitted by chemical vapors, repellents and pheromones are commonly considered together.

A widely used repellent is N,N-diethyl-m-toluamide, which is effective against mosquitoes, fleas, gnats, and many other insects. This substance is present in Off! as a 14% solution and in Cutter in various forms. This author can testify to the effectiveness of both brands against very hungry mosquitoes.

The antennae of mosquitoes are covered with receptors that detect the currents of carbon dioxide and water vapor rising from warm-blooded hosts. They can track these currents, much like a heat-seeking missile, to their source and proceed to feed. However, when the mosquito encounters an atmosphere filled with the repellent, the signals from its receptors are distorted in some fashion and it has difficulty recognizing or finding the host.

In this preparation[1] you will make the active ingredient of the commercial repellent Off!. The starting material is m-toluic acid, which is converted to the acid chloride with thionyl chloride.

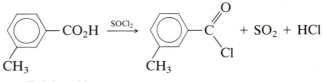

m-Toluic acid

The acid chloride is converted to the amide by addition of diethylamine.

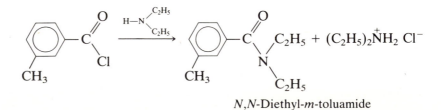

N,N-Diethyl-m-toluamide

[1] The preparation is patterned after the one described by Wang, *J. Chem. Educ.*, **51**, 631 (1974).

50.3 Preparation of *N,N*-Diethyl-*m*-Toluamide

(A) *m*-Toluyl Chloride

Assemble the apparatus shown in Figure 8.1(b) using thoroughly dried glassware and a 50-mL round-bottomed flask. The top of the reflux condenser should be connected to a water aspirator by means of a T-tube, as shown in Figure 8.2(b). In the flask place 1.35 g (0.010 mole) of *m*-toluic acid, 1.5 mL (2.5 g, 0.021 mole) of thionyl chloride ($SOC1_2$), and *one* drop of dimethylformamide (DMF).[2] Add a boiling chip, and reflux the mixture gently onto a steam bath in the hood for 30 min. During the heating, the toluic acid goes into solution and sulfur dioxide and hydrogen chlorie are evolved. In some procedures for preparing acid chlorides the excess thionyl chloride would be removed at this point by evaporation under reduced pressure, but that is not necessary in this case because the excess will be destroyed in the next step.

⫸ *CAUTION* Handle thionyl chloride with great care. The liquid burns the skin, and the vapor is an irritant and harmful to breathe. If any is spilled on the skin, wash the affected area thoroughly with soap and water.

(B) *N,N*-Diethyl-*m*-toluamide (Off!)

Remove the steam bath from the reflux setup and replace it with a pan of ice water. While the reaction mixture cools, prepare a solution of 3.5 mL of diethylamine (2.5 g, 0.034 mole) (**Caution**—*stench*! Use hood) in 15 mL of diethyl ether and place the solution in the dropping funnel.

⫸ *CAUTION* Ether is extremely volatile and highly flammable. Extreme care must be taken to avoid fire hazards.

Add the amine solution, drop by drop, to the acid chloride with *constant agitation*. At first, a vigorous reaction occurs with each drop accompanied by the evolution of heat and the formation of a white cloud of the amine hydrochloride that fills the apparatus. After a few milliliters of the amine solution have been added, however, the reaction will become more moderate. When all of the amine solution has been added, remove the ice bath and allow the reaction mixture to stand for about 10 min to ensure complete reaction. Transfer the mixture to a separatory funnel with the aid of 5–10 mL of dichloromethane, and wash it successively with 15-mL portions of water, 5% aqueous sodium hydroxide, 3 *M* hydrochloric acid, and finally with water. Dry the light brown dichloromethane–ether solution with a little anhydrous magnesium sulfate and filter it through a filter tube containing a small wad of glass wool into an Erlenmeyer flask. Rinse the reaction flask with

---

[2] DMF serves as a catalyst in the conversion of the toluic acid to the acid chloride with thionyl chloride. See discussion in Section 9.8 (H).

1–2 mL of methylene chloride and add the rinse to the Erlenmeyer flask. Remove most of the solvent on a steam bath in the hood.

▸ *CAUTION*   Prolonged exposure to high concentrations of methylene chloride vapors may induce cancer. As with all organic solvents, work in a well-ventilated area when using it.

The crude amide can be purified by column chromatography using about 15 g of alumina or silica gel and dichloromethane as the elution solvent. The first compound to come off the column is *N*,*N*-diethyl-*m*-toluamide. Evaporation of the solvent gives about 1.0–1.5 g of the purified amide as a colorless to light tan oil (bp 160–165° at 20 mm pressure).

If an IR or NMR instrument is available, obtain the spectrum of the amide and verify that the assigned structure is correct. The purity of the sample can be verified by gas chromatography on a Carbowax column at 220°.

Questions   1. Propose syntheses for the gypsy moth and beagle sex pheromones from any starting material containing six or fewer carbon atoms.

2. What is the mechanism for conversion of carboxylic acids to acid chlorides using thionyl chloride?

3. What is the mechanism for the DMF catalysis of the conversion of carboxylic acids to acid chlorides using thionyl chlorides?

4. What is the mechanism for conversion of the *m*-toluyl chloride to the amide using diethylamine?

5. In the reaction of diethylamine with *m*-toluyl chloride, a small amount of *m*-methylbenzaldehyde is formed. How can you account for this? What might the other products be?

51 Antimicrobial Agents: Sulfanilamide

51.1 Introduction

The development and introduction of sulfa drugs as antimicrobial agents represents one of the genuine "miracles" of chemistry. Before 1936 a bacterial infection did not just mean an inconvenient visit to the doctor, it was a serious life-threatening situation. Sulfanilamide and its congeners, the "sulfanilamides," were the first effective antimicrobial chemotherapeutic agents and their introduction produced a sharp drop in mortality from infection.

The discovery of sulfanilamide, the first sulfa drug, grew out of the studies of Gerhard Domagk[1] and co-workers at I. G. Farbenindustrie (the leading German dye firm of the time) in the 1930s and was, in a sense, an accident. It was known that Gram-positive bacteria (i.e., bacteria that adsorb Gram's stain) could be destroyed by certain azo dyes; by trial and error it was found that the azo dye Prontosil was particularly effective. The first clinical trial was on a 10-month-old baby girl who was dying from a septicemia infection; the drug produced a dramatic cure. In 1935 French scientists discovered that Prontosil was itself

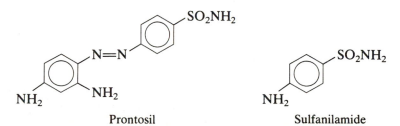

Prontosil   Sulfanilamide

[1] Domagk was awarded the 1938 Nobel prize in medicine for his discovery of sulfanilamide chemotherapy.

445

ineffective but that a natural tissue enzyme split the azo linkage to produce sulfanilamide, which was quite active. Sulfanilamide concentrations of only micrograms per milliliter are required.

Since then hundreds of compounds related to sulfanilamide have been prepared. Structure-activity studies have shown that the benzene ring and the para NH_2 group are essential but that the SO_2 group can be replaced by heterocyclic groups to produce sulfa drugs of even greater therapeutic value. Of these, sulfathiazole and sulfadiazine are particularly effective. The use of sulfa drugs fell off after the discovery of penicillin and related antibiotics, but they are still used for certain urinary infections.

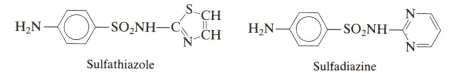

Sulfathiazole   Sulfadiazine

It has been found that the effectiveness of sulfa drugs results from their interference with the synthesis of folic acid (vitamin B_{10}), an essential metabolite in certain microorganisms. Sites on the enzyme involved with folic acid synthesis that are usually occupied by p-aminobenzoic acid (PABA) are occupied by the structurally related p-aminobenzenesulfonic derivative and thereby prevent folic acid formation. Compounds that act in this way are called antimetabolites and are an example of competitive inhibition. Since we humans acquire our folic acid by dietary intake, the sulfanilamides do not interfere with our metabolism. If PABA is administered along with the sulfa drug, the bacteriostasis is counteracted.

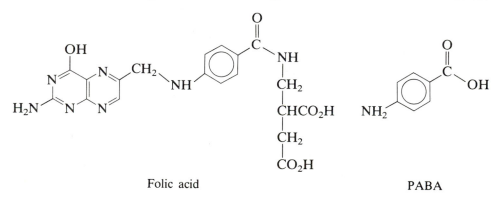

Folic acid   PABA

51.2 Synthesis of Sulfanilamide

A convenient synthesis of sulfanilamide (p-aminobenzenesulfonamide) and related aminoarylsulfonamides make use of p-acetamidobenzenesulfonyl chloride as shown in the following scheme.

It is not feasible to convert p-aminobenzenesulfonic acid (sulfanilic acid) to the sulfonyl chloride since the amino group of the p-aminobenzenesulfonyl chlo-

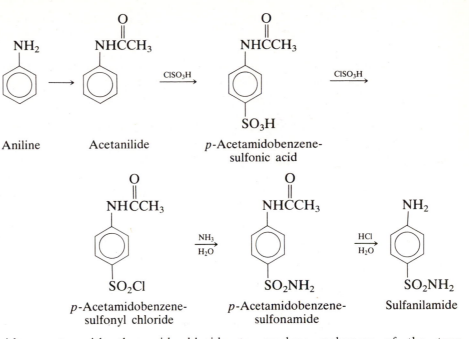

ride reacts with the acid chloride to produce polymers of the type $+NH—C_6H_4—SO_2NH—C_6H_4—SO_2+$.

Protecting the amino group of aniline by conversion to acetanilide permits direct chlorosulfonation with chlorosulfonic acid to obtain *p*-acetamidobenzenesulfonyl chloride. A slight excess over 2 moles of chlorosulfonic acid is used, since chlorosulfonation is a two-step process. The sulfonic acid formed in the first stage is converted to the sulfonyl chloride by reaction with a second molecule of chlorosulfonic acid.

Reaction of *p*-acetamidobenzenesulfonyl chloride with excess aqueous ammonia produces corresponding sulfonamide. Other sulfonamides can be synthesized by substituting the appropriate amino compound. For reaction with aminothiazole, aminodiazole, and similar compounds, a tertiary base such as pyridine is added to combine with the hydrogen chloride formed. Because carboxylic amides are hydrolyzed more easily than sulfonic acid amides, it is a simple matter to deacetylate the acetamido compound by selective hydrolysis under controlled conditions and secure the aminobenzenesulfonamide.

51.3 Preparation

(A) *p*-Acetamidobenzenesulfonyl Chloride[2]

In a dry 20 × 150-mm test tube, that is clamped securely in such a way that it can be immersed in a removable ice bath, place 2.0 mL (3.4 g, 0.03 mole) of

[2] Since hydrogen chloride is evolved, it is desirable to conduct the reaction in a hood or to use an inverted large funnel connected to an aspirator pump.

chlorosulfonic acid, Cl—SO_3H. Insert a thermometer into the test tube. Cool the acid to 10–15°, but *not below* 10°, in a water bath containing a few pieces of ice.

▶ CAUTION   Chlorosulfonic acid must be handled with extreme care. It causes severe burns if dropped on the skin. Your laboratory instructor may wish to dispense it directly into your test tube from a buret.

Then add 0.68 g (0.005 mole) of finely powdered, dry acetanilide, in about seven portions. These additions are best accomplished by raising the thermometer with one hand (*but do not remove the thermometer from the test tube*) and adding the portion with a spatula held in the other hand. Stir the mixture with the thermometer; when the temperature returns to 15°, add the next portion. After all of the acetanilide has been added (a few small particles may remain undissolved), remove the ice bath and replace it with a hot-water heating bath. Heat the reaction mixture to 60–70° and maintain this temperature for an hour.

Pour the reaction mixture *slowly and carefully* onto about 20 g of ice chips. *It is important to carry out this step cautiously to minimize spattering of any unreacted chlorosulfonic acid.* To transfer the remaining drops of reaction mixture from the test tube, add dropwise 2 mL of cold water to the tube (**Caution**—*there may be some spattering!*) and add the rinse to the beaker. This process may need to be repeated to complete the transfer. The reaction product separates as an off-white gummy mass that soon becomes hard. Break up any lumps that have formed, collect the product on a Hirsch funnel, and wash it with several portions of cold water. The crude product, sucked as dry as possible, is used directly in the next step.[3]

(B) *p*-Acetamidobenzenesulfonamide

Place the crude, damp product from the first step in a 125-mL Erlenmeyer flask and add 3 mL of concentrated aqueous ammonia (28%) diluted with 2 mL of water. An immediate exothermic reaction ensues. Rub the mixture with a glass stirring rod until a smooth, thin paste is obtained, and then heat it at 70° for 30 min. Cool the flask in an ice bath and add dropwise, with stirring, dilute (1 : 1) sulfuric acid until the mixture falls to pH 3–5 (measured with wide-range pH paper). After chilling the mixture in an ice bath, collect the product on a Hirsch funnel, wash with cold water, and dry. The yield of crude, dry product is 0.6–0.7 g. This material is sufficiently pure for the next step.

Reserve 0.50 g of dry product for the next step and recrystallize the remainder from hot water (about 60 mL per gram) and take the melting point of the purified material. Test the solubility of the product in dilute acid and dilute alkali. Explain the result. The recorded melting point for pure *p*-acetamidobenzensulfonamide is 219°.

---

[3] Moist *p*-acetamidobenzenesulfonyl chloride can be purified by dissolving it in a mixture of acetone and toluene (1:1), separating any water, and allowing the solvent to evaporate until crystallization takes place. In a purified form, the product keeps well; the crude product tends to decompose on standing.

(C) *p*-Aminobenzenesulfonamide (Sulfanilamide)

Place the crude, dry *p*-acetamidobenzenesulfonamide obtained in the previous preparation in a 25-mL round-bottomed flask and add an amount of dilute hydrochloric acid (1 volume of concentrated acid to 2 volumes of water) equal to 4 mL of dilute acid per gram of substance. Boil the mixture gently under reflux for 30–40 min, which is sufficient time to hydrolyze the amide bond.

Remove the heat source and the condenser, and add in small portions to the cooled reaction mixture a total of about 0.7 g of powdered sodium bicarbonate (**Caution**—*frothing!*) until the mixture is just alkaline against pH paper (pH 7–8). Be sure to stir the mixture between additions in order to be sure that the bicarbonate has reacted. During the neutralization, the free amine separates as a white, crystalline precipitate. Cool the mixture in an ice bath, collect the product by suction filtration, and wash it with a little cold water. The yield is about 0.70–0.72 g/g of *p*-acetamidobenzenesulfonamide.

Recrystallize the product from water, using about 12 mL of water per gram of sulfanilamide. If the solution is colored, add a little decolorizing charcoal before the hot solution is filtered through a fluted filter. The sulfanilamide[4] separates on cooling as long white silky needles. If too much water is used in the recrystallization the yield will be quite low. Typical yields are 0.5–0.8 g. Save a sample to turn in and test the remainder for solubility in dilute acid and dilute alkali. Explain the results.

Determine the melting point of a dry sample of sulfanilamide; the recorded melting point of pure *p*-aminobenzenesulfonamide is 163°.

Questions
1. Chlorosulfonic acid attacks acetanilide in the para position. Explain with the aid of resonance structures why this is so.

2. Show equations for the synthesis of sulfathiazole starting from acetanilide and any other required reagents. Would you expect sulfathiazole to be soluble in dilute acid or dilute base?

3. Show the role of aqueous acid in the hydrolysis of *p*-aminobenzenesulfonamide.

4. Why is it impractical to synthesize sulfanilamide by any process that would require *p*-aminobenzenesulfonyl chloride as an intermediate? What reaction would occur if *p*-aminobenzenesulfonyl chloride were formed momentarily in a synthetic process?

5. The solubility of sulfanilamide in water at room temperature is about 8 g/L. If you recrystallized sulfanilamide from 12 mL of water per gram of solid, what fraction of the material would be lost in the mother liquor? How could you recover it?

---

[4] Sulfanilamide is a powerful therapeutic agent and must be taken only with the advice and supervision of a physician.

52 Artificial Sweeteners

52.1 Introduction

Artificial sweetening agents are extremely important to the food industry because of the national concern with diet and weight control. They are important medicinally, as well, as food additives for diabetics who must limit their sugar intake. Unfortunately, there is evidence that artificial sweeteners can have other physiological effects. For example, even though Dulcin is extremely sweet and so could be used in very small amounts, prolonged use can lead to toxic effects resulting from its conversion in the body to *p*-aminophenol. There are indications that saccharin and cyclamates are mild carcinogens (but weak enough so that some argue that their use saves far more lives than they take). Although chemists have been trying for a long time (saccharin was discovered in 1879 by Professor Ira Remsen at the Johns Hopkins University) to develop a powerful sweetener with no side effects, it is only recently that a new agent, "aspartame" (NutraSweet), has appeared that apparently fills the bill.

One of the problems with developing new sweetening agents is that there is no apparent relationship between structure and activity. Some of the strongest sweeteners have structures bearing no obvious relationship to sugars; in some cases even minor changes in structure can completely destroy a compound's sweetness. In Table 52.1 are collected a few examples of sweet compounds and their close relatives along with their relative sweetness. The values should not be taken too literally because perceived relative sweetness varies from individual to individual, and even for the same taster it varies with concentration. The structures of a few of these substances are also shown.

TABLE 52.1
Relative Sweetness

Compound	Relative sweetness
Ethanol	Very low
Glycerol	Low
Lactose	0.4
Glucose	0.7
Sucrose	1.0
Fructose	1.1
Cyclamates	30
Dulcin	100
Methyl L-aspartyl-L-phenylalanine (L,L-aspartame)	100
Saccharin	600
2-Amino-4-nitro-1-propoxybenzene	4000
2-Nitro-4-amino-1-propoxybenzene	0
Phenylthiourea	0[a]
L,D-Aspartame, D,L-aspartame, and D,D-aspartame	Bitter

[a] Bitter to some.

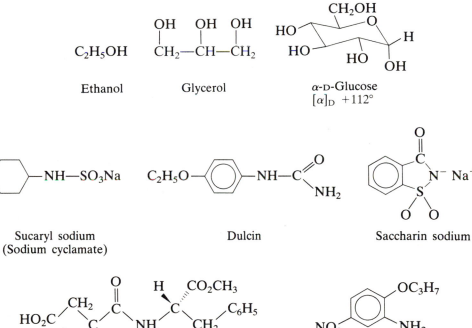

Ethanol   Glycerol   α-D-Glucose $[\alpha]_D$ +112°

Sucaryl sodium (Sodium cyclamate)   Dulcin   Saccharin sodium

Methyl L-aspartyl-L-phenylalanine

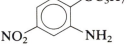

2-Amino-4-nitro-1-propoxybenzene

52.2 p-Ethoxyphenylurea (Dulcin)

Dulcin was first prepared in Germany in 1883. Two methods for preparing it are given here. The first involves the conversion, with acid, of potassium cyanate (KCNO)[1] to isocyanic acid, which then undergoes a typical nucleophilic addition reaction with p-ethoxyaniline to yield Dulcin. The addition reaction is analogous to the derivatization of amines with phenyl isocyanate, $(C_6H_5—N═C═O)$ (see Chapter 9).

$$NCO^- + H^+ \rightleftharpoons H—N═C═O$$
Cyanate

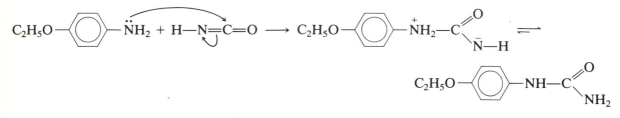

The second method is similar except that the necessary isocyanic acid is generated by acid-catalyzed hydrolysis of urea. Under the conditions of this experiment, urea is in equilibrium with ammonium cyanate, which in turn is in equilibrium with isocyanic acid. The formation of ammonium cyanate from urea is the reverse of the famous 1828 experiment of Wöhler that led to the overthrow of the "vital force" theory of the origin of organic compounds.

$$H_2N—CO—NH_2 \overset{H^+}{\rightleftharpoons} NH_4^+ \ NCO^- \rightleftharpoons NH_3 + H—N═C═O$$

(A) Preparation by Cyanate Method

In a 50-mL Erlenmeyer flask, dissolve 1.4 mL (1.4 g, 0.01 mole) of p-ethoxyaniline (p-phenetidine) in 10 mL of water and 2 mL of glacial acetic acid. While swirling the solution vigorously, add, drop by drop, a solution of 1.6 g (0.02 mole) of potassium cyanate (KNCO) (or 1.3 g of sodium cyanate) dissolved in 5 mL of water (**Caution**—*be certain that you are using a cyanate, not a cyanide!*). As soon as a precipitate of the arylurea begins to appear, add the remainder of the cyanate solution at once and mix the contents thoroughly. Allow the mixture to stand with occasional shaking for an hour or longer, add 5 mL of

---

[1] Potassium cyanate must not be confused with the extremely poisonous potassium cyanide (KCN).

water, and cool the mixture in an ice bath. Collect the product on a suction filter, wash it with a little cold water, and recrystallize it from hot water (30 mL/g) with the addition of decolorizing carbon. Avoid prolonged boiling since this leads to the formation of N,N'-di(ethoxyphenyl) urea. The yield is about 1 g; the recorded melting point of Dulcin is 173–174°.

(B) Preparation by Urea Method[2]

In a 50-mL round-bottomed flask place 1.4 mL (1.4 g, 0.01 mole) of p-ethoxy-aniline (p-phenetidine), 2.4 g (0.02 mole) of urea, and 5 mL of water. To this mixture add 1 mL of concentrated hydrochloric acid and 5 or 6 drops of glacial acetic acid. Attach a reflux condenser, swirl the material well, and boil the mixture vigorously for 30 min or until the reaction is complete. At first the dark-colored solution remains clear, but in 15–20 min the product begins to separate rapidly. When the mixture sets to a semisolid crystalline mass, stop the heating *at once*.

After cooling the flask to room temperature, add 3–4 mL of cold water, stopper the flask tightly, and shake it thoroughly to make a slurry of the crystals. Collect the product on a suction filter, wash it with cold water, and press the crystals firmly. Recrystallize the crude product from boiling water (30 mL/g) with the addition of decolorizing carbon. Chill the material in an ice bath before collecting the purified crystals. The yield is about 1 g (literature, mp 173–174°).

Questions

1. What is produced when potassium cyanide is treated with acid?
2. Write a mechanism for the acid catalysed conversion of urea into ammonium cyanate.
3. On prolonged boiling with aqueous acid Dulcin is converted into N,N'-di(ethoxyphenyl) urea. Write a mechanism for this process.
4. Propose a synthesis of phenylthiourea.
5. With the acid of the table in Chapter 10 (or the Appendix) sketch the NMR expected for Dulcin.

[2] Kurzer, *Organic Syntheses*, Collective Volume IV, **51**, 52 (1963).

53 Sugars

53.1 Introduction

Carbohydrates are one of the large and important groups of compounds found in nature. They include wood and cotton (cellulose), potato and corn starches, and honey and maple syrup (sugars). The simple sugars or monosaccharides such as D-glucose, D-fructose, and D-galactose, are the units from which the more complex carbohydrates are built up by elimination of water (condensation polymerization). Low-molecular-weight polymers containing 2–10 units are called oligosaccharides (*oligo* means few). Examples of disaccharides are sucrose, maltose, cellobiose, and lactose. Raffinose is a trisaccharide with three units: galactose–glucose–fructose. Starches and cellulose are polysaccharides of high molecular weight, built up from thousands of glucose units.

The monosaccharides are polyhydroxy aldehydes (aldoses) of polyhydroxy ketones (ketoses). In solution, they exist in an equilibrium involving an open-

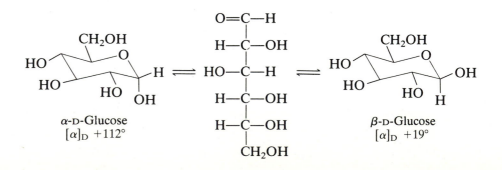

α-D-Glucose
$[\alpha]_D$ +112°

β-D-Glucose
$[\alpha]_D$ +19°

chain structure and cyclic hemiacetal forms. The latter may be five-membered rings (furanose) or six-membered rings (pyranose). The cyclic forms exist in two stereoisomeric configurations (called anomers), designated as α- and β-forms, which are not mirror images of each other. They arise because a new chiral center is created when the ring structure is formed, and it can exist in either of two configurations. In aqueous solution either one of the individual α- or β-anomers will reach the same equilibrium mixture (mutarotation). For D-glucose, this equilibrium value is $+52°$. The anomers can be isolated in the form of derivatives such as α- and β-D-glucose pentaacetates, or α- and β-methyl-D-glucosides, when sugars are subjected to acylation or alkylation.

53.2 Monosaccharide and Disaccharide Tests

Carry out tests (A), (B), and (C) on 2% solutions[1] of the monosaccharides D-glucose (hydrate) ($C_6H_{12}O_6 \cdot H_2O$) and D-fructose ($C_6H_{12}O_6$) and of the disaccharides sucrose ($C_{12}H_{22}O_{11}$), maltose (hydrate) ($C_{12}H_{22}O_{11} \cdot H_2O$), and lactose (hydrate) ($C_{12}H_{22}O_{11} \cdot H_2O$).

Briefly describe and explain each result, using equations where possible. For comparison of the different sugars it is advantageous to record the results in a tabular form.

(A) Test for Reducing Sugars

Sugars that have a free aldehyde group (or can form one under the reaction conditions) are oxidized by cupric ion (see Chapter 48) to produce the carboxylic acid and cuprous ion.

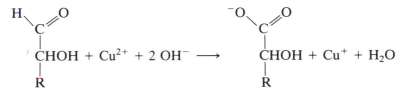

Two reagents commonly used for this purpose are Fehling's solution[2] and

Benedict's solution.[3] These are basic reagents in which the cupric ion is prevented from precipitating as the hydroxide by being complexed as either tartrate or citrate (both are blue). The cuprous ion does not form a tight complex and precipitates as the brick-red cuprous oxide Cu_2O.

In a test tube place 5 mL of freshly mixed Fehling's solution (equal volumes of No. 1 and No. 2) or 5 mL of Benedict's solution.

Heat the solution to gentle boiling and add 2–3 drops of the glucose solution. Continue to boil gently and observe the results after a minute or two. Continue to add the glucose solution, 2–3 drops at a time, and heat for a short while after each addition, until the deep blue color just disappears.

Repeat the test with the other sugar solutions. Discontinue the test if no reduction is observed after 5 or 7 drops of the sugar solution has been added.

Write a structural equation for each sugar that gives a positive test. For each negative test indicate what structural feature is absent.

Hydrolysis of Sucrose. To 10 mL of the 2% solution of sucrose, add 1–2 mL of dilute hydrochloric acid and heat on a steam bath for 0.5 hr. Carefully neutralize with 10% aqueous sodium hydroxide and apply a test for reducing sugars.

(B) Osazone Test

α-Hydroxy carbonyl compounds, such as the reducing sugars, react with phenylhydrazine to produce biphenylhydrazones, called osazones.

$$\begin{array}{c} H \\ | \\ C=O \\ | \\ CHOH \\ | \\ R \end{array} + 3\ C_6H_5NHNH_2 \longrightarrow \begin{array}{c} H \\ | \\ C=N-NHC_6H_5 \\ | \\ C=N-NHC_6H_5 \\ | \\ R \end{array} + C_6H_5NH_2 + NH_3 + H_2O$$

The mechanism of this curious process involves initial formation of the monophenylhydrazone, which undergoes a series of proton exchanges to yield a keto compound (**I**).

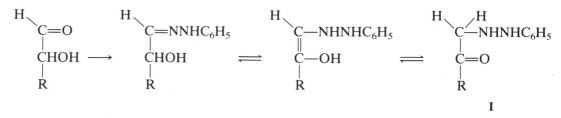

I

[3] **Preparation of Benedict's Solution**. Benedict modified Fehling's solution and devised a single test solution that does not deteriorate on standing. For qualitative work this solution may be prepared as follows: Dissolve 20 g of sodium citrate and 11.5 of anhydrous sodium carbonate in 100 mL of hot water in a 400-mL beaker. Add slowly to the citrate–carbonate solution, stirring continually, a solution of 2 g of copper sulfate crystals in 20 mL of water. The mixed solution should be perfectly clear (if not, pour it through a fluted filter).

The keto compound reacts with a second molecule of phenylhydrazine to give a new hydrazone (**II**).

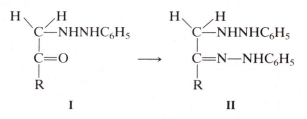

I　　　　　　　　　　　II

This hydrazone then undergoes an oxidative proton abstraction and N—N bond cleavage to produce the imine hydrazone (**III**) and aniline. The imine hydrazone (**III**) then undergoes an addition–elimination reaction with a third molecule of phenylhydrazone to form the observed osazone.

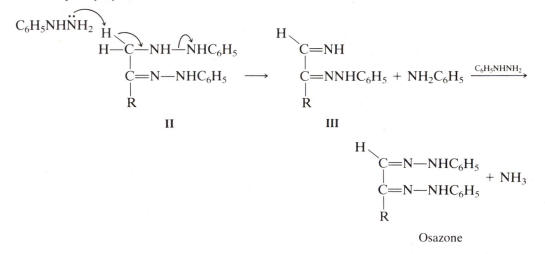

II　　　　　　　　　　　III

Osazone

In applying the osazone test for comparison of various sugars, it is advisable to perform all the tests simultaneously. Place 5 mL of each sugar solution separately in large test tubes, add to each 3 mL of freshly prepared phenylhydrazine reagent,[4] and 2 or 3 drops of saturated aqueous sodium bisulfite (to avoid

---

[4] **Preparation of Phenylhydrazine Reagent**. The directions for preparing the phenylhydrazine reagent are given here for your information only. Pure phenylhydrazine should be handled only by an experienced chemist.

　Dissolve 4 g of phenylhydrazine hydrochloride, C_6H_5—NH—$NH_2 \cdot HCl$, in 36 mL of water, and add 6 g of sodium acetate crystals and a drop of glacial acetic acid. If the resulting solution is turbid, add a small pinch of decolorizing charcoal, shake vigorously, and filter the solution. The reagent deteriorates rapidly.

oxidation). Mix thoroughly, and heat in a beaker of boiling water. Note the time of immersion and record the time in which the osazones[5] precipitate. Shake the tubes from time to time to avoid forming supersaturated solutions of the osazones. Continue heating them in the bath of boiling water for 15–20 min and allow them to cool slowly.

If a microscope is available, it is instructive to examine the osazones. From each tube in which crystals have formed, transfer a small quantity of material to a microscope slide, and examine it under the microscope. Filter the remainder of the crystals at once with suction, wash with a little water, and let them dry. The osazones discolor on standing. The melting points of the common osazones lie close together (200–206°) and are not useful for identification.

(C) Acetylation of Glucose

In a mortar, pulverize 0.9 g (0.003 mole) of anhydrous glucose with 0.5 g of anhydrous sodium acetate. Transfer the mixture to a 25-mL round-bottomed flask and add 5 mL (5 g, 0.15 mole) of acetic anhydride. Heat the mixture under a reflux condenser on a steam bath for 1.5 hr with occasional vigorous agitation. Pour the warm solution in a thin stream, with vigorous stirring, into 50 mL of ice-cold water in a beaker. Disintegrate any lumps of the crystalline precipitate and allow the finely divided material to stand in contact with the water, occasionally stirring it, until the excess acetic anhydride has been hydrolyzed. This will require about 1 hr. Collect the crystals with suction and press them as dry as possible with a clean cork or flat stopper. Transfer the crystals to a beaker, mix thoroughly with about 50 mL of water, and allow to stand with occasional stirring for about 2 hr longer. Collect the crystals with suction and press them as dry as possible on the filter. Recrystallize the crude β-D-glucose pentaacetate from about 5 mL of methanol. The yield of purified product is about 1 g. The recorded melting point of the purified compound is 135°.

α-D-Glucose pentaacetate can also be prepared by acetylation with acidic catalysts. Acids interconvert α- and β-D-glucose and the pentaacetates, and at equilibrium in acetic anhydride, the α-pentaacetate constitutes 90% of the product. With sodium acetate, a basis catalyst, the β-pentaacetate predominates because sodium acetate catalyzes interconversion of α- and β-D-glucose but not the pentaacetates, and the rate of acetylation of β-D-glucose is much faster than that of the α-isomer.

(D) Benzoylation of Glucose

In a 50-mL Erlenmeyer flask, place 5 mL of the 2% glucose solution, 3 mL of 10% aqueous sodium hydroxide, and 0.4 mL of benzoyl chloride (**Caution—**

---

[5] Consult Shriner, Fuson, Curtin, and Morrill, *The Systematic Identification of Organic Compounds*, 6th ed. (New York: Wiley, 1980).

The precipitate formed with D-mannose (0.5–1 min) is the sparingly soluble phenylhydrazone. Sucrose must undergo hydrolysis before precipitation of glucosazone occurs.

irritating vapor! Work in hood). Shake the flask vigorously at frequent intervals until the odor of benzoyl chloride no longer can be detected (test *very cautiously*); at least 15 min will be required. When the product has crystallized, collect the crystals by suction filtration and wash them thoroughly with water. If the product tends to remain gummy, it may be allowed to stand in contact with the alkaline solution until the following laboratory period.

Recrystallize the crude product from ethanol and determine the melting point. The recorded melting point of D-glucose pentabenzoate is 179°.

Questions

1. What is the precipitate that forms in 52.2(A)? Compare these results with those obtained in previous tests of simple aldehydes and ketones in Chapter 9.

2. What is meant by each term?
   (a) aldohexose   (b) ketopentose

3. Of what value is Fehling's test in classifying an unknown sugar?

4. Compare the melting points and rates of precipitation of the osazones of the more common sugars and explain how these values might be used in the identification of an unknown sugar.[5] (*Hint:* Consult the book cited in footnote 5.)

5. Explain the fact that glucose and fructose give the same osazone.

6. What is the significance of the fact that five acetyl or benzoyl groups can be introduced into the molecule of glucose or fructose?

7. What is the significance of the fact that only eight acetyl or benzoyl groups (and not ten) can be introduced into the molecule of the disaccharide sucrose?

8. Give an accurate definition of the term *carbohydrate*.

9. What sugars are present in the solution after hydrolysis of sucrose? Will the sugars form an osazone with phenylhydrazine?

10. Why is the hydrolysis of sucrose sometimes referred to as the inversion of sucrose? What is an invert sugar?

11. What agents other than acids promote the hydrolysis of complex carbohydrates?

12. Fehling's solution contains a known amount of copper per milliliter of solution; suggest a method for determining the amount of glucose in a solution of unknown strength.

54 Biosynthesis of Alcohols

54.1 Fermentation of Sugars

The fermentation processes involved in bread making, wine making, and brewing are among the oldest chemical arts. For many years it was believed that the transformation of sugar by yeasts into ethanol and carbon dioxide was inseparably connected with the life processes of the yeast cell. This view was abandoned when Eduard Buchner[1] (Nobel laureate, 1907) demonstrated that yeast juice will bring about alcoholic fermentation in the absence of any live yeast cells. He proposed that the fermenting activity of yeast was due to the presence of remarkably active catalysts of biochemical origin, now called enzymes. It was later shown that enzymes are complex but nonliving polypeptides, that in some instances can be synthesized in the laboratory. It is now recognized that most of the chemical transformations that go on in living cells of plants and animals are brought about by enzymes. The overall equation for the fermentation of sucrose is typical.

$$\text{Sucrose } + \text{ H}_2\text{O} \longrightarrow \text{ glucose } + \text{ fructose } \longrightarrow 4 \text{ CH}_3\text{CH}_2\text{OH} + 4 \text{ CO}_2$$
$$\text{C}_{12}\text{H}_{22}\text{O}_{11} \qquad\qquad \text{C}_6\text{H}_{12}\text{O}_6 \quad \text{C}_6\text{H}_{12}\text{O}_6$$

The first step in the fermentation of disaccharides, such as sucrose or maltose, is a simple hydrolysis to monosaccharides: hexoses like glucose and fructose. These are then converted to their 6-phosphate esters, then to fructose 1,6-diphosphate, and to phosphate esters of the trioses dihydroxyacetone (a ketotriose) and

---

[1] Eduard Buchner is widely confused with Ernst Büchner (note the difference in spelling) who invented (1888) the funnel bearing his name.

glyceraldehyde (an aldotriose). The dihydroxyacetone and glyceraldehyde are readily interconverted by enolization. The next step involves phosphorylation of the glyceraldehyde and oxidation of the resulting glyceraldehyde-3-phosphate to the phosphoglyceric acid, which lead to pyruvic acid. The pyruvic acid loses carbon dioxide to form acetaldehyde, and the aldehyde is reduced to ethanol (Figure 54.1). Enzymes show an extraordinary selectivity—each step of the fermentation requires a particular enzyme as catalyst. Inorganic salts are also important: Pasteur found that salts such as magnesium sulfate and various phosphates were needed to promote yeast growth and fermentation. The chief industrial sources of sugars for fermentation are molasses residues and starches from grains (corn, rye) and potatoes. The starches are high-molecular-weight polysaccharides that are converted by the enzyme diastase into the disaccharide maltose, which furnishes two molecules of glucose (see Chapter 52). After fermentation has been completed, fractional distillation produces an ethanol–water azeotrope (bp 78.15°) containing 95.6% alcohol by weight (97.2% by volume). To obtain anhydrous (absolute) ethanol commercially, water is removed from the azeotrope by distillation with benzene to furnish the pure product (bp 78.37°). For tax purposes, beverage alcohol is rated in terms of proof spirit; 100 proof (U.S.A) is 50% ethanol by volume, 42.5% by weight. To avoid payment of the beverage tax, alcohol can be denatured by the addition of substances that make it unfit to drink. Higher alcohols (C_3–C_5), called fusel oil, are obtained in small amounts by fractionation of the fermented liquor. These alcohols do not come from the sugars, but arise from enzyme action on amino

FIGURE 54.1 *General Outline for Steps in Alcoholic Fermentation (P* = phosphate ester)*

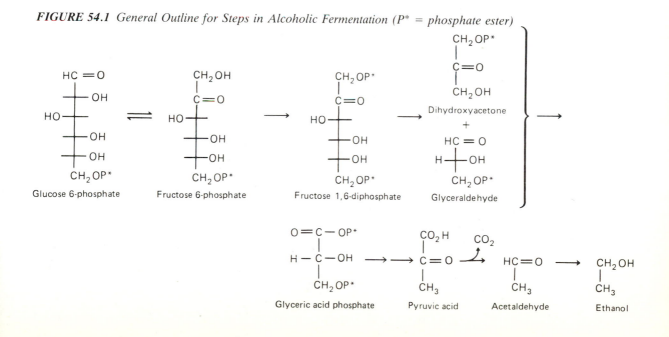

acids derived from the raw materials used and from yeast cells. The fusel alcohols are all primary, mainly *n*-propyl, isobutyl, isopentyl (3-methyl-1-butanol), and optically active 2-methyl-1-butanol. In the United States, industrial ethanol is manufactured mainly from ethylene, a product of the "cracking" of petroleum hydrocarbons. By reaction with concentrated sulfuric acid, ethylene is converted to ethyl hydrogen sulfate, which is hydrolyzed to ethanol by dilution with water. Isopropyl, *sec*-butyl, *t*-butyl, and higher secondary and tertiary alcohols also are produced on a large scale from alkenes derived from the cracking process.

54.2 Ethanol by Fermentation

Place 20 g (0.06 mole) of sucrose (common granulated sugar) in a 250-mL Erlenmeyer flask, add 175 mL of water, 20 mL of Pasteur's salts solution,[2] and one-quarter envelope of dry yeast or one-quarter cake of compressed yeast, rubbed to a thin paste with about 10 mL of water. Shake vigorously and close the flask with a rubber stopper holding a delivery tube arranged so that any gas evolved must bubble through about 10 mL of water in a test tube. Allow the mixture to stand at a temperature of 25–35° until fermentation is complete, as indicated by the cessation of gas evolution (about a week is required). Without stirring up the yeast any more than is necessary, decant the liquid through a plug of cotton or glass wool into a 250-mL round-bottomed flask. Attach a fractionating column and attach a distillation head to the top of the column. Add a boiling chip and distill about 50 mL of liquid into a weighed 100-mL graduated cyclinder (discard the yeast residue in the distilling flask). Note and record the temperatures at every 5-mL interval. Record the total volume of the distillate and determine its weight by the increase from the starting weight. From the weight and volume calculate the specific gravity of the distillate. If a hydrometer is available, a more accurate measure of the specific gravity of the distillate can be obtained.[3] From the table of densities of aqueous solutions of ethanol given in the Appendix, calculate the weight of alcohol in the distillate.[4] Calculate the percentage yield of ethanol produced in the fermentation.

One can obtain 95% ethanol from the dilute alcohol obtained above by slow redistillation through the same fractionating column, except this time collect only the fraction with a boiling point of 78–82°; if too little distillate is obtained in this range continue the distillation and collect the fraction boiling at 82–88°. Discard the residue, which contains mostly water and the fusel oil.

[2] Pasteur's salts solution consists of 2.0 g potassium phosphate, 0.20 g calcium phosphate, 0.20 g magnesium sulfate, and 10.0 g ammonium tartrate in 860 mL of water. If Pasteur's salts solution is not available, 0.25 g of disodium hydrogen phosphate may be substituted. The fermentation can be carried on without any added salts but is slower and yields a more difficultly separated yeast residue.

[3] The hydrometer should be the short form for use with 50 mL or less of liquid, with a range of 0.900–1.000 in graduations of 0.002 or less.

[4] The density, d_4^{20}, of a liquid is the measured specific gravity multiplied by the density of water at 20°.

It is instructive to carry out the iodoform test and esterification test with small portions of the distillate as described in Chapter 9. The NMR spectrum of ethanol is shown in Figure 10.24.

Questions

1. When benzene is used to dry 95% (by weight) ethanol, two azeotropes are involved: a ternary azeotrope (bp 65°) containing 74% by weight benzene, 18.5% ethanol, and 7.5% water, and a binary azeotrope (bp 68°) consisting of 68% benzene, and 32% ethanol. Assuming a safety factor of 10% excess, calculate the amount of benzene needed to dry 1 kg of 95% ethanol. How much absolute ethanol would be obtained ?

2. How could one ascertain whether the alcohol or the carboxylic acid furnishes the hydroxyl group that appears in the molecule of water produced in formation of an ester ?

55 Peptides–Biopolymers

55.1   Structure

Peptides are polymers derived from amino acids and have molecular weights less than 5000–10,000; similar polymers with molecular weights greater than this are called proteins. The component amino acids are joined by amide bonds from the amino group of one unit to the carboxyl group of the next.

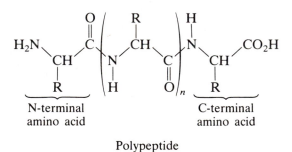

Polypeptide

In writing structures of polypeptides, the convention is to place the N-terminal amino acid on the left and the C-terminal amino acid on the right. Then the amino acids are named left to right using either their common names of their standardized three-letter codes.

464

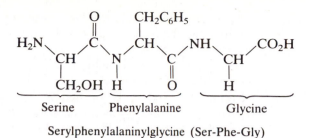

Serylphenylalaninylglycine (Ser-Phe-Gly)

55.2 Laboratory Synthesis of Polypeptides

The standard strategy for forming amide bonds is to prepare a chemically active component and allow it to react with an amine. For example, a carboxylic acid can be converted to an acyl chloride, which, on treatment with an amine, gives an amide. This reaction is the basis for both carboxylic acid and amine identification (see Chapter 9).

$$R-\overset{\overset{\text{O}}{\|}}{C}-OH \xrightarrow{\text{activation}} R-\overset{\overset{\text{O}}{\|}}{C}-Cl \xrightarrow{NH_2R} R-\overset{\overset{\text{O}}{\|}}{C}-NHR + HCl$$

This simple strategy, so effective for monofunctional group components, fails miserably when applied to making amide bonds between amino acids. With such bifunctional group compounds, the initial activation step produces an intermediate that reacts with itself to produce a polymer before the amine component is added. For example, the simple amino acid glycine, on conversion to glycyl chloride, polymerizes to polyglycine and other condensation products.

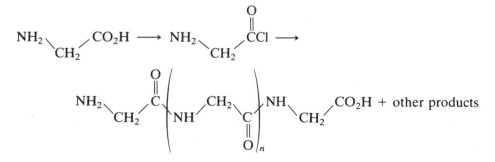

To avoid this unwanted reaction it is necessary to "protect" the amino group, temporarily, that is, to convert the NH_2 to some other unreactive functional group. After the desired coupling has occurred, the protecting group can be removed.

The repeated sequence of protection, activation, and reaction terminated by a final deprotection step is time-consuming and fraught with many subtle difficulties. However, the ability to synthesize any arbitrary amino acid polymer is so important to modern biochemical research that a great effort has been invested in making the process practical. A brilliant achievement in this area was Merrifield's[1] step-by-step synthesis of the enzyme bovine pancreatic ribonuclease, a protein with a chain of 124 amino acid units. A total of 369 consecutive chemical reactions was achieved with an overall yield of 17% which corresponds to an average yield of better than 99% for each step.

In the peptide synthesis described here, we will tackle a much simpler example, glycylglycine. The chemical sequence to be followed is outlined below.

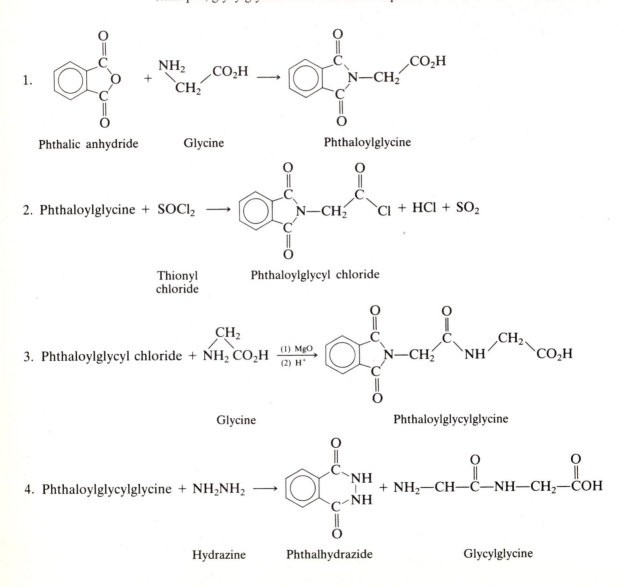

1. Phthalic anhydride   Glycine   Phthaloylglycine

2. Phthaloylglycine + SOCl₂ →   + HCl + SO₂

   Thionyl chloride   Phthaloylglycyl chloride

3. Phthaloylglycyl chloride + NH₂ CO₂H →

   Glycine   Phthaloylglycylglycine

4. Phthaloylglycylglycine + NH₂NH₂ →

   Hydrazine   Phthalhydrazide   Glycylglycine

[1] Merrifield was awarded the Nobel prize in 1984.

In step 1 the amino group of the glycine is protected by reaction with phthalic anhydride to produce pththaloylglycine, in which the original NH_2 group has been transformed into an imide. Because the nitrogen of an imide is conjugated with a pair of carbonyl groups, it is no longer basic or nucleophilic. Because the nitrogen bears no hydrogens, it cannot even hydrogen-bond to other bases. The amino group has been chemically deactivated and one can now proceed to activate the carboxylic acid without fear of having it react with the amino group of a glycine unit.

In step 2 the amine-protected glycine is converted into an acid chloride with thionyl chloride. In this reaction an equivalent of hydrogen chloride is liberated. If the amino group had not been protected it would have been converted into a hydrochloride salt, which would have slowed the reaction with thionyl chloride.

In step 3 the acid chloride is allowed to react with another equivalent of glycine to produce the phthaloylglycylglycine. In principle, one could convert this material into an acid chloride, as in step 2, and use that to add a third amino acid and so on. We will stop at the dipeptide stage.

The final step, after all the desired amino acids have been joined, is to remove the protecting phthaloyl unit. This can be accomplished by heating the phthaloylglycylglycine with hydrazine, which does two consecutive nucleophilic displacements on the imide to yield glycylglycine and the very stable phthalhydrazide. Directions for this last step are not given here because hydrazine is quite toxic and hazardous for beginners to work with.

55.3 Preparation of Phthaloylglycylglycine

(A) Phathaloylglycine

Place a well-pulverized mixture of 0.6 g (0.004 mole) of phthalic anhydride and 0.3 g (0.004 mole) of glycine in a 20 × 150-mm test tube. Clamp the tube vertically and insert a 250-degree thermometer into the mixed solids so that the bulb is covered. Heat the tube gently until the solids melt, stir them with the thermometer so that they mix well, and then heat the molten mass at 150–190° for 15 min. Allow the reaction mixture to cool and then recrystallize it in the test tube from 10 mL of water. The yield of phthaloylglycine is about 0.7–0.8 g, mp 192–193° (literature, mp 196–198°).

If IR and NMR instruments are available, determine the spectra of the product.

(B) Phthaloylglycyl Chloride

In a small round-bottomed flask place 0.4 g (0.002 mole) of recrystallized and dried phthaloylglycine (the product must be dry before adding the next reagent). Add 2.0 mL (3.3 g, 0.028 mole) of thionyl chloride (**Caution**—*vapors and liquid*

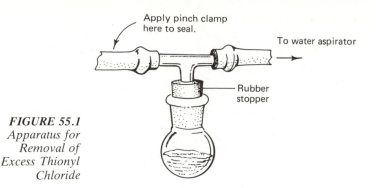

Apply pinch clamp
here to seal.

To water aspirator

Rubber
stopper

FIGURE 55.1
*Apparatus for
Removal of
Excess Thionyl
Chloride*

are strongly irritating to the skin, nose, and eyes!) and 1 drop of dimethylformamide (DMF). Attach a condenser bearing a T-tube leading to a water aspirator
(see Figure 8.2) and heat the mixture at a gentle reflux rate for 1 hr.

Remove the condenser, add a new boiling chip, and evaporate the excess
thionyl chloride from the reaction mixture under reduced pressure using the
vacuum adapter illustrated in Figure 55.1. In carrying out the evaporation apply
heat gently and continue to pull a vacuum until the residue solidifies. The yield is
about 0.44 g, mp 74–82° (literature, mp 85–86°). The crude product is used
directly in the next step.

(C) Phthaloylglycylglycine

In a large test tube prepare a suspension of 0.14 g of glycine and 0.1 g of
magnesium oxide in 6 mL of water and cool it to 5° in an ice bath. Add to the
cooled suspension, drop by drop, a solution of 0.35 g (0.0016 mole) of phthaloylglycyl chloride in 2 mL of tetrahydrofuran. The reaction mixture should be
maintained at 5° during the addition. When the addition is complete, stir the
reaction mixture for 10–15 min at room temperature to complete the reaction.
Acidify the mixture (pH paper) with hydrochloric acid. Chill the acidic mixture
in an ice bath for 20 min and collect on a Hirsch funnel the crystalline product
that separates. Recrystallize the product from 6–7 mL of water. The yield is
0.2–0.3 g, mp 224–226°.

If IR and NMR instruments are available, determine the spectra of the
product and compare them with the spectra of phthaloylglycine.

Questions   **1.** Explain why the preparative procedure used to synthesize glycylglycine would
fail if you attempted to introduce threonine, tyrosine, or glutamic acid.

2. The pK_1 values for glycine, glycylglycine, and glycylglycylglycine are 2.35,
3.14, and 3.23, respectively. How can one account for the large difference
between glycine and the peptides derived from it?

3. Explain why a benzoyl group could not be used as an N-protecting group for
peptide synthesis.

56 Metalloporphyrins

56.1 Introduction

The complexes between metal ions and porphyrins represent one of the more important classes of naturally occurring metal-containing compounds (e.g., hemoglobin in blood). Porphyrins are macrocyclic tetrapyrrole systems with conjugated double bonds and various groups attached to the ring perimeter. The unsubstituted system is called *porphine*; the substituted porphines are called porphyrins (commonly abbreviated as PH_2 regardless of any substituents).

A neutral PH_2 can accept protons on the nitrogens to form a monoprotonated species (PH_3^+) and diprotonated species (PH_4^{2+}). The neutral PH_2 species can

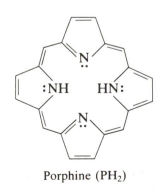

Porphine (PH_2)

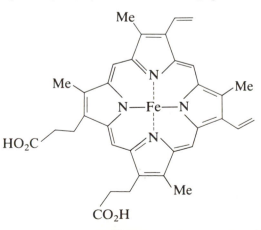

Heme
(a metalloporphyrin
occurring in hemoglobin)

also lose one or two protons to form the monoanion (PH^-) or the dianion (PH^{2-}). It is usually the dianion that complexes with a metal ion to form a *metalloporphyrin* (MP). Biologically active porphyrins invariably contain a metal ion. Hemoglobin and myoglobin contain iron porphyrins as do the cytochromes and the enzyme catalase. Many of the metal salts of porphyrins are semiconductors.

$$PH_2 + H^+ \rightleftharpoons PH_3^+$$

$$PH_3^+ + H^+ \rightleftharpoons PH_4^{2+}$$

$$PH_2 \rightleftharpoons PH^- + H^+$$

$$PH^- \rightleftharpoons P^{2-} + H^+$$

$$P^{2-} + M^{2+} \rightleftharpoons PM \qquad \text{(metalloporphyrin)}$$

In the first part of this experiment you will synthesize $\alpha,\beta,\gamma,\delta$-tetraphenylporphyrin by the condensation of pyrrole with benzaldehyde in propionic acid as a solvent. The porphyrin is produced in low yield, but it is very easy to isolate and purify because it is virtually insoluble in propionic acid. In the second part of the experiment, the porphyrin will be converted into its copper complex by heating a dimethylformamide (DMF) solution of copper acetate with the tetraphenylporphyrin. The metalloporhyrins with other metals can be prepared by similar methods.

$$C_6H_5CHO + H^+ \rightleftharpoons C_6H_5\overset{+}{C}HOH$$

The mechanism of the reaction between pyrrole and benzaldehyde is not clear. The principal reaction steps are presumably acid-catalyzed condensations of benzyl cations onto a pyrrole ring (pyrrole undergoes aromatic substitution reactions with particular ease), which could lead to a cyclic species containing four pyrroles and four benzaldehydes (see above). However, the conversion of this tetrahydrotetraphenylporhyrin to the observed tetraphenylporphyrin requires an oxidation and it is not clear what is being reduced. Typical overall yields in this reaction based on the weight of benzaldehyde and pyrrole used is about 25%. There are many candidates for partners for the final oxidation–reduction step; molecular oxygen has been proposed as the oxidizing agent.

56.2 Absorption Spectroscopy of Porphyrins

The porphyrins contain an immense π-network that spans the four pyrrole rings and the central 16-membered ring. The porphyrins usually absorb intensely (molar absorptivity constant, ε, of about 10^5; see Section 10.4) in the visible region near 400 mm. This band, called the Soret band, is named after the biochemist who first observed the band in hemoglobin. The position of the Soret band is sensitive to pH, solvent, and the presence of complexed metal ions as in the metalloporphyrins.

In addition to the Soret band, the spectral region between 500–700 nm, the visible region, contains a number of weaker bands with an appearance that is also dependent on the pH, solvent, and complexed metal ion. For most simple porphyrins in neutral solvents, the visible spectrum consists of four bands.

56.3 Preparations

(A) $\alpha,\beta,\gamma,\delta$-Tetraphenylporphyrin

In a 100-mL round-bottomed flask, equipped with a Claisen adapter, a dropping funnel, and a reflux condenser, place 30 mL of propionic acid (**Caution—** *corrosive liquid!*) and 2.0 mL (2.1 g, 0.020 mole) of benzaldehyde.[1] Heat the solution until it boils gently and then add dropwise over a 3-min period a solution of 1.5 mL of freshly purified pyrrole[2] in 3 mL of propionic acid. After all of the pyrrole has been added, continue to boil the mixture for 30 min and then cool it to room temperature. Collect the crude tetraphenylporphyrin on a Büchner or

---

[1] The benzaldehyde should be from a freshly opened bottle. Benzaldehyde is slowly oxidized to benzoic acid on standing in contact with air. Small amounts of benzoic acid are easily removed from benzaldehyde by passing it down a short column of alumina (2–3 cm long).

[2] On standing, pyrrole is oxidized to dark-colored polymeric impurities that interfere with the preparation. If present, these are readily removed by passing the dark pyrrole through a short column of alumina. A simple method is to prepare a pencil column (Section 7.6) using a Pasteur pipet packed with 2–3 cm of alumina. A 2-mL sample of crude pyrrole passed through this column will emerge light tan and is suitable for this experiment.

Hirsch funnel (it is advisable to first wet the filter paper with a little water to prevent the solid from flowing around the edges). Wash the solid with about 15 mL of methanol, then with 15 mL of hot water, and finally with 3 mL of cold methanol (to remove excess water). Continue to draw air through the funnel in order to dry the sample. The yield of dark purple tetraphenylporphyrin is about 0.5 g; the purity is reported to be better than 97%.

If a spectrometer is available, prepare 100 mL of a stock solution of 40 mg of tetraphenylporphyrin in methylene chloride in a 100-mL volumetric flask. Dilute 1.00 mL of this solution to 10 mL in a 10 mL volumetric flask and determine the spectrum over the range of 350–700 nm. Record the wavelength of each band maximum and calculate each extinction coefficient.

▐▶ *CAUTION*   Prolonged exposure to high concentrations of methylene chloride vapors may induce cancer. As with all organic solvents, work in a well-ventilated area when using it.

(B) Copper(II) Tetraphenylporphyrin Complex

In a 50-mL round-bottomed flask, place 200 mg (0.33 mmole) of tetraphenylporphyrin, 20 mL of dimethylformamide (DMF), and 65 mg (0.33 mmole) of copper(II) acetate monohydrate. Add a boiling chip, attach a reflux condenser, and boil the mixture gently for 10 min. Cool the mixture to room temperature and add 20 mL of water. Collect the product on a Büchner funnel and wash it with about 25 mL of water, followed by 5 mL of cold methanol to speed the drying process. Continue to draw air through the filter until the product is dry. The yield of the purple copper porphyrin is about 200 mg.

If a spectrometer is available, prepare 10 mL of a solution of the copper(II) tetraphenylporphyrin in methylene chloride as described above and determine the spectrum over the range of 350–700 nm. Record the wavelength of each band maximum and calculate each extinction coefficient. Compare the spectrum with that of tetraphenylporphyrin.

Questions   **1.** How many resonance structures can be written for porphine?

2. It is not possible to prepare the porphyrin analog in which pyridine is substituted for pyrrole. In light of your answer to question 1, suggest a reason for this failure.

3. It is possible to prepare a cyclic ether analogous to tetrahydrotetraphenylporphyrin in which oxygens replace the nitrogens of the pyrrole rings. Write a structure for this ether and equations for its formation from furan and benzaldehyde.

4. What are some of the possible impurities in the preparation of tetraphenylporphyrin and how does the purification scheme remove most of them?

Appendix

A Tables of Physical Data

	(In millimeters of mercury)					
T, °C	Water	Benzene	Bromo-benzene	*p*-Dibromo-benzene	Toluene	Ethanol
30	31.8	118	6	—	37	78
40	55.3	181	10	—	60	135
50	92.5	269	17	—	93	223
60	149.2	389	28	—	138	353
70	233.8	547	43	4	203	543
80	355.5	754	66	7	288	813
90	526.0	1016	98	12	402	1187
95	634.0	1171	118	15	472	1415
100	760.0	1344	141	18	557	1695
110	1074.0	1748	199	28	747	2364
120	1489.0	2238	275	42	965	—

Vapor Pressures of Organic Substances and of Water at 30–120°C

Vapor Pressures of Organic Substances[a] at 100°C

(In millimeters of mercury)

Acetophenone	26.5	p-Dichlorobenzene	73.1
Aniline	45.7	Dimethylaniline	37.9
Benzaldehyde	60.5	Ethyl acetoacetate	80
Benzoic acid	1.8	Ethylbenzene	307
Bromobenzene	141	Ethyl benzoate	17.4
Biphenyl	9	2-Furaldehyde	110
n-Butyl acetate	325	Nitrobenzene	20.8
n-Butyl alcohol	390	p-Nitrobenzoic acid	0.006
o-Chlorotoluene	133	p-Nitrotoluene	7.9
p-Chlorotoluene	123	Salicylic acid	0.86
o-Cresol	13.6	Toluene	557
p-Dibromobenzene	18	Triphenylmethanol	0.01

[a] The vapor pressures of salts, such as aniline hydrochloride and sodium acetate, are generally so low at 100°C that they can be considered to be nil.

Vapor Pressures of Water and Organic Substances at 90–100°C[a]

(In millimeters of mercury)

T, °C	Water	α-Pinene	p-Dichloro-benzene	o-Chloro-toluene
90.0	525.8	99.6	49.5	84.1
90.5	525.8	101.5	50.5	85.6
91.0	546.0	103.3	51.5	87.3
91.5	556.4	105.2	52.5	88.9
92.0	566.9	107.2	53.6	90.6
92.5	577.6	109.1	54.6	92.2
93.0	588.5	111.1	55.7	94.0
93.5	599.5	113.1	56.8	95.7
94.0	610.7	115.2	58.0	97.5
94.5	622.1	117.3	59.1	99.3
95.0	633.7	119.4	60.3	101.1
95.5	645.5	121.5	61.5	103.0
96.0	657.4	123.7	62.7	104.8
96.5	669.6	125.9	63.9	106.7
97.0	681.9	128.2	65.2	108.7
97.5	694.4	130.5	66.5	110.7
98.0	707.1	132.8	67.7	112.7
98.5	720.1	135.1	69.1	114.7
99.0	733.2	137.5	70.4	116.8
99.5	746.5	139.9	71.8	118.9
100.0	760.0	142.4	73.1	121.0

[a] Based on data presented in Weast, Ed., *Handbook of Chemistry and Physics* (Boca Raton, FL: CRC, Inc., 1983).

Density and Vapor Pressure of Water at 0–35°C

T, °C	Vapor pressure, Hg	Density $d_{4°}^{t°}$	T, °C	Vapor pressure, Hg	Density $d_{4°}^{t°}$	T, °C	Vapor pressure, Hg	Density $d_{4°}^{t°}$
0	4.58	0.99987	12	10.48	0.99952	24	22.18	0.99733
1	4.92	0.99993	13	11.19	0.99940	25	23.54	0.99708
2	5.29	0.99997	14	11.94	0.99927	26	24.99	0.99682
3	5.68	0.99999	15	12.73	0.99913	27	26.50	0.99655
4	6.09	1.00000	16	13.56	0.99897	28	28.10	0.99627
5	6.53	0.99999	17	14.45	0.99880	29	29.78	0.99597
6	7.00	0.99997	18	15.38	0.99862	30	31.55	0.99568
7	7.49	0.99993	19	16.37	0.99843	31	33.42	0.99537
8	8.02	0.99988	20	17.41	0.99823	32	35.37	0.99505
9	8.58	0.99981	21	18.50	0.99802	33	37.43	0.99473
10	9.18	0.99973	22	19.66	0.99780	34	39.59	0.99440
11	9.81	0.99963	23	20.88	0.99757	35	41.85	0.99406

Aqueous Ethanol

Density $d_{4°}^{20°}$	C_2H_5OH			Density $d_{4°}^{20°}$	C_2H_5OH		
	% by weight	% by volume	g/100 mL		% by weight	% by volume	g/100 mL
0.98938	5	6.2	4.9	0.85564	75	81.3	64.2
0.98187	10	12.4	9.8	0.84344	80	85.5	67.5
0.97514	15	18.5	14.6	0.83095	85	89.5	70.6
0.96864	20	24.5	19.4	0.81797	90	93.3	73.6
0.96168	25	30.4	24.0	0.81529	91	94.0	74.2
0.95382	30	36.2	28.6	0.81257	92	94.7	74.8
0.94494	35	41.8	33.1	0.80983	93	95.4	75.4
0.93518	40	47.3	37.4	0.80705	94	96.1	75.9
0.92472	45	52.7	41.6	0.80424	95	96.8	76.4
0.91384	50	57.8	45.7	0.80138	96	97.5	76.9
0.90258	55	62.8	49.6	0.79846	97	98.1	77.4
0.89113	60	67.7	53.5	0.79547	98	98.8	77.9
0.87948	65	72.4	57.1	0.79243	99	99.4	78.4
0.86766	70	76.9	60.7	0.78934	100	100	78.9

B Preparation of Gas Chromatographic Columns

Of the several ways to prepare gas chromatography columns, we have found that the following procedure works well and consistently gives $\frac{1}{4}''$ columns providing about 1200 plates/ft, quite good enough for the student laboratories.

Step 1: Prepare the Empty Column. Cut off the desired length of $\frac{1}{4}''$ copper tubing (6 ft is a practical length) and bend it to the desired shape to fit the GC instrument. We prefer to bend the column before it has been filled. Place the column fittings on the empty column, attach the column to the instrument (we find it convenient here to substitute a jig that has fittings in the same locations as the instrument), and bind the fittings to the column by tightening them in accordance with the manufacturer's directions. Remove the column from the instrument.

Step 2: Coat the Solid Support. The directions given below are for preparing a Chromosorb G column coated with 3% (w/w) DC 710. The general approach is to prepare a solution of the desired liquid phase in an appropriate solvent and soak the solid support in the solution until it adsorbs the liquid phase. It is necessary to make the solution about 15% more concentrated than would be required for total adsorption.

In a graduated cylinder measure 50 mL of sieved 100/120 Mesh Chromosorb G (50 mL = 23.4 g) and prepare a solution of 0.81 g ($23.4 \times 0.03 \times 1.15$) of DC 710 in 125 mL of methylene chloride. In a 250-mL (or larger) filter flask place the methylene chloride solution and add the solid support. Swirl the solution for about 15 sec and then *gradually* apply a water aspirator vacuum to the flask. Be careful in applying the vacuum at first because there is a large amount of air trapped in the solid support and the solution foams badly. Swirl the flask for about 30 sec after the full vacuum has been applied to remove most of the trapped air. Release the vacuum and let the mixture stand for about 15 min with occasional swirling. Collect the coated solid by vacuum filtration through a large sintered glass funnel; continue to draw air through the solid for 15 min. Transfer the wet solid to a large evaporating dish and allow any remaining solvent to evaporate. Finally, heat the solid in a drying oven at 80–100° until it is thoroughly dry and free flowing. Do *not* sieve the coated support.

Step 3: Fill the Empty Column. Insert into the exit end (the end nearest the detector) of the empty column a small twist of glass wool and press it into place with a pair of tweezers. This glass-wool twist keeps the packing from being blown out by the gas flow, so be sure it is large enough to be held securely in place. On the other hand, if too large a wad of glass wool is used, there may be too much of a pressure drop.

Attach the exit end of the column to a water aspirator and attach a small funnel to the entrance end of the column. Turn on the aspirator and pour a small portion of the coated support into the funnel. Repeatedly tap the column as the solid flows in. Thorough tapping is the key to a well-packed column free of efficiency-destroying voids. Repeat the additions of support with tapping until no more solid support can be added.

Remove the funnel and aspirator connections from the filled column and remove about $\frac{1}{2}''$ of packing from the *entrance end* of column. Press a small twist of glass wool securely into the entrance end of the column.

Step 4: Bake the Column. The column may have accumulated impurities in the packing process, and it is good practice to bake out the column at the same temperature that will be used in the measurements. This is conveniently done by attaching the entrance end of the column to the GC, heating the GC oven to the desired temperature, and then turning on the gas flow. Since the exit end is *not* connected, any vaporized materials that are blown out will not contaminate the detector. After several hours of this treatment, the column is ready for use.

C Expanded IR Correlations

The table of characteristic infrared frequencies given below is divided into five parts, according to the general region in which the absorptions occur.

Frequency, cm^{-1}	Origin, Intensity, Comments
I. Carbon–Hydrogen Stretching Region:[a] 3600–2500 cm^{-1}	
3500–2500	carboxylic acid O—H, single strong, broad band
3400–3200	polymeric O—H stretch, strong, broad in high dilution band sharpens and moves toward 3650 cm^{-1}
near 3400	nonhydrogen-bonded N—H, medium intensity, sharp
near 3200	hydrogen-bonded N—H, medium intensity, sharp NH_2 group usually a doublet with ca. 50 cm^{-1} separation
near 3300	alkyne C—H, strong
3095–3010	alkene C—H, medium intensity
near 3050	aromatic C—H, variable
2980–2850	alkane C—H, medium to strong
2850–2750	aldehyde C—H, weak; one or two bands
II. Triple-Bond Stretching Region: 2300–2100 cm^{-1}	
2260–2240	unconjugated C≡N, strong
2240–2210	conjugated C≡N, strong (aryl and vinyl)
2240–2100	—C≡C—, strong for terminal acetylenes, medium to absent for disubstituted acetylenes
III. Deuterium Stretch: 2130–1940 cm^{-1}	
2130–1940	C—D, essentially 0.7 of corresponding C—H stretch
IV. Double-Bond Stretching Region:[b] 1900–1550 cm^{-1}	
1850–1800 and 1790–1720	anhydride C=O, both strong
1815–1750	acid chloride C=O, strong
1800–1770	vinyl ester C=O, strong
1750–1735	saturated ester C=O, very strong
1740–1720	saturated aldehyde C=O, strong
1730–1715	α, β-unsaturated and aryl ester C=O, strong
1725–1700	saturated carboxylic acid C=O, strong
1715–1680	α, β-unsaturated and aryl aldehyde C=O, strong
1715–1680	α, β-unsaturated and aryl carboxylic acid C=O, strong
1690–1630	amide C=O, strong
1690–1630	oxime C=N, variable intensity
1680–1620	unconjugated C=C, variable intensity
1630–1575	—N=N—, variable intensity

[a] This region includes not only carbon–hydrogen stretches but also nitrogen–hydrogen and oxygen–hydrogen stretches.

[b] The lower part of this region overlaps slightly with the upper part of the next, fingerprint region.

Frequency, cm^{-1}	Origin, Intensity, Comments

V. Fingerprint Region:[c] 1650–500 cm^{-1}

Frequency, cm^{-1}	Origin, Intensity, Comments
1650–1550	N—H bending, variable intensity
1600–1585	aromatic skeletal vibration, medium intensity
1500–1400	aromatic skeletal vibration, medium intensity
1485–1445	CH$_2$ bending, variable intensity
1470–1430 and 1380–1370	CH$_3$ bending, strong
1385–1380 and 1370–1365	*gem*-dimethyl, strong
1410–1310 and near 1200	phenol, strong
1410–1310 and near 1150	tertiary ROH and ROR, strong
1370–1335 and 1170–1155	sulfonamides, strong
1350–1260 and near 1150	secondary ROH and ROR, strong
1350–1260 and near 1100	primary ROH and ROR, strong
1350–1250	aliphatic amine C—N, strong
near 1410 and 1220–1020	aromatic amine C—N, weak
965	*trans* —CH=CH—, strong
985 and 910	CH$_2$=CH—, strong
890	CH$_2$=C$\diagdown^{\diagup}$, strong
840–810	—CH=C$\diagdown^{\diagup}$, strong
700	*cis* —CH=CH—, variable intensity
750 and 690	C$_6$H$_5$—, strong
750	*ortho* C$_6$H$_4\diagdown^{\diagup}$, very strong
780–700	*meta* C$_6$H$_4\diagdown^{\diagup}$, very strong
825	*para* C$_6$H$_4\diagdown^{\diagup}$, very strong
800–600	C—Cl, strong
700–500	C—Br, strong

[c] This region includes the many possible C—H, O—H, and N—H bends as well as the various single-bond stretches. The number of possible frequencies is so large and their interactions so strong that it is meaningless to identify a particular absorption with any simple vibrational motion.

D Expanded NMR Correlations

Chemical shifts of $X—CH_3$, $X—CH_2—Y$, and $X—CH—Y$ can be estimated
$$\overset{|}{Z}$$

with acceptable accuracy by using the data in the following table.[1] Shifts are computed relative to a reference value that differs depending on whether it is 1°, 2°, or 3°: $CH_3 = 0.87$ ppm, $CH_2 = 1.20$ ppm, and $CH = 1.55$ ppm. A distinction is also made between substituents that are attached directly (α-substituents) and those with an intervening carbon aton (β-substituents).

Examples of the application of this method are given at the end of the table.

Substituent	Type of Hydrogen	Alpha Shift ppm	Beta Shift ppm
—Aryl	CH_3	1.40	0.35
	CH_2	1.45	0.53
	CH	1.33	—
—C=C—	CH_3	0.78	—
	CH_2	0.75	0.10
—Cl	CH_3	2.43	0.63
	CH_2	2.30	0.53
	CH	2.55	0.03
—Br	CH_3	1.80	0.83
	CH_2	2.18	0.60
	CH	2.68	0.25
—OH	CH_3	2.50	0.33
	CH_2	2.30	0.13
	CH	2.20	—
—NRR′	CH_3	1.30	0.13
	CH_2	1.33	0.13
	CH	1.33	—
$—O\overset{O}{\overset{\|}{C}}R$ and $—O\overset{O}{\overset{\|}{C}}OR$	CH_3	2.88	0.38
	CH_2	2.98	0.43

[1] Based on the table and data given in Chapter 4, Appendix B, of Silverstein, Bassler, and Morel, *Spectrometric Identification of Organic Compounds*, 4th ed. (New York: Wiley, 1981).

Substituent	Type of Hydrogen	Alpha Shift ppm	Beta Shift ppm
$\overset{\displaystyle O}{\underset{\displaystyle \|}{}}$			
—CR, where R is	CH_3	1.23	0.18
alkyl, aryl, OH,	CH_2	1.05	0.31
OR′, H, or N	CH	1.05	—

Example

$$CH_3\text{—}CH_2\text{—}CH_2\text{—}OH$$

$$\underset{1}{\uparrow} \quad \underset{2}{\uparrow} \quad \underset{3}{\uparrow}$$

1. reference CH_3 0.87
   no α 0.00
   no β <u>0.00</u>
   total   0.87 ppm

2. reference CH   1.20
   no α 0.00
   β—OH   <u>0.13</u>
   total   1.33 ppm

3. reference CH_2 1.20
   α—OH   2.50
   no β <u>0.00</u>
   total   3.70 ppm

E Laboratory Reports

Technical reports are an important part of scientific research. Even the most exciting research result is of little value until it has been communicated to other researchers. In chemistry, oral communication is valuable and one can impress colleagues by brilliant lectures on one's research. However, ultimately, the spoken word is not considered an acceptable substitute for a written report; the chemist's attitude is that until it is published, it hasn't really been done. The written report is just as important in industry although, because of the need to preserve trade secrets, the emphasis there is on internally circulated documents rather than journal articles. Because of this emphasis on writing, the early development of technical writing skills will help you both in your academic work and later in your professional career.

Format. Reports and publications generally adhere to the format outlined below. This particular pattern is not the only way that the results and conclusions could be presented, but it is logical and it is the organization that chemists have come to expect. The basic idea is to answer the questions: "Why was this research undertaken?", "What did you do?", "What results were obtained?", "What do the results mean?", and "What conclusions are to be drawn?" Sometimes, particularly in presenting organic research containing many independent experiments that are largely self-descriptive, the Results section is combined with the Discussion section. Another frequent practice of organic chemists is placing the Experimental section after the Discussion section. Nevertheless unless you have a sound reason for change, it is suggested that you follow the outline given here.

1. Introductory Material

(a) *Title*—balance between brevity and descriptive accuracy; usually 5–10 words.

(b) *Author(s) names(s).*

(c) *Address where the study was conducted.*

(d) *Abstract*—a brief 3–10 lines summary of the main findings of the work reported.

2. Introduction

(a) Tell the reader what the report is about.

(b) When appropriate, give a *brief* historical background to the study.

(c) Describe the chemical and physical process studied.

(d) Include a statement of purpose describing what you intend to prove.

3. Experimental

Outline the major steps and procedures used to prepare your compounds or make your measurements. You should write for your audience. In general you should assume that the reader is familiar with typical laboratory apparatus and procedures, and your experiments should be detailed only to the level required to let your reader understand how you did the experiment. Finding the appropriate depth requires considerable judgment and is one of the most difficult aspects of technical writing.

When you are describing experiments that repeat work already published (by you or someone else), you should give a reference to the published procedure and then describe in detail only the deviations from the published work. If you followed the published procedure exactly, that portion of your experimental section could be as brief as "Dimethyl terephthalate was prepared according to the procedure of Robbins[3]"—the complete citation to Robbins' work being given in footnote 3.

Any *unexpected* hazards encountered in the work should be described and emphasized.

4. Results

(a) In most cases the "Results" section of your report will be very different from your laboratory notebook. Your notebook is a daily record of everything you did; your report is a distillation of your notebook. The most obvious difference is that the data are not usually presented in the order observed but rather in the most concise form possible. This generally means presentation of numerical data in simple tables. Strive to make the tables self-explanatory so that only a minimum amount of text is required. Graphs and figures also can take the place of many words.

(b) If the preparation of a compound is presented, be sure to include sufficient data to demonstrate the identity and purity of the compound. In this section present only the observations. Any discussion of what the facts mean should be postponed to the "Discussion" section.

(c) If you collected numerical data and reduced the raw results to some simpler form, use sample calculations to show how you did it.

(d) When numerical data are presented, some form of error analysis should be included and the number of significant figures made clear.

5. Discussion

(a) Repeat your main findings. All of these should have been included in the Experimental section, and only the most important aspects of the experiments are repeated here.

(b) State your conclusions based on these findings.

(c) Discuss briefly the significance of any difference between your results and the expected ones.

6. Conclusions

Summarize as succinctly as possible the major deductions of your work. If your discussion section has already presented the conclusions, this section may be skipped.

7. Acknowledgments

In research situations many people contribute to the success of the work. The rule is that the authors cited in the introductory material should have made major *intellectual* contributions to the conception and execution of the study. Supporting staff such as draftsmen, machinists, glassblowers, repair people, secretaries, and other paid professionals are customarily *not* cited even though everyone knows that their contributions are essential. If one of them has made a particularly important contribution, this Acknowledgments section is the place to recognize that effort. It is customary to cite nonroutine sources of money used to support the work. Thus a grant from the National Science Foundation would be cited, but regular departmental funds would not be cited.

8. References and Footnotes

Some journals publish references and footnotes collected at the end of the text; others include them at the bottom of the page on which they are first cited. Whichever way they are to be printed, it is probably best if your manuscript collects them at the end.

Grammar and Style. There are many excellent texts that deal with grammar and style. A particularly valuable source is *The ACS Style Guide*,[2] which is an inexpensive manual for authors of papers to be published in American Chemical Society journals. Scientific grammar is the same as ordinary grammar, but there are some special conventions that should be obeyed if your paper is to be well received.

Foremost among these is to avoid the first person. The goal of science is to be objective as possible, and a convention has evolved that interprets statements such as "I heated the flask for 10 hr" as being too personal and insufficiently objective. When describing laboratory experiments, it is best to stick to the passive voice: "The flask was heated for 10 hr." There is a growing tendency to relax this style in the Conclusions section, where you will frequently find phrases such as: "We have demonstrated that . . . ."

The ACS Style Guide presents a long, 23-page list of abbreviations and symbols accepted in American Chemical Society journals. The choice of units between the traditional metric system or the newer International System of Units (SI) remains controversial. Whichever you prefer, do not use a mixture of units or the reader will be hopelessly confused.

A grammatical misdeed that causes much trouble in scientific writing is the "dangling modifier" or other forms of indefinite antecedents. Because of the desire for tight logic in scientific writing, it is important that verbal phrases refer unambiguously to another—appropriate—word or phrase in the sentence.

[2] Janet S. Dodd (Ed.), *The ACS Style Guide* (Washington, DC: American Chemical Society, 1986).

Ambiguous

Knowing the concentration and volume of acid consumed, the amount of base present could be calculated. (The amount of base doesn't know anything!)

Clear

The amount of base present could be calculated from the concentration and volume of acid consumed.

Writing. Many students when asked to write a scientific paper have trouble getting started. If you have this problem, it is a good idea to start with the Experimental section, move next to Results and then Discussion, and save the Abstract, Introduction, and Conclusions for last. If at all possible, use a word processor so that you do not have to be concerned with getting it right the first time. Scientific writers, even experienced ones, usually go through many drafts.

F In Case of Accident

Always call or notify a laboratory instructor as soon as possible.

Fire   **Burning Reagents:** Immediately extinguish any gas burners in the vicinity. Fire extinguishers, charged with carbon dioxide or monoammonium phosphate powder under pressure, are available in various parts of the laboratory.

For burning oil use powdered sodium bicarbonate.

Burning Clothing: Avoid running (which fans the flame) and take great care not to inhale the flame. Rolling on the floor is often the quickest and best method for extinguishing a fire on one's own clothing.

Smother the fire as quickly as possible using wet towels, laboratory coats, heavy (fire) blankets, or carbon dioxide extinguisher.

Treatment of Small Burns: Submerge the burned area in cold water until the pain subsides. Blot the area dry, gently, with a sterile gauze and apply a dry gauze as a protective bandage. In small second or third degree burns in which blisters have formed or broken, or in which deep burns are encountered, see a physician as soon as possible.

Extensive Burns: These require special treatment to avoid serious or fatal outcome—*summon medical treatment at once.* Combat the effects of shock by keeping the patient warm and quiet.

Injuries and Chemical Burns

Reagents in the Eye: Wash *immediately* with a large amount of water, using the ordinary sink hose, eye-wash fountain, or eye-wash bottle—*do not touch the eye.* After the eye has been washed thoroughly for 15 min, if *any* discomfort remains, see a physician.

Reagents on the Skin: *Acids*—Wash immediately with a large amount of water, then soak the burned part in sodium bicarbonate solution. Cover the burned area with a dressing bandage and see a physician.

Alkali—Wash immediately with a large amount of water, then soak the burned area in 1% boric acid solution to neutralize the alkali. Cover the burned area with a dressing and see a physician.

Bromine—Wash *immediately* with a large amount of water, then soak the burned area in 10% sodium thiosulfate, or cover with a *wet* sodium thiosulfate dressing, for at least 3 hr and see a physician.

Organic Substances—Most organic substances can be removed from the skin by washing immediately with ordinary ethanol, followed by washing with soap and warm water. If the skin is burned (as by phenol), soak the injured part in water for at least 3 hr and see a physician.

Cuts: Wash the wound with sterile gauze, soap, and water. Cover with a sterile dressing and keep dry.

Index

Numbers in **boldface** refer to preparations or procedures.

Atomic Weights

Aluminum	Al	26.98	Magnesium	Mg	24.31
Antimony	Sb	121.75	Manganese	Mn	54.94
Arsenic	As	74.92	Mercury	Hg	200.59
Barium	Ba	137.34	Nickel	Ni	58.71
Boron	B	10.81	Nitrogen	N	14.01
Bromine	Br	79.91	Oxygen	O	16.00
Calcium	Ca	40.08	Palladium	Pd	106.40
Carbon	C	12.01	Phosphorus	P	30.97
Chlorine	Cl	35.45	Platinum	Pt	195.09
Chromium	Cr	52.00	Potassium	K	39.10
Copper	Cu	63.55	Silicon	Si	28.09
Fluorine	F	19.00	Silver	Ag	107.87
Hydrogen	H	1.01	Sodium	Na	22.99
Iodine	I	126.90	Sulfur	S	32.06
Iron	Fe	55.85	Tin	Sn	118.69
Lead	Pb	207.19	Vanadium	V	50.94
Lithium	Li	6.94	Zinc	Zn	65.37